HIGH TEMPERATURE FATIGUE

Properties and Prediction

FRONTISPIECE

The upper photograph (original magnification $\times$ 15 000), courtesy of Dr J. K. L. Lai, shows the cell structure induced in 316 austenitic steel after 713 cycles of high strain fatigue at 625°C, the test lasting 1 day. The lower micrograph (original magnification $\times$ 15 000) shows the structure in a specimen subjected to the identical fatigue test (i.e. at a plastic strain range of 0·2%) but which was afterwards solution annealed at 1050°C for 1 h. The cell structure has survived the heat treatment, though companion tests have shown that only about 30% of the cyclic hardening previously introduced by fatigue is retained.

HIGH TEMPERATURE FATIGUE

Properties and Prediction

Edited by

R. P. SKELTON

Technology Planning and Research Division,
Central Electricity Research Laboratories,
Central Electricity Generating Board,
Leatherhead, Surrey, UK

ELSEVIER APPLIED SCIENCE
LONDON and NEW YORK

ELSEVIER APPLIED SCIENCE PUBLISHERS LTD
Crown House, Linton Road, Barking, Essex IG11 8JU, England

Sole Distributor in the USA and Canada
ELSEVIER SCIENCE PUBLISHING CO., INC.
52 Vanderbilt Avenue, New York, NY 10017, USA

WITH 20 TABLES AND 133 ILLUSTRATIONS

Softcover reprint of the hardcover 1st edition 1987

British Library Cataloguing in Publication Data

High temperature fatigue: properties and
prediction.
1. Materials—Fatigue. 2. Materials
at high temperatures
I. Skelton, R. P.
620.1′123 TA418.38

Library of Congress Cataloging in Publication Data

High temperature fatigue.

Bibliography: p.
Includes index.
1. Metals—Fatigue. 2. Metals at high temperatures.
I. Skelton, R. P.
TA460.H5 1987 620.1′1217 87-15460

ISBN-13: 978-94-010-8046-0 e-ISBN-13: 978-94-009-3453-5
DOI: 10.1007/978-94-009-3453-5

Preface

About 35 years ago, thermal fatigue was identified as an important phenomenon which limited the lifetime of high temperature plant. In the intervening years many investigations have been carried out, primarily to give guidance on likely endurance (especially in the presence of time-dependent deformation) but latterly, with the introduction of sophisticated testing machines, to provide knowledge of the underlying mechanisms of failure.

A previous edited book (*Fatigue at High Temperature*, Elsevier Applied Science Publishers, 1983) summarised the state-of-the-art of high temperature fatigue testing and examined the factors influencing life, such as stress state, environment and microstructural effects. It also considered, in some detail, cyclic crack growth as a more rigorous approach to life limitation.

The aim of the present volume (which in style and format follows exactly the same lines as its predecessor) is once again to pursue the desire to translate detailed laboratory knowledge into engineering design and assessment. There is, for example, a need to consider the limitations of the laboratory specimen and its relationship with engineering features. Many design procedures still rely on a simple endurance approach based on failure of a smooth specimen, and this is taken to indicate crack initiation in the component. In this volume, therefore, crack propagation is covered only incidentally, emphasis being placed instead on basic cyclic stress–strain properties, non-isothermal behaviour, metallography, failure criteria and the need for agreed testing procedures.

As was the case for the previous volume, I have approached authors who are known to each other, and who were appraised of the individual contributions. My only intervention as editor has been to supply cross references in order to forge the separate chapters into a unit whole. However, when all the contributions had been received, I realised that a rounding-off chapter (7) was required. This I attempted myself; it contains a minimum number of references, and it is hoped that it will perhaps be useful in setting the scene for those relatively new to the subject of high temperature fatigue.

The plan of the book is as follows. The introductory chapter arose almost by accident, when the author was surprised to learn that the earliest reference to non-linear deformation behaviour was in a paper dated 1729 (and written in Latin!). A separate interest in the background to constant strain testing and, indeed, the history of mechanical testing in general led to several lines of enquiry and, ultimately, to Chapter 1. The next chapter, in a more orthodox vein, looks in some detail at a topic which, in papers on high strain fatigue, has always seemed to take second place to the endurance results, namely, cyclic stress–strain behaviour. This information is of course vital to the stress analyst who has to process it before arriving at an appropriate strain range on the endurance curves. Another question often asked is: how far do isothermal laboratory tests reflect the situation in service where temperatures vary throughout the cycle? It is hoped that Chapter 3 provides at least some of the answers.

From the microstructural point of view it is known that high temperature fatigue wrings many changes in the substructure. These are summarised in Chapter 4, and the author states the role metallography has to play in the overall strategy of lifetime assessment. In Chapter 5 the background to high temperature design codes is discussed and a possible way forward suggested which takes the metallurgy of failure mechanisms into account. Finally, in Chapter 6 we see that having come of age, high temperature high strain fatigue may be required to obey certain standards in order to achieve full respectability.

As was the case in the earlier volume, the materials dealt with are, of necessity, those of interest to the power generating industry, that is, the low-alloy ferritic and austenitic steels. The higher-alloyed ferritic (9Cr–12Cr) steels are currently under active consideration although, at the time of writing, relatively little fatigue data exist. However, superalloys are treated to some extent, and Chapter 4 includes a section on Ti and Al alloys of interest to the aircraft/automotive industries respectively.

It will become clear that the cyclic stress–strain relation (expressed, for example, as the Ramberg–Osgood law) permeates much of the book. Indeed, it is one medium through which engineers, metallurgists, experimentalists and materials scientists usefully communicate: so that as well as providing the final endurance data upon which lifetime assessment of components is based, the experimentalist can, if so desired, become involved along the way, for example in finite element or other simplified analyses.

The editor bears responsibility for the overall choice of topics. Inevitably, many of the contributions were solicited from colleagues within his own industry, and so in fields where there is continued healthy debate, the views expressed are personal and not necessarily those of the CEGB. I would like to thank my fellow authors for their ready acceptance to write a chapter, despite other commitments. Once again, regarding my own work, I wish to thank the CEGB for providing typing and drawing facilities. It is finally hoped that the combined information in this and the earlier volume will be helpful to the engineering and metallurgical fraternity who are responsible for designing and assessing the lifetimes of structures operating at high temperature.

R. P. SKELTON

Contents

Preface v

List of Contributors x

1. Historical Introduction: Stresses, Strains and Material Behaviour 1
   R. P. Skelton

2. Cyclic Stress–Strain Properties During High Strain Fatigue . 27
   R. P. Skelton

3. Materials Response to Thermal–Mechanical Strain Cycling . 113
   D. A. Miller and R. H. Priest

4. Metallography of High Temperature Fatigue 177
   W. J. Plumbridge

5. The Physical Metallurgy of Failure Criteria 229
   R. Hales

6. The Case for Standards in High Temperature Fatigue . . 261
   G. B. Thomas

7. The Relation Between Laboratory Specimen and the Practical Case 301
   R. P. Skelton

Index 321

List of Contributors

R. Hales

Technology Planning and Research Division, Central Electricity Generating Board, Berkeley Nuclear Laboratories, Berkeley, Gloucestershire GL13 9PB, UK

D. A. Miller

Operational Engineering Division (Midlands Area), Central Electricity Generating Board, Scientific and Technical Branch, Bedminster Down, Bridgwater Road, Bristol BS13 8AN, UK

W. J. Plumbridge

Department of Metallurgy and Materials Engineering, University of Wollongong, New South Wales 2500, Australia

R. H. Priest

Operational Engineering Division (Midlands Area), Central Electricity Generating Board, Scientific and Technical Branch, Bedminster Down, Bridgwater Road, Bristol BS13 8AN, UK

R. P. Skelton

Technology Planning and Research Division, Central Electricity Generating Board, Central Electricity Research Laboratories, Kelvin Avenue, Leatherhead, Surrey KT22 7SE, UK

G. B. Thomas

Engineering Materials and Metallurgy, ERA Technology Ltd, Cleeve Road, Leatherhead, Surrey KT22 7SA, UK

1

Historical Introduction: Stresses, Strains and Material Behaviour

R. P. SKELTON

Central Electricity Research Laboratories, Leatherhead, Surrey, UK

SUMMARY

'For from all tempers he could service draw
The worth of each, with its alloy, he knew;
And, as the confident of nature, saw
How she complections did divide and brew.'
JOHN DRYDEN (1631–1700)

In this short introduction it is shown how adequate prediction of material properties could not be made until accurate methods were available for measuring displacements, and hence strains, on suitable specimens. Over the years, laboratory testpiece sizes have shrunk dramatically, because the sensitivity of measuring techniques has increased many fold. This fact, combined with modern knowledge of microstructural events, should lead to an improvement in our ability to predict materials behaviour and hence the lifetime of components in service.

1.1. BACKGROUND

1.1.1. Specimens

The present-day high strain fatigue test in direct stressing at elevated temperature is descended from a long line of experiments in mechanical testing. This goes back a hundred years or so in terms of high temperature [1], but in terms of uniaxiality began with the tensile tests on iron wires conducted by Leonardo da Vinci (1452–1519) [2]. The first known fatigue test (on mining chain) was actually conducted in 1829 [3] and, in briefly tracing the development of the sophisticated strain-controlled experiment of today, it is of course necessary to realise, first, that not all fatigue tests have been conducted on simple specimens and, secondly, that deformation modes are not necessarily uniaxial. Indeed, the theory of elasticity has its origins in the study of a bending beam and the correct placing of the neutral axis [2], and both the bending and torsion modes were able to give measurable deflections long before accurate means were available to record the corresponding parameters in direct loading.

In the following discussions it will be helpful to have the definitions below.

(i) A *specimen test* employs a piece of the material in question whose gauge length and uniform cross section (usually circular) are specified. It is assumed, not always correctly, that mechanical properties are uniform over the gauge. On this definition of a specimen, anisotropic welds and heat-affected zones, for example, or (pedantically) any testpiece containing a crack are excluded; so too are testpieces where extensometry cannot be mounted directly on the gauge. However, the gauge length may be allowed to reduce to zero, as is the case with an hourglass specimen.

(ii) Any test assembly not conforming to the above is thus a *structure* which can be of varying complexity. A *features test* may highlight the crack initiation properties of a weld or stress concentrator in a simple geometry (see Chapter 7). A fatigue test may be conducted on a *subassembly* or indeed even the whole *component*. It will be seen later that specimen and component testing are not alternatives but are complementary in any analysis of failure.

1.1.2. Disciplines

In the history of technology, there are many accounts of the design and construction of bridges, machinery and so forth, but relatively few on the strength of their materials of construction and the method of testing [2, 4–6]. The ultimate aim in traditional engineering design is one of 'optimisation in loads for use of least material' and the achievement of that aim over the years has involved the combined contributions of empirical testing, elasticity, plasticity and time-dependent deformation theories, and the efforts of metallurgists in identifying the elements responsible for producing strong alloys. Further, once the structure has been designed, the 'engineer' (who will be identified later) is faced with the task of predicting how long it will last, especially in high temperature environments. In examining the background of the modern fatigue test this short review will keep in mind:

(i) The concept of small specimen testing.
(ii) The development of simple stress–strain laws for predicting strength and fracture properties.
(iii) The observation of time-dependent mechanical deformation and its use in predicting future behaviour.

The engineering stress analyst and the materials scientist are both concerned with mechanical properties, principally materials behaviour under external stress. The former deals directly with external properties while his colleague attempts to explain observed behaviour in terms of internal microstructural effects. With regard to *internal stresses*, a stress analyst would have to deal, for example, with long-range residual stresses arising from welding or thermal transients, whereas a materials scientist calculates (say) forces between dislocations or back-stresses generated during creep. One of the main tasks of materials studies for high temperature applications is to derive relations estimating the failure time of the material, and hence of the component, which are of practical use and which are based securely on the microstructural behaviour of the alloy concerned. 'Materials Science' is officially about 50 years old and, as McLean has pointed out [7], has had a fairly rapid development largely because the scientific laws on which it is founded were readily available from other disciplines.

Despite these considerations, there have been isolated investigations in the past [2] which strangely anticipate this new Science. Thus, between 1720 and 1722, we find Réamur conducting tests on wires to check their heat treatment and also measuring hardness by indentation. In a paper of 1784, Coulomb showed that heat treatment (annealing) affected the limit of elasticity but did not alter the basic elastic properties of metals. Small deflections were supposed not to disturb the molecular arrangement, but beyond the elastic limit permanent sliding occurred as the material hardened, causing an increase in the 'cohesive forces'. During torsional oscillations, more damping occurred with the onset of plasticity. Young in 1807 had given an explanation of ductility, brittleness and toughness in terms of 'cohesive powers' and had provided an estimate of the molecular size as $\sim 2{\cdot}5 \times 10^{-10}$ in (10^{-9} mm) [2]. Much earlier, Hooke had made a no less remarkable speculation [8]: 'the particles therefore that compose all bodies... owe the greatest part of their sensible or potential extension to a vibrative motion'. He proposed a basic frequency of one million pulses per second which decreased in extension and increased during compression.

1.2. MECHANICAL BEHAVIOUR AND THE STRESS–STRAIN LAWS

Early tests on the strength of materials were concerned solely with establishing the load to cause failure of a given specimen or structure, and observing the point at which fracture took place. No means were available for measuring the displacements which occurred and, excepting experiments on wires, compression and bending tests on fairly rigid solids preceded those in tension because of difficulties arising with gripping. The numerical law of elasticity for a slender wire was published by Hooke in 1678 [8], but it was not until the nineteenth century that stress–strain laws were established which included a contribution from plasticity. These have been reviewed by Osgood [9].

1.2.1. Ultimate Tensile Strength

It should be noted that early investigators did not possess the concept of stress as force divided by an area, and so the breaking loads observed were specific to the testpiece under examination. Further, the quality of material could confuse the results in unexpected ways: da Vinci's experiments showed that, for the same load, longer wires broke but shorter wires of the same cross section did not. Nowadays we would argue [10] that the longer wire had a greater probability of containing defects, and it is remarkable that specimen-to-specimen scatter must have been apparent to da Vinci, who urged that the test should be repeated several times to check the results [2]. Galileo (1564–1642) gave some indication of the ultimate strength of copper [2] and realised that the strength of a column was proportional to its cross-sectional area and not to its length.

Mersenne (1588–1648) performed experiments on vibrating wires of musical instruments to determine the effect of tension and other parameters upon pitch [11], but he was also interested in their breaking point [12]. This is an early example of a specimen and component test actually coinciding. Mariotte (1620–1684) was able to state that fracture occurs at a certain limit of elongation [2]. His dumb-bell-shaped tensile specimens (of wood) are a clear precursor of the modern short test specimen. Much later, Vicat in 1834 reported [13] the fracture behaviour of iron wires used in the cable of suspension bridges. The diameter was unspecified but the testpiece, although not machined, must clearly be regarded as a *specimen test*.

1.2.2. Elastic Laws

The justly celebrated law of Robert Hooke (1635–1702) that '... the Rule or Law of Nature in every springing body is, that the force of power thereof to restore itself to its natural position is always proportionate to the distance or space it is removed therefrom...' [8] was a landmark in the difficult birth of elasticity which has been admirably traced elsewhere [2, 6, 7]. Hooke realised that the material was internally stretched in due proportion down to a fine scale, but his law could equally well apply to a structure. Not until 1807 did Thomas Young conceive that the ratio of stress (σ) to strain (ε) (as we now know them) characterised a material property as measured in a specimen. Young arrived at this result [14] by

TABLE 1
SIGNIFICANT EVENTS IN ELASTICITY STUDIES

Investigator	*Date(s)*	*Comments*
da Vinci	1452–1519	Columns: strength varies directly as some ratio of cross section
Galileo	1564–1642	Strength $\propto$ cross-sectional area
Mariotte	1620–1684	Elongation $\propto$ force
Hooke	1678	'Ut tensio, sic vis'
Jacob Bernoulli	1654–1705	Curvature of elastic beam at a point $\propto$ bending moment there
Euler	1707–1783	Introduced an 'elastic constant' in bending beam theory
Coulomb	1773	Concept of shear stress
Young	1807	Modulus—a material property
Cauchy	1822	Concept of stress and strain at a point
Navier	1826	Modern definition of E
Poisson	1830	Poisson's ratio

considering the original height of a steel column that would halve its length under its own weight (1500 miles!). The practical definition of Young's modulus E was provided in 1826 by Navier [2]:

$$E = \sigma/\varepsilon \tag{1}$$

Table 1 lists in chronological order the concepts ultimately responsible for eqn (1). Poisson's ratio was not discovered until some years later. Young was uncertain about lateral dimensional changes and it is doubtful whether they could have been detected at that time.

1.2.3. Elastic–Plastic Laws

1.2.3.1. Measurements in Tension

Hooke was concerned with the nature of the elastic restoring force [8], and he did not record evidence of permanent set although the sensitivity of his tests on 40 ft (12·2 m) wires would have been adequate. The first indication that tension might be proportional to the load raised to some power was that of Bulfinger in 1729 [15] who considered the equilibrium of the tension fibres in a bent beam. A similar non-uniform relation was also derived by Riccati in 1731 [16]. This author also refers to some earlier experiments of s'Gravesande (1688–1742) [17], in which the original specimen length was not restored after the application of large loads (elasticity having been 'given up').

In a paper otherwise concerned with the correct placing of the neutral axis in a beam, Hodgkinson in 1822 [18] subjected a 7 ft (2·13 m) iron wire weighing 2 dwts. 17 grs. (5·12 g) to ever-increasing loads and noted a permanent set of 0·03 in (0·76 mm). His tabulated results are shown in Fig. 1 in terms of strain. Taking the density of wrought iron as $7{\cdot}7 \times 10^3\ \mathrm{kg\ m^{-3}}$, the cross-sectional area of the wire is calculated to be $0{\cdot}31\ \mathrm{mm}^2$ and so, assuming elastic unloading, eqn (1) predicts a value $E = 1{\cdot}66 \times 10^5$ MPa from this early data of Hodgkinson.

In 1827 Lagerhjelm stated [4] that if C be the extension of a bar at the elastic limit and Δ the extension when it breaks, then the product $C\sqrt{\Delta}$ is a constant and, further, that permanent set produces an increase in volume. A similar deformation law was produced several years later by Gerstner [2, 4], who also reported strain hardening, deduced from experiments in loading and unloading. Some of these early relations are of questionable validity but they do indicate that attempts were being made to measure deviations from Hooke's law.

Stress–strain formulae for elastic–plastic deformation derived since 1830 have been surveyed by Osgood [9], who included his own relation:

$$\varepsilon_t = \frac{\sigma}{E} + \left(\frac{\sigma}{A}\right)^{1/\beta} \tag{2}$$

where ε_t is total strain, σ is the stress, and A and β are constants. It is one of the few expressions which recognises that elastic deformation continues to act beyond yield, a point appreciated by Young [14]. In eqn (2) a plot of

$$\log\left(\varepsilon_t - \frac{\sigma}{E}\right)$$

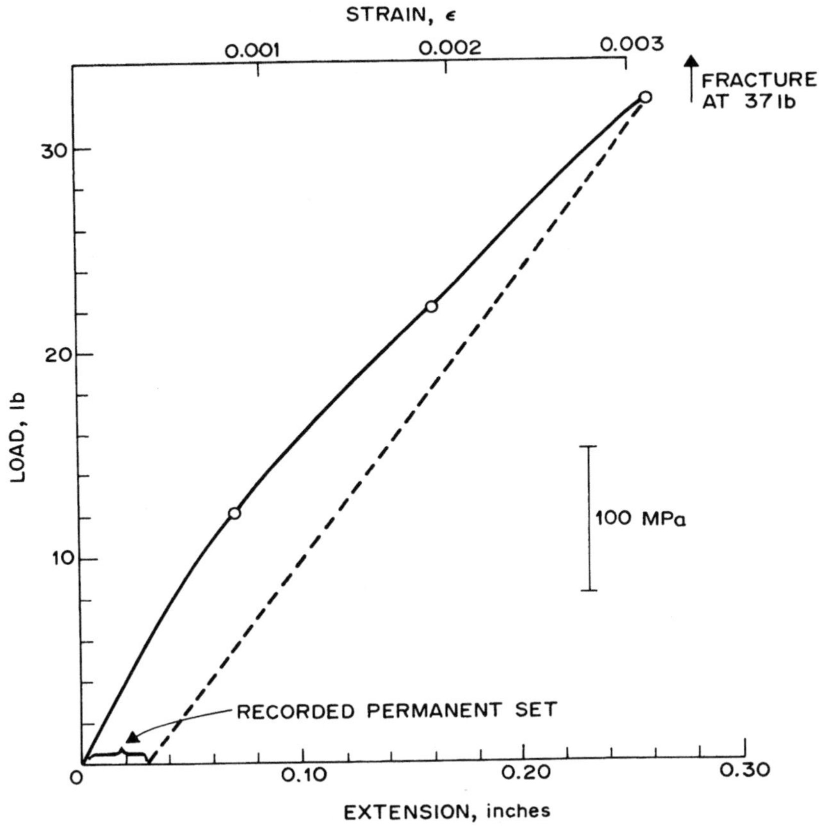

FIG. 1. Hodgkinson [18] gave sufficient information for a calculation of Young's modulus, a parameter not in use at the time.

i.e. the plastic strain, against log σ gives a straight-line fit for most metals and alloys. This formula is examined extensively for fatigue deformation in Chapter 2.

By the end of the first quarter of the nineteenth century, investigators were also aware that the elastic modulus appeared to be relatively insensitive to alloying, heat treatment and plastic deformation.

1.2.3.2. Measurements in Compression

Early theories of strength supposed that solids were incompressible. In any case, given the likelihood of buckling, it is improbable that uniaxial

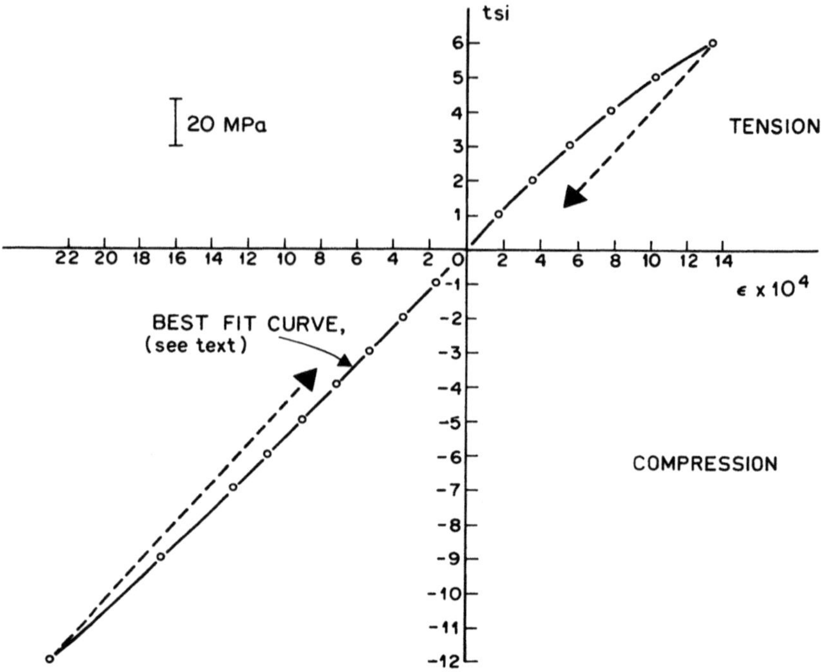

FIG. 2. Only by testing very long bars could Hodgkinson obtain measurable strains in compression.

specimens could have been made long enough for realistic measurements to have been taken. In considering bending, Parent (1666–1716) supposed that elastic properties in tension differed from their counterparts in compression. It was again left to Hodgkinson in about 1840 to determine a stress–strain curve for 10 ft (3 m) long bars of cast iron [19]. His results, quoted by Unwin [5, 20] are plotted in Fig. 2 together with the relation deduced by Hodgkinson:

for tension: $\sigma = 6220\varepsilon - 1\,298\,000\varepsilon^2$
for compression: $\sigma = 5773\varepsilon - 233\,500\varepsilon^2$

where σ is expressed in tons/sq. in.

If desired, Fig. 2 may be regarded as a forerunner of a single-cycle hysteresis loop: gauge lengths in the meantime have tended to become somewhat shorter!

1.3. EXPERIMENT AND DESIGN

Some of the investigations listed in Table 1, for example, were undertaken as a scientific study of the properties of solids. A history of the testing of the materials of construction would no doubt begin with the work of Musschenbroek (1692–1761), who invented a tensile lever machine for deforming small samples [2]. By this time ($\sim$1729) the idea that the properties of a small specimen could reflect those of larger structures seems to have been firmly implanted, despite the unavailability of the stress concept. Apparently Buffon, who tested iron bars and was a contemporary of Musschenbroek, felt that small specimens (of wood) were unrepresentative and preferred to test beams up to 28 ft (8·5 m) long [2].

With the invention of the screw cutting lathe in 1797 leading to ease of specimen manufacture and better surface finish, the way was clear for more accurate experimentation and reproducibility of results. Once the difference between elastic limit and ultimate tensile strength was appreciated, a more rational approach to design could then be made, as Navier was urging in 1826 [2]. The traditional method of avoiding failure in service was to multiply known breaking loads (latterly stresses) in laboratory tests by a factor of safety. Starting with a factor of one tenth in 'prehistory' [7] the joint requirements of avoiding cumbersome appearances and of using materials more economically reduced this to one half of the elastic limit (Poncelet), and then to one third of the UTS (Telford) [2]. Where dynamic action was involved, the breaking weight was to be divided by 'six times the live load added to three times the dead load' [5]. In 1858 the Board of Trade actually adopted an absolute value of 5 tons in^{-2} (77 MPa) for wrought iron structures. Some of these factors are very conservative.

Because of such uncertainties, large-scale static and dynamic tests of model parts of structures were undertaken to aid design, principally in the field of railway engineering [21]. The notion of exactly simulating service loads on structures (or even specimens) of course has persisted to this day under the term 'spectrum loading'.

None of these measures employed strain as a design criterion. The accurate measurement of strain on more reasonably sized specimens has occurred fairly late in materials testing. By 1860, for example, Kirkaldy was only able to state that total elongation in typical tensile tests was uniform in the plastic region (that is, until necking occurred) but the measuring technique was too insensitive for a determination of modulus. However, the passage of some 20 years saw the invention of the mirror extensometer by

Bauschinger, subsequently modified by Martens so that deflections of 10^{-5} mm were measurable [5]. From this time onwards, accurate determination of the laws of deformation was possible both in tensile and reversed-stress testing.

1.4. FATIGUE TESTING

The sequence of events in fatigue testing has taken a similar path to that in unidirectional testing, though over a shorter time period. First, there is the observation that repeated loading causes fracture. Then, experiments of an apparently elastic nature are carried out, resulting in an endurance law due to Basquin [22]:

$$N_f^{\alpha_1} S = \text{constant} \tag{3}$$

where N_f is the cycles to failure, S is the semi-stress range and α_1 is a constant. Finally, with the advent of accurate extensometry, the analogous case for elastic–plastic fatigue (i.e. where hysteresis is involved) was provided by the Coffin–Manson [23, 24] relation:

$$N_f^{\alpha_2} \varepsilon_p = \text{constant} \tag{4}$$

where ε_p is the plastic strain range and α_2 is a constant.

It should be noted that eqns (3) and (4) are *predictive* laws, i.e. they specify failure at some future time (or cycles). The *deformation* relation in fatigue,

$$\Delta\sigma = A\varepsilon_p^{\beta} \tag{5}$$

where $\Delta\sigma$ and ε_p are the *ranges* of stress and plastic strain is the cyclic equivalent of eqn (2), and is now about 35 years old.

High temperature high-strain fatigue testing has also taken place within this period, as compared to about 100 years of high temperature unidirectional testing [1], though an early direct stress machine in 1925 could operate at temperatures up to 700°C [1].

1.4.1. Traditional Methods

1.4.1.1. Elastic Regime

The work of Wöhler, undertaken between 1850 and 1870, is generally recognised as the first which paid strict attention to the magnitude of stress ranges and which was sufficiently accurate to form the first basis for the scientific design of structures against fatigue. A good account of the development of the series of tests is provided in books by Unwin [5] and

Gough [25]. Usually, reversed stresses were obtained by torsional or rotating bending methods, the latter simulating the behaviour of railway axles in service. These specimens were some 35 mm in diameter. Direct stressing was also employed, but only in the repeated tension mode. A contemporary machine of Osborne Reynolds was able to deliver both thrust and tension (but in unequal ratios) on specimens of $\frac{1}{2}$ in (12·5 mm) gauge length and $\frac{1}{4}$ in (6·3 mm) diameter.

It is important to note that, at this period, the control mode in these machines was stress (strictly, load). Wöhler began his investigations with full-scale tests [26], proceeded to model axles and concluded with laboratory specimens [3], all of which produced the first S–N failure curves. Wöhler attempted to relate fatigue strength with the elastic limit but could not, of course, measure such small deflections directly. In resorting to a bend method he became the first to deduce the presence of residual stresses [2], and he concluded that the initial plastic deformation did not affect the subsequent stress range during fatigue.

1.4.1.2. Elastic–Plastic Regime

The development of the fatigue *failure* law thus waited upon accurate testing equipment provided by Wöhler. Further progress on *deformation* behaviour in fatigue was in turn impossible without the accurate extensometry (in both axial and transverse directions) which Bauschinger, for example, employed in his experiments. Further, by about 1880 (a period of rapid growth in testing laboratories), uniaxial specimens could be rigidly gripped and thus taken into compression, and Bauschinger was the first investigator to observe metal deformation behaviour under cyclic conditions [2, 5, 25]. He found that the tensile and compressive limits of elasticity ('primitive' values) could be raised by cycling to their 'natural' values or, conversely, lowered if the material had seen significant working during manufacture. Nowadays, we would recognise these processes as cyclic hardening or softening, respectively, and would be more interested in the variation of peak stresses rather than the elastic limits. Bauschinger also observed that the compressive yield point was lowered with respect to the initial loading in tension, and this effect is discussed more fully in Chapter 2. For many years after these tests there was much experimentation and discussion on whether, as Bauschinger suggested, the 'natural' elastic limits represented a safe fatigue range.

Some 30 years elapsed before Bairstow conducted a classic series of tests on the hysteresis of commercial iron and steel. The author claimed [27] that 'the measurement of the length of the specimen under the actual conditions of test has constituted an important new feature in the history of fatigue testing'. It must be noted that load was still the control parameter, so that the progressive length changes observed during the experiment would nowadays be classed as ratchetting. Some of Bairstow's hysteresis loops were therefore not quite closed, but his technique of gradually increasing the load range on a single specimen is immediately recognisable as an incremental step test (see Chapter 2). As a result, Bairstow was able to deduce a stress–strain law:

> 'If the widths of the hysteresis loops are plotted as ordinates [sic], on a range of stress base, it will be seen that the curve obtained is approximately a straight line.... As the ordinate is zero at the intersection, the hysteresis loop must disappear... [i.e.] the specimen must be completely elastic.'

In fact, Bairstow's results obey eqn (5), as Morrow has shown [28].

Bairstow's specimens were of $\frac{1}{2}$ in (12·5 mm) gauge length with a $\frac{3}{8}$ in (9·5 mm) radius at the shoulders, and were of 0·17–0·25 in (4·3–6·5 mm) diameter. Some of his loops, with the axes plotted conventionally, are shown in Fig. 3. It is assumed that the extensometer is attached directly to the gauge length (a description of the machine is provided in Gough's book [25]). Amongst other observations, Bairstow noted that any permanent extensions were to be ascribed to stress *range* (not maximum), that there was evidence of an attenuation in ratchetting (i.e. concomitant hardening), that the plastic strain range (referred to as 'cyclical permanent set') decreased with cycles for an axle steel but increased for a Bessemer steel and, finally, that the hysteresis loop was not necessarily a sign of failure, an early reference to the 'limited life' philosophy adopted in today's design approach. Bairstow's paper does not date with re-reading.

Other contemporary papers include those of Smith and Wedgewood [29], who defined the cyclic state as when the cyclic yield stresses in tension and compression became equal, and Hopkinson and Williams [30], who by calorimetric measurements showed that the energy dissipated during fatigue increased approximately as the fourth power of the stress range. From the arguments of Chapter 2, it is possible to show that this predicts a value of $\beta = 0{\cdot}33$ in eqn (5).

From this period until the beginning of constant strain range experiments in the early 1950s, there seems to have been a lull in high strain fatigue testing (but see Appendix). Simple laboratory tests were undertaken for assessing component performance and there was a return to testing

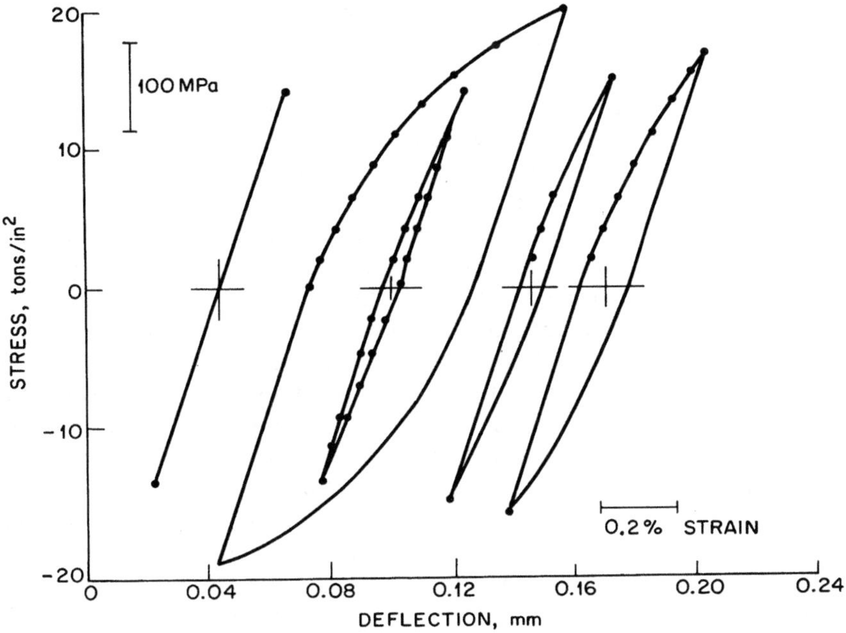

FIG. 3. Some of Bairstow's results on axle steel [27] replotted conventionally with modern units added.

scale- and full-sized models where the simple methods were found wanting [31]. Physicists and metallurgists concentrated on internal structural behaviour during fatigue, but much time was apparently wasted in exploring accelerated methods for determining a fatigue limit [3]. Valuable empirical endurance data were gathered on new or improved materials, but only after the Second World War was there a move to test simple specimens or structures under exactly simulated service loads in the high temperature regime, in order to design to a finite life.

1.4.2. The Beginnings of Constant Strain Testing

When extensometers can be attached directly to the specimen gauge and the resulting test carried out between limits of constant displacement, these special conditions give rise to the *strain-controlled test.* If for any reason extensometers have to be attached remote from the gauge, then the same *displacement-controlled* conditions do not necessarily produce a constant strain range at the gauge because of strain redistribution in the shoulders of the specimen and grips etc. Nevertheless, freedom from gross ratchetting is guaranteed in this latter, less exacting test.

In his book of 1924 Gough [25] mentioned a 'specific strain' testing machine, but did not quote any experimental results. The first steps in modern displacement control were taken in the USA by Townsend [32] who, in simulating the behaviour of lead cable in service, performed tests in U-shaped bending, vibration and cantilever bending of a specially tapered strip. In the latter two tests, a variable throw cam was employed. Bending was in one direction only, i.e. the compressive region was not deliberately explored. In the subsequent discussion Whittemore [32] stated that 'several years ago' he tested the fatigue endurance of sheet duralumin by bending the specimen around a cylinder in one direction, then unwrapping it and bending it around a cylinder of the same diameter in the opposite direction. Shortly afterwards, Dunsheath and Tunstall in England [33] vibrated lead strips 'backwards and forwards' and were able to produce a plot of displacement amplitude against time to failure.

In these tests, therefore, a conscious effort was being made to reproduce the service condition as one of repeated displacement rather than stresses. However, it may well be that this task had already been unwittingly performed in a much earlier era, resulting from the activities of a commission set up in 1847 inquiring into the suitability of iron as a structural material for railway applications [21]. James and Galton tested an iron bar in three-point bend [2], deflection at the centre being caused by a rotating eccentric. Fairbairn's celebrated test (1860–61) on a built-up wrought iron girder some 22 ft (6·7 m) long, again in bending, was also actuated in this way [2, 25]. The conclusion 'that wrought-iron girders, when loaded to the extent of a tensile strain of 7 tons per in^2 [sic], are not safe...' shows that the terms stress and strain were still being applied somewhat loosely.

In modern times, strain-controlled tests are performed to simulate the repeated thermal strains which arise when power plant is started and stopped. The first tests in 1954 employed stainless steel tubes which were constrained while the temperature was cycled between given limits [23] but, since then, the preferred mode is to test isothermally and simulate thermal strains by mechanical means, since this makes the recording of the hysteresis loop much easier (but see Chapter 3). The first creep–fatigue results appeared in about 1963 [34] and it is interesting to note that these were displacement-controlled rather than strain-controlled tests. Progress over the past 25 years will be documented in other pages of this book, especially Chapter 6.

1.5. PREDICTION

1.5.1. Empirical Approach

Hooke's law is an early example of a simple deterministic relation derived from experimental data. For many years, properties of constructional materials were considered to be time-independent, although Marriotte supposed the discrepancy between his calculated fracture load in bend, and the experimental tensile value as due to a 'time effect', and Girard in 1798 found that deflection of wooden struts increased with time under load [2]. The first numerical prediction came in 1834 from Vicat's work on wires [13]: he showed that 'the speeds of subsequent extension are very approximately proportional to the times' so that 'a suspension bridge, the cables of which were loaded above $\frac{1}{4}$ [of their UTS] could... sink continually from one year to the next'. In the field of fatigue, following Hodgkinson's experiments, an early theory assumed that stress reversals caused the UTS to deteriorate with time [25]. Erroneous then, it has

TABLE 2
SIGNIFICANT EVENTS IN CREEP STUDIES [35, 36]

Investigator(s)	*Date(s)*	*Comments*
Vicat	1834	Creep at finite stress (limited life design?)
Le Chatelier	1890	Decrease of strength as strain rate decreased
Thurston	1895	First to present creep curve; stress relaxation
Hopkinson & Rogers	1905	Studied creep at high temperature
Phillips	1905	
Andrade	1910–1914	Primary, secondary, tertiary creep
Chevenard	1919	Characterised creep for service materials in chem-
Dickenson	1922	ical and power industry
Bailey	1926	Hardening and recovery creep
McVetty	1934	Analysis of creep–time curve
Barr–Badgett	1935	Attempt to predict long-term strength from short term tests
—	1945 onwards	Stress rupture plots; curves for 0·1%, 0·2%, etc., strain
Larson–Miller	1952	Extrapolation based on rate process theory (activ-
Sherby–Dorn	1954	ation energy for creep)
Manson–Haferd	1953	Logarithmic extrapolation
Monkman–Grant	1956	Empirical ductility–strain rate relation (may also be derived from microstructural considerations)
—	1959 onwards	Cavity nucleation and growth theories and effect upon creep curve

ironically become correct now, for we know that the cyclic yield stress of certain alloys decreases during exposure to high temperature fatigue.

With the advent of high temperature uniaxial creep testing, empirical laws capable of extrapolation were developed, so that specimen times to rupture or elongation could be predicted [35, 36] and applied to service. A chronological summary is provided in Table 2. It is considered [7] that parametric methods of extrapolation have probably reached their limit, so that future predictions of remanant life must be based on the relations which govern microstructural events.

The introduction of a creep dwell period into the high strain fatigue cycle also causes great difficulties in empirical extrapolation. First, testing is expensive and more complex than creep testing alone, so data are not readily available. Secondly, laboratory testing of specimens at low strains and long dwells approaching the service condition leads to impossibly long testing times. Thirdly, the time-dependent failure mechanism means that, in eqn (4), cycles to failure generally decrease non-linearly as dwell times are increased, the variation also depending upon strain level.

When one service cycle can be identified with a laboratory cycle, then accelerated testing in fatigue provides a means of predicting a failure time if material properties are time-independent. Fairbairn's beam test was

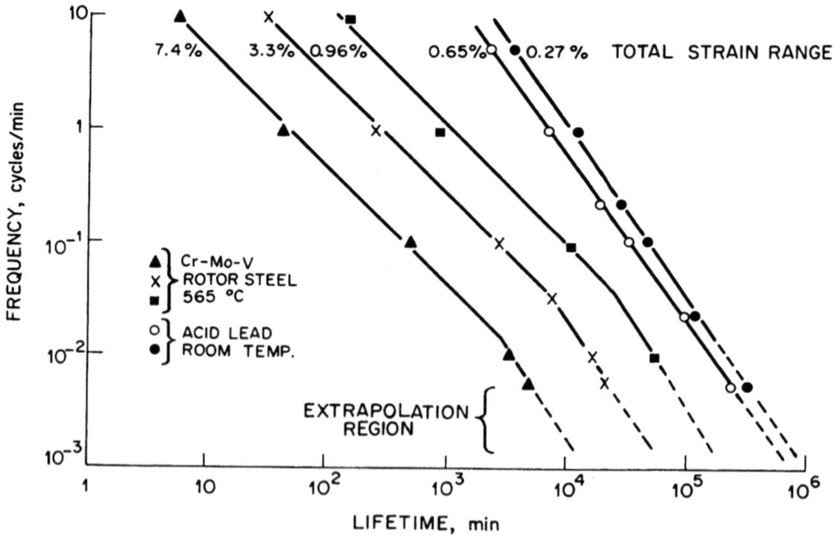

FIG. 4. Attempts at empirical extrapolation may be complicated by changes in slope.

conducted at 8 cycles per minute and endured $1{\cdot}2 \times 10^7$ reversals. For 100 service loadings per day he thus predicted a lifespan of 328 years [25]. Relations such as eqn (3) or (4) are therefore predictive but do not include a frequency effect. This was provided by Eckel, who was concerned with the lifetime of slowly cycled lead cables [37]:

$$\nu^m t_f = N_f/\nu^{1-m} = \text{constant} \tag{6}$$

In this relation ν is the frequency of testing, t_f is the failure time and m is a constant so that, at a given strain level, lifetimes at lower frequencies can be predicted. Some of Eckel's results are shown in Fig. 4. For high temperature alloys, attempts at extrapolation are hampered by a change in the slope m (see Fig. 4) which is due to a change in the fracture mechanism [38].

The limitations of empirical extrapolation methods have thus forced metallurgical engineers to adopt measures based on simple physical arguments coupled with careful metallography, and models based on microstructural changes which may occur in service. These matters are pursued in Chapters 4 and 5.

1.5.2. A New Field for Materials Science

1.5.2.1. Application of Microstructural Theory

In 1902 Ewing and Humfrey [39] published the first paper in which the microscope was directed towards the surface of a specimen which had been deformed in fatigue. They observed that slip activity was present below the conventional elastic limit and that slip bands were the sites of embryo cracks.

This investigation at once showed the importance of metallography in interpreting results or in deciding the future course of a test programme. A modern equivalent would be the polishing and etching of a test piece in order to verify that a given period of creep/fatigue had produced intergranular cracking rather than a transgranular mode. Metallography can thus help trace boundaries between different mechanisms, but a complete predictive theory requires modelling using time-dependent relations upon which this particular branch of Materials Science is founded. Examples are the Arrhenius relation for thermally activated events (rate process theory), the equations of diffusion, point defect theory, laws of oxidation and precipitation, and nucleation theory. All of these may be used in assessing microstructural 'damage' in which grain boundary cavitation and particle growth play a large part.

It may be noted that 'modelling' of material behaviour (in terms of a fracture damage parameter) can be based either on continuum concepts or on those of metal science [40]. Ashby and Dyson have recently remarked [41] on the similarities which exist in the expressions, for example where lifetime is defined by the growth of cavities.

The title for this section is borrowed from a very percipient study by McLean [7] who has outlined the direction Materials Science should be taking in order to reduce the safety factors still applied in modern engineering design. He argues that 'the present state of Materials Science resembles that of elasticity early in the 19th century, when a theory had been evolved but had scarcely reached designers', so that by analogy it should now be a powerful ally for design and life prediction. The understanding that has now been acquired between microstructure and mechanical properties in simple laboratory tests must now be applied in earnest to the following technological problems [7].

(a) Reducing the amount of expensive laboratory testing in which combinations of properties are critical, e.g. creep–fatigue.
(b) Predicting future life from known events on the microstructural scale with assistance from quantitative metallography.
(c) Extrapolating from simple laboratory tests to the fluctuating multiaxial stresses in service.

Following McLean's depiction, the progress achieved in relevant fields is summarised in Fig. 5. Although the selection is somewhat arbitrary (a full diagram would include thermodynamic studies, alloy systems and materials development) the figure shows how the barriers between the various disciplines must disappear if progress is to be made on a unified front. We have now entered that borderland of physics and engineering where 'should a Boundary Commission be appointed, I for one would decline to serve' [42].

The flowchart in Fig. 6 suggests a typical route that may be followed from the design of a component or feature to an assessment of its probable life. The traditional role of a metallurgical engineer in this scheme is suggested at point A, where tests are conducted if necessary and advice is given on material properties. His new role is suggested at points B, where a complete life calculation is required to be carried out. For plant which has seen service, a remanent life calculation for creep deformation, for example, would include such techniques as monitoring of structural degradation (e.g. dislocation mesh growth) and creep cavitation on the plant item, together with mechanical property tests on small samples of service-exposed material [43]. Techniques for a creep–fatigue assessment are described

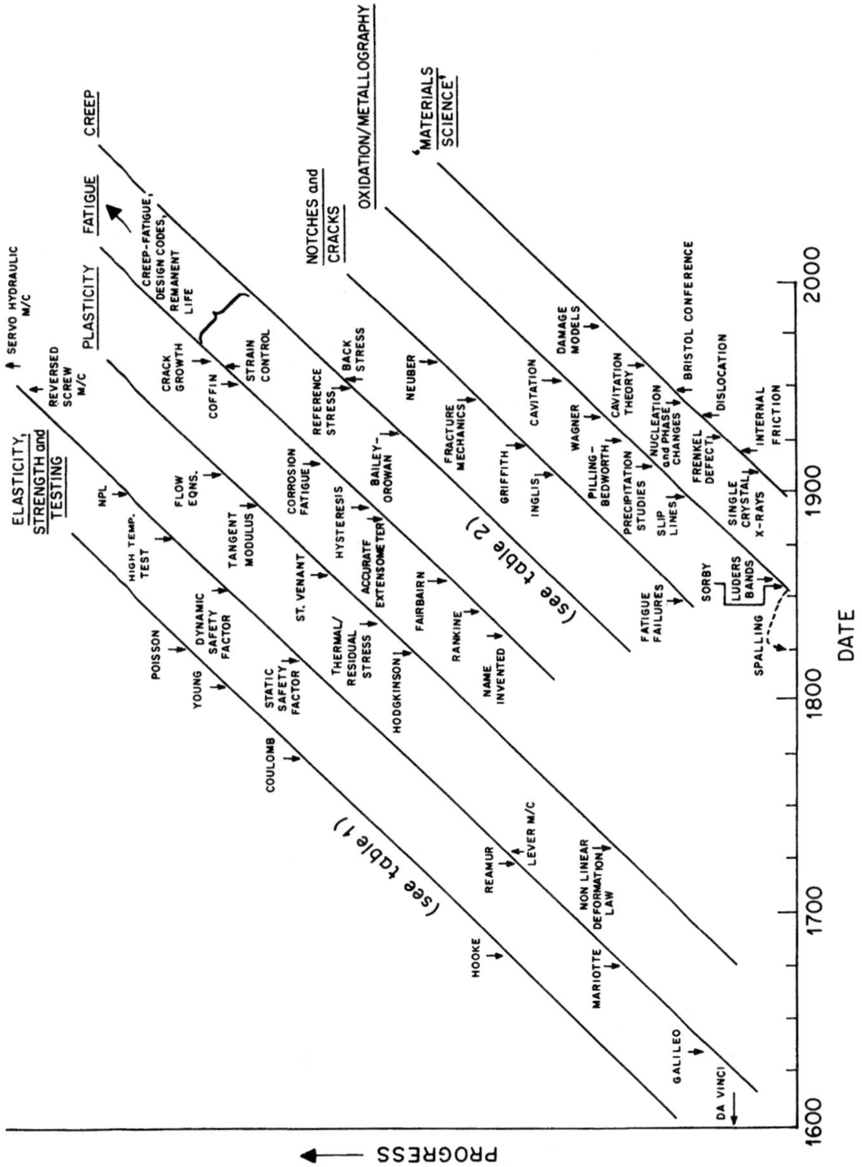

FIG. 5. Materials Science considered in relation to other disciplines.

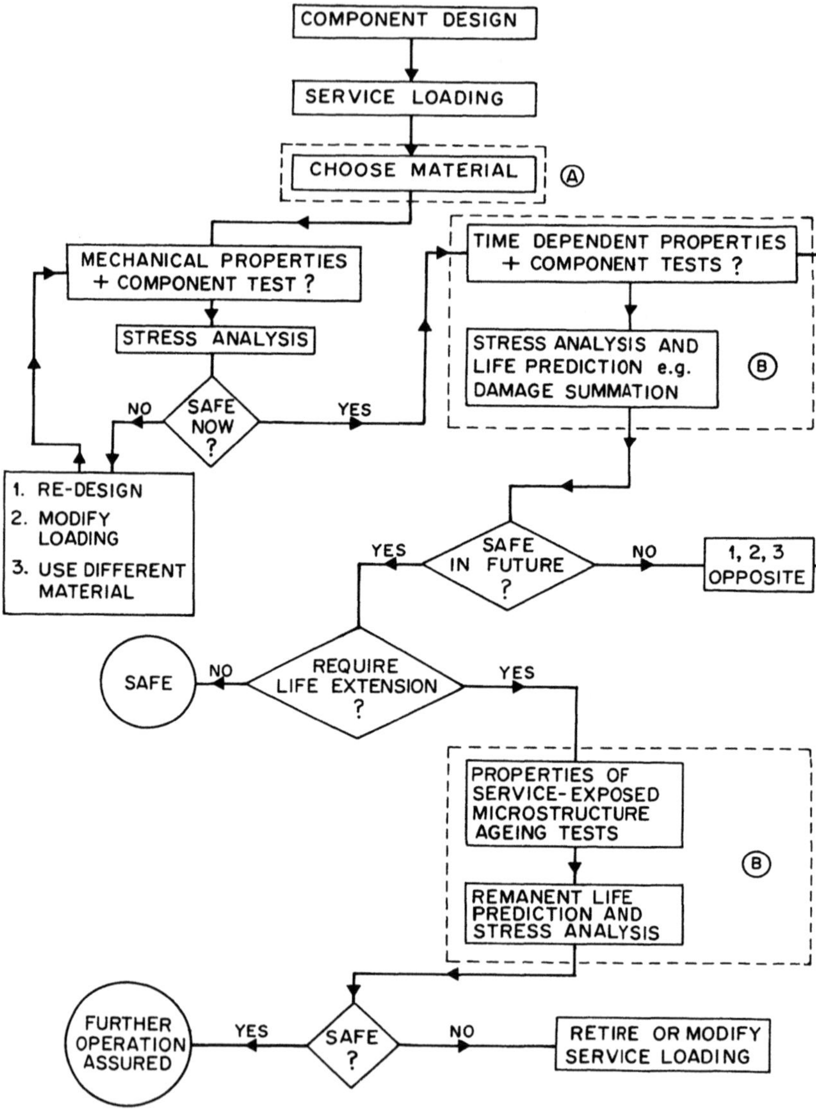

FIG. 6. The requirements of metallurgical engineering now extend beyond advising on materials.

elsewhere in this book. All such information may, of course, be used retrospectively to explain a service failure.

1.5.2.2. Reduction of Safety Factors

During the past three centuries, the accuracy of dimension measurement has increased from 1 part in 10^4 (with Hooke) through 1 part in 10^6 (Whitworth) to 1 part in 10^9 (interferometric methods) [7]. Best workshop practice has lagged several orders of magnitude behind these values [7]. In the modern materials laboratory, a typical load measurement in a uniaxial application would be accurate to 0·1%, total strain can be detected in microstrain (10^{-6}), while cyclic plastic strains are measurable down to 10^{-4} [44]. It can be appreciated that although specimen lengths have drastically reduced, the relative accuracy of measurement has still increased.

When applying carefully measured laboratory data to the practical situation further uncertainties come into play. These arise from (a) the accuracy to which stresses and strains can be calculated or measured on the plant item, (b) inherent scatter in the laboratory data themselves (for example, in cyclic stress–strain properties or in fatigue endurances), and (c) other factors, e.g. environmental effects in service. For these reasons, factors of safety are traditionally introduced, as discussed in Section 1.3.

It is significant that although the degree of precision has increased both in measuring and component manufacture, current factors of safety are no less onerous than those used in days gone by. In some applications the '2 and 20' rule is employed (a design curve which factors high strain fatigue data, i.e. strain ranges by 2 or cycles by 20, whichever gives the least endurance [45]). If welds are involved in high temperature service, a reduction factor of 4 on strain is not uncommon. It is therefore the duty of Engineering Science to attain the design factor of unity implied in the quotation at the beginning of this chapter.

REFERENCES

1. Loveday, M. S., 1982, High temperature and time dependent mechanical testing: an historical introduction, *Measurement of high temperature mechanical properties of materials*, London, HMSO, pp. 1–9.
2. Timoshenko, S. P., 1953, *History of strength of materials*, New York, McGraw-Hill.
3. Watson, P. and Hill, S. J., 1982, Fatigue life assessment of ground vehicle components, *Design of fatigue and fracture resistant structures*, STP761, Philadelphia, ASTM, pp. 5–27.

4. Todhunter, I. and Pearson, K., 1886, *History of the theory of elasticity and of the strength of materials*, Vols 1 and 2, Cambridge University Press.
5. Unwin, W. C., 1910, *The testing of materials of construction*, 3rd edn, London, Longmans Green & Co.
6. Gordon, J. E., 1978, *Structures*, Harmondsworth, Penguin.
7. McLean, D., 1976, A new field for materials science, *Mat. Sci. Engng*, **26**, 141–52.
8. Hooke, R., 1678, De potentia restitutiva or of Spring: explaining the power of springing bodies, *Cutlerian lectures*, John Martyn, Printer to the Royal Society, pp. 1–24.
9. Osgood, W. R., 1946, Stress–strain formulas, *Proc. J. Aeronaut. Soc.*, **13**, 45–8.
10. Irwin, G. R., 1964, Structural aspects of brittle fracture, *Appl. Mat. Res.*, **3**, 65–81.
11. Hall, A. R., 1970, *From Galileo to Newton 1630–1720*, London, Fontana-Collins.
12. Gordon, J. E., 1968, *The new science of strong materials*, Harmondsworth, Penguin.
13. Vicat, M., 1834, Sur l'allongement progressif du fil de fer soumis à diverses tensions, *Annales Ponts et Chaussees*, **CLXV**, 40–4. For an English translation, see Ref. 1, pp. 9–12.
14. Young, T., 1807, Lecture XIII: On passive strength and friction, *A course of lectures on natural philosophy and the mechanical arts*, Vol. 1, London, The Royal Society, pp. 135–56. See also Vol. 2, p. 46.
15. Bulfinger, G. B., 1729, De solidorum resistentia specimen, *Commentarii Academiae Scientiarum Imperialis Petropolitanae*, Vol. 4, pp. 164–81.
16. Riccati, J., 1731, Verae et germanae virium elasticarum leges ex phaenominis demonstratae, *De Bonienci Scientarum et Artium Instituto Atque Academia Commentarii*, Vol. 1, pp. 524–44.
17. Forbes, R. J. and Dijksterhuis, E. J., 1963, *A history of science and technology*, Harmondsworth, Penguin.
18. Hodgkinson, E., 1822, On the transverse strain and strength of materials, *Mem. Proc. Manchester Lit. Phil. Soc.*, **4**, 225–89.
19. Hodgkinson, E., 1846, Experimental researches on the strength and other properties of cast iron, *Defect of elasticity*, British Association meeting in Cork, 1843, pp. 407–11.
20. Unwin, W. C., 1886, An old puzzle about the strength of beams, *The Engineer*, 29 October, p. 351.
21. *Report of the Commissioners Appointed to Enquire into the Applications of Iron to Railway Structures*, 1848, London, William Clowes and Sons Ltd.
22. Basquin, O., 1910, The exponential law of endurance tests, *Proc. ASTM*, **10**, 625–30.
23. Coffin, L. F., 1954, A study of the effects of cyclic thermal stresses on a ductile metal, *Trans. ASME* (*Ser. A*), **76**, 931–50.
24. Manson, S. S., 1954, Behaviour of materials under conditions of thermal stress, NACA Tech. Note 2933.
25. Gough, H. J., 1924, *The fatigue of metals*, London, Scott, Greenwood & Son.
26. Petersen, R. E., 1963, Fatigue of Metals, Part 3—Engineering and design aspects, *Mat. Res. Standards*, **3**, 122–39.

27. Bairstow, L., 1909–1910, The elastic limits of iron and steel under cyclical variations of stress, *Phil. Trans. R. Soc. A*, **210**, 35–55.
28. Morrow, J. D., 1965, Cyclic plastic strain energy and fatigue of metals, *Internal friction, damping and cyclic plasticity*, STP 378, Philadelphia, ASTM, pp. 45–87.
29. Smith, J. H. and Wedgewood, G. A., 1915, Stress–strain loops for steel in the cyclic state, *J. Iron Steel Inst.*, **XCI**, 365–97.
30. Hopkinson, B. and Williams, G. T., 1912, The elastic hysteresis of steel, *Proc. R. Soc.*, **87A**, 502–11.
31. Mann, J. Y., 1967, *Fatigue of materials*, Melbourne University Press.
32. Townsend, J. R., 1927, Fatigue studies of telephone cable sheath alloys, *Proc. ASTM*, **27**, 153–72.
33. Dunsheath, P. and Tunstall, H. A., 1928, The Physical properties of lead cable sheaths, *J. Inst. Elect. Engrs*, **66**, 280–9.
34. Walker, C. D., 1965, Strain-fatigue properties of some steels at 950°F (510°C) with a hold in the tension part of the cycle, *Joint Int. Conf. on Creep*, Inst. Mech. Engrs, London, Vol. 178, pp. 3.49–3.56.
35. Finnie, I. and Heller, W. R., 1959, *Creep of engineering materials*, New York, McGraw-Hill.
36. Smith, G. V., 1950, *Properties of metals at elevated temperatures*, New York, McGraw-Hill.
37. Eckel, J. F., 1951, The influence of frequency on the repeated bending life of acid lead, *Proc. ASTM*, **51**, 745–60.
38. Coles, A. and Skinner, D., 1965, Assessment of thermal fatigue resistance of high temperature alloys, *J. R. Aeronaut. Soc.*, **69**, 53–5.
39. Ewing, J. A. and Humfrey, J. C. W., 1902, The failure of metals under repeated alterations of stress, *Phil. Trans. R. Soc. A*, **200**, 241–50.
40. High Temperature Materials—Modelling Service Behaviour, NPL Conference, 18–19 June 1985.
41. Ashby, M. F. and Dyson, B. F., 1984, Creep damage mechanisms and micromechanisms, NPL Report DMA(A), March 1984.
42. Ewing, J. A., 1933, *An engineer's outlook*, London, Methuen.
43. Cane, B. J., 1983, Estimating the remanent creep life of power plant components, *High Temp. Technol.*, **4**, 215–28.
44. Lukas, P. and Klesnil, M., 1973, Cyclic stress–strain response and fatigue life of metals in low amplitude region, *Mat. Sci. Engng*, **11**, 345–56.
45. Langer, B. F., 1962, Design of pressure vessels for low cycle fatigue, *J. Basic Engng*, **84**, 389–402.

APPENDIX

It is remarkable that though the technology was available in the eighteenth century to measure quite small expansions of solids, none of it was transferred to the measurement of mechanical deformation. Réamur in 1722 noted a 1·4% expansion of a 300 mm iron bar from ambient temperature to red heat [A1]. Derham in 1736 [A2] from work done 20 years earlier had deduced a contraction of 0·08% from the change in frequency of a seconds pendulum ($\sim$1 m long) when cooled in a freezing mixture. Musschenbroek in 1731 could measure, using a dial gauge, the expansion of a 147 mm long bar to within 0·002 mm, i.e. to a strain of 0·0014% [A1]. Similar dilatometers due to Ellicot [A3] in 1735, Smeaton [A4] in 1754 and Ramsden (in a subsection of Ref. A5) in 1785 could detect movements of 0·0035 mm, 0·0013 mm and 0·0035 mm, respectively.

The first hand-held screw micrometer has been attributed to James Watt in 1772 [A6]. Henry Maudslay (1771–1831) subsequently developed a similar instrument accurate to 0·0025 mm while, in the early nineteenth century, Joseph Whitworth achieved a sensitivity one hundred times better than this [A6]. Not until the work of Bauschinger do we see the benefits of metrology applied to the mechanical testing of materials.

The first testing machine ever has been accorded to Réamur in 1722 [A1]. A quenched iron wire was clamped in a vice and was gradually deflected by a screw device until the wire broke. The advance of the screw was noted on a linear graduated plate as a measure of the 'flexibility' of the wire, i.e. the more the advance, the more the 'body' of the testpiece. No dial gauge appears to have been used to increase the sensitivity.

Réamur also noted that flexibility of the wire could also be determined by means of added weights, via a cord and pulley arrangement. This method would clearly measure the bending strength of the material and so, even in this earliest of testing machines, one had a choice of load or displacement control!

A notable mid-period paper (Section 1.4.1.2) appeared in 1929 [A7]. Using hysteresis loop measurements, the authors tested sections from a 150 mm diameter railway axle which had been in service some 35 years (cycles unspecified). The outer layers, which had seen greater stress reversals during life, gave a closed hysteresis loop, the plastic strain range of which was less than material near the centre, tested under the same conditions. In the latter case, the hysteresis loop took several cycles to close. It was thus verified that cyclic hardening had occurred in the outer layers of the axle during service.

REFERENCES

A1. Sisco, A. G., 1956, *Réamur's memoirs on steel and iron* (English translation of original printed in 1722), Chicago, University of Chicago Press.

A2. Derham, W., 1735–1736, Experiments concerning the vibrations of pendulums, *Phil. Trans. R. Soc.*, **39**, 201–3.
A3. Ellicot, J., 1735–1736, The description and manner of using an instrument for measuring the degrees of the expansion of metals by heat, *Phil. Trans. R. Soc.*, **39**, 297–9.
A4. Smeaton, J., 1754, Description of a new pyrometer with a table of experiments made therewith, *Phil. Trans. R. Soc.*, **48**(II), 598–613.
A5. Roy, W., 1785, An account of the measurement of a base on Hounslow Heath, *Phil. Trans. R. Soc.*, **75**, 385–481.
A6. Hume, K. J., 1980, *A history of engineering metrology*, London, MEP Publications.
A7. Smith, J. H., Connor, C. A. and Armstrong, F. H., 1929, The correlation of fatigue and overstress, *J. Iron Steel Inst.*, **120**, 267–95.

2

Cyclic Stress–Strain Properties During High Strain Fatigue

R. P. SKELTON

Central Electricity Research Laboratories, Leatherhead, Surrey, UK

SUMMARY

This chapter reviews an important aspect of high strain fatigue testing, namely the cyclic stresses generated when experiments are conducted at a given total strain range. It has in mind the design engineer, who may wish to use more sophisticated data in an analysis, and the materials scientist who has before him a wide open experimental field at elevated temperature. After an introduction to the basic concepts of cyclic hardening or softening, there follows a section on definitions of parameters used throughout the chapter, particularly the Ramberg–Osgood relation as modified for the cyclic case. Testing methods are then briefly examined and alternatives such as constant plastic strain range experiments considered. The multifarious ways of presenting cyclic stress–strain data are next investigated and a 'best' (i.e. least confusing) method recommended.

The properties of the hysteresis loop itself are then analysed in some detail and, amongst other relations, it is shown how the cyclic hardening exponent is related to energy changes, friction stress and back stress, and the Bauschinger effect.

In the next two sections, typical high temperature alloys are classed either as hardening or softening, and several empirical relations of these changes with cycles are reviewed. The saturation stress–strain values for these materials are then summarised most efficiently in a series of tables from which the cyclic stress–strain curve for each material can be generated for several temperatures and strain rates.

After reviewing other effects which are relevant to the situation in service (strain rate, temperature, sequential effects, indentation hardness changes etc.) there follows a very short list of the microstructural causes responsible for many of the observed features (this topic is dealt with in another chapter). Passing fairly swiftly over typical engineering representations of actual behaviour, it is then shown how laboratory data are applied in calculating the strain concentration at a notch root. Three methods are examined.

The chapter concludes with some general observations from laboratory experiments that could profitably be transferred to design and analysis.

2.1. INTRODUCTION

About 30 years ago, following the original experiments of Coffin [1], endurance curves at high temperature began to appear with strain range replacing the stress range which had characterised the fatigue tests of the previous 100 years. This trend arose because the cycles of interest spanned from 10^2 to 10^5, and they reflected the more arduous duty which high temperature plant was expected to undertake. In order to achieve failure at these relatively low endurances, the metal must yield during each reversal and it was realised that stress is a very insensitive parameter for describing fatigue behaviour. Further, to avoid changes in the testpiece length, total strain limits were used to *control* the fatigue experiment, and it is such a common technique nowadays that only the briefest of descriptions is necessary in papers on high strain fatigue (HSF) investigations.

Data are either plotted in terms of *plastic strain range* (width of the hysteresis loop at zero stress) against number of cycles or *total strain range* versus those cycles. Very generally, the former method would be useful for a research investigation and, indeed, led to the original endurance law [1], while the second method has more application in design and assessment. Both have been the mainstay of many hundreds of papers in the elevated temperature, HSF regime in the intervening years. An example (the only one to appear in this chapter) is given in Fig. 1(a) [2].

The cyclic stress–strain data which are generated in the course of a strain-controlled fatigue test are not so often emphasised, and this is unfortunate in many ways. Figure 1(a), which demonstrates the near-independence of endurance on plastic strain in the case of copper over the range −195°C to 650°C, might suggest that the material is equally strong at all temperatures. However, from the corresponding cyclic stress–strain properties given in Fig. 1(b) [3], it is seen that Fig. 1(a) disguises a sharp drop in stress range above 300°C, so that a given plastic strain range is achieved at much lower stress levels.

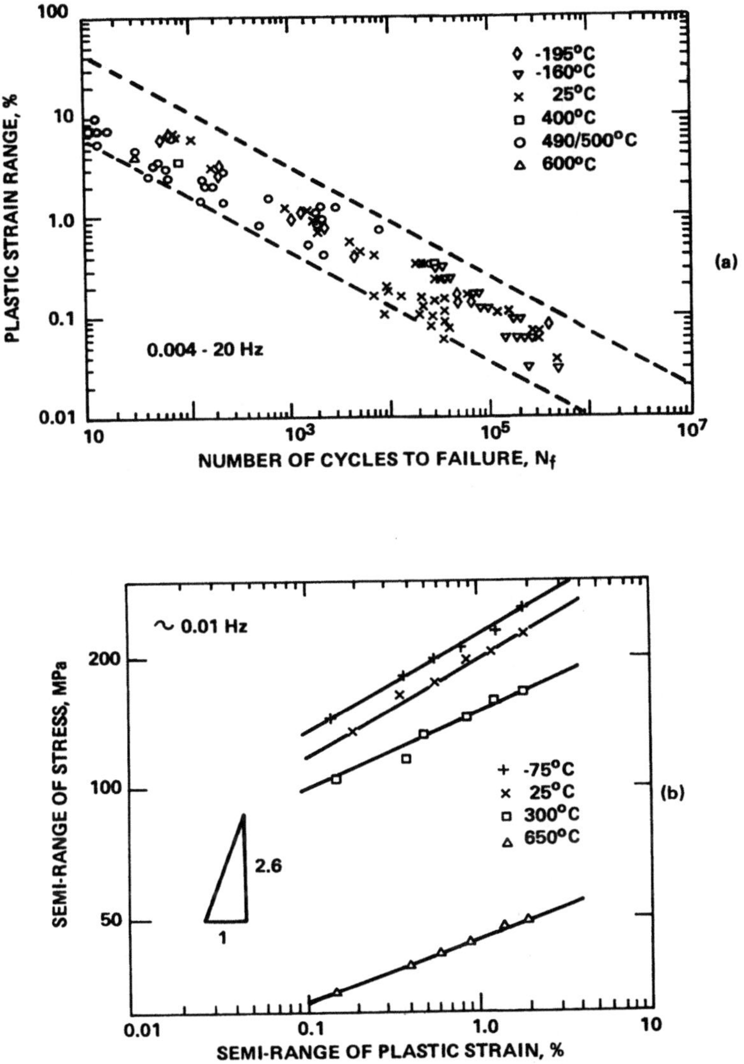

FIG. 1. Similarities in the endurance of OFHC copper at several temperatures (a), mask a much larger variation in cyclic strength (b). (After Refs 2 and 3 reproduced with permission of ASM and Fatigue of Engineering Structures, Ltd.)

Cyclic stress–strain data are thus not as familiar as endurance plots, but their importance should not be underestimated. To the research worker, who observes that well annealed simple alloys normally cyclically harden while cold-worked materials usually soften (these terms to be defined in Section 2.2), such behaviour provides valuable pointers to the microstructural processes responsible. Some alloys indeed first harden then soften, for example annealed $2\frac{1}{4}$Cr–1Mo steel in the range 480–540°C [4, 5], depending on the strain range. Again, long-term fatigue tests at high temperature sometimes show metallurgical instabilities, accelerated ageing, and degradation of strength [6], which can all occur before the onset of a fatigue crack. Similarly, the designer needs stress–strain data when performing an initial lifetime assessment for critical locations in high temperature plant: he too should be aware of long-term strength behaviour when contemplating calculations in support of extending the life of plant. Application of the data is discussed further in Section 2.11 and also in Chapter 7.

One of the aims of this chapter is to trace the typical cyclic stress–strain response of alloys at present in use at elevated temperature. Room temperature behaviour has already been reviewed by Klesnil and Lukas [7], and also by Landgraf [8] who compared the variation of monotonic (unidirectional) yield strengths and strain hardening exponent (q.v.) with their cyclic counterparts, and discussed their variation with hardening/softening and the control the metallurgist can exert in producing the required material behaviour. It will not be possible to provide a complete source of cyclic stress–strain data, but it may be noted that tabulated values appear in the NRIM data sheets (referenced separately) and also in a book largely dealing with other matters on HSF [9].

The estimation of the likely cycles-to-failure at a specific location in a component is only the final stage of an assessment. This must be preceded by a stress analysis which calculates, among other things, the surface strain range. Its value in service is likely to change cycle-by-cycle and so, at all stages, there must be an adequate input from the experimental results of laboratory investigations. The data appearing in this survey, therefore, are obtained on simple specimens deformed in direct stressing HSF before the appearance of a crack. Results are sometimes quoted at the 'half-life' stage where it is assumed that the material has stabilised. For some alloys, cracking at this stage may well have caused a drop in stress; Ellison and Paterson reported [10] a crack depth of 0·1 mm at 10% of the endurance in a 1Cr–Mo–V steel at 565°C.

Typical strain rates during continuous cycling tests lie in the range

4×10^{-3}–10^{-6} s^{-1} (see Appendix 1). During a dwell period, however, when the stress relaxes, much lower rates can occur.

2.2. DEFINITION

In practical applications, as when installations are run up to full power and subsequently shut down, reversed deformation occurs at varying temperatures and the surface layers of components are thermally *contained.* In laboratory modelling of these layers, certain simplifications are frequently made so that testing is undertaken isothermally at the highest temperature (but see Chapter 3) between limits of constant strain ε_t, and the plastic strain per cycle ε_p is measured. The peak stresses are left to adopt their own levels and, for simple continuous cycling at least, the tension and compression values are very nearly equal at each stage. Values of plastic strain range ε_p of interest lie between 0·0002 (about the limit of detection) to 0·02 (which would be regarded as a very severe strain reversal in a component).

In a push–pull laboratory test on a smooth cylindrical specimen, the load cell output is usually plotted on the vertical (Y) axis and the extensometer (displacement) signal on the horizontal (X) axis of the recorder. The resulting figure, suitably converted to units of stress and strain, is a closed *hysteresis loop.*

2.2.1. Continuous Cycling

When a push–pull HSF test between total strain limits is commenced, it is usually found that the tension–compression stress range $\Delta\sigma$ does not remain constant, but gradually increases or decreases towards a saturation value. This process is termed *cyclic hardening* or *cyclic softening* respectively and if, in Fig. 2(a), an early loop is characterised by the tips AA, then hardening is represented by a loop such as HH. Conversely, softening reduces the original stress range to SS (Fig. 2(a)).

Hardening is usually complete within 5–10% of the life, whereas for most metals and alloys, softening is generally much more prolonged. At the saturation point the hysteresis loop is said to be *stable*, a schematic illustration being given in Fig. 2(b). Some obvious parameters connected with this loop are as follows:

(a) The total strain range ε_t, the plastic strain range ε_p and the total stress range $\Delta\sigma$, already encountered. Sometimes the semi-range

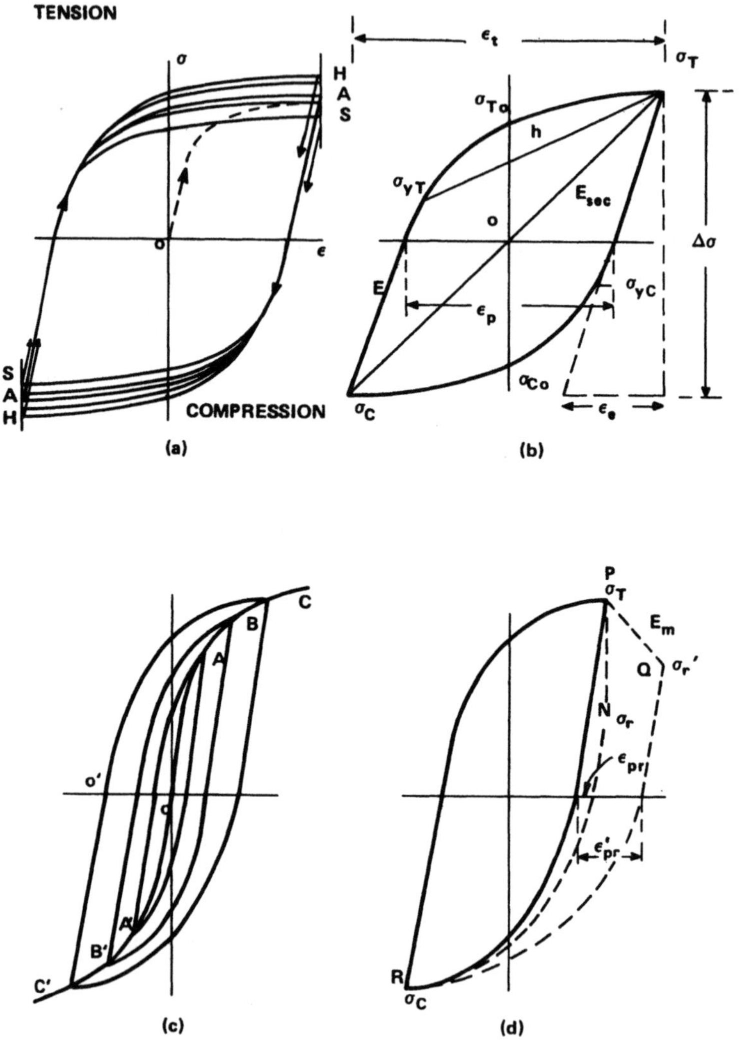

FIG. 2. (a) Hardening and softening behaviour. (b) Definitions associated with stable loop. (c) Establishment of cyclic stress–strain relation. (d) Stress relaxation behaviour.

values (i.e. *amplitudes*) of these quantities are quoted, but these can cause confusion if the peaks σ_T, σ_C cease to be symmetrical.

(b) A *yield stress* in tension, σ_{yT} and in compression, σ_{yC}.

(c) Points σ_{TO} and σ_{CO} may be regarded as residual tension and compressive stresses, respectively, when the specimen is forced to return to its original length. The span σ_{TO} to σ_{CO} is thus a *residual stress range*.

(d) *Young's modulus, E*, which is recorded during the tension and compression unloading phases. Effective moduli are defined by a *tangent modulus* E_{tan} (average value h), and a *secant modulus*, E_{sec}. The latter, for example, is used in defining a strain concentration factor [11].

Less obvious parameters associated with the loop of Fig. 2(b) may also be cited.

(e) The energy dissipated in each cycle (area of loop) together with the energy stored each half cycle.

(f) The Bauschinger strain, related to a lowering of σ_{yC} below the value for the first tensile value, σ_{yT}.

(g) Internal stresses which are, respectively, a *friction stress*, σ_f, and a *back-stress*, σ_b. These latter quantities will be considered further in Section 2.5.

2.2.2. A Cyclic Stress–Strain Relation

From Fig. 2(b) it is apparent that the total strain range is the sum of the *elastic* strain range ε_e and the plastic strain range. Since $\varepsilon_e = \Delta\sigma/E$ we have:

$$\varepsilon_t = \varepsilon_p + \frac{\Delta\sigma}{E} \tag{1}$$

Equation (1) applies for any complete cycle and so, referring to the evolutionary stages of Fig. 2(a), it follows that since ε_t is constant an increase in $\Delta\sigma$ during hardening with each cycle is accompanied by a decrease in ε_p, and conversely for softening. It is manifest that the hysteresis loop can never quite be closed until the point of stability but, in practice, the stress increments after the first ten cycles or so are quite small.

In eqn (1) therefore, the relation between ε_p and $\Delta\sigma$ depends on the exact stage of hardening or softening. In the saturation stage, however, it is possible to characterise each alloy. Suppose a stable loop AA′ has been obtained at a given ε_t (Fig. 2(c)). Upon increasing this value symmetrically about O, it is found that a new stable loop BB′ very quickly establishes itself within several cycles, and so on for further loops CC′, DD′,.... If the material is truly stable, these stages are reproduced upon repeating the strain ranges in descending order. It is then found that if the locus of stress ranges, given by the tips AA′, BB′ etc. is plotted against each corresponding total strain range, the Ramberg–Osgood relation [12, 13] is obeyed:

$$\varepsilon_t = \frac{\Delta\sigma}{E} + \left(\frac{\Delta\sigma}{A}\right)^{1/\beta} \tag{2}$$

where A and β are constants. Substituting for ε_t from eqn (1):

$$\Delta\sigma = A\varepsilon_p^{\beta} \tag{3}$$

where A is defined as the *cyclic strength coefficient* and β is the *cyclic hardening exponent*. Thus a logarithmic plot of $\Delta\sigma$ vs. ε_p should yield a straight line of slope β.

The properties of eqn (3) will be referred to many times in this chapter. For the present it may be noted that:

(a) The constant A is the hypothetical strength at 100% plastic strain range ($\varepsilon_p = 1$).
(b) The term β reflects the shape of each individual loop (see Appendix 2 and Section 2.4), and it is related to at least two other physical properties (see Section 2.5).
(c) It is possible that eqn (3) may apply even in the evolutionary stages of hardening/softening, provided that the rate of change of β and A with number of cycles can be defined adequately (see Section 2.6).
(d) Equations (2) and (3) are the most popular of many which have been proposed between stress and strain, originally derived for monotonic loading [13]. There is some evidence of a slight discontinuity in slope β below a certain plastic strain, so it is important to specify the range over which eqn (3) is to be applied (see Section 2.7).
(e) Strictly, eqn (2) predicts a finite value of ε_p, and hence yield, for any stress, but noted departure from elastic loading is found to be in good agreement with experiment.

2.2.3. Effect of Strain Rate, Dwell and Offset Limits

For symmetrical cycling, eqn (1) applies for any strain rate $\dot{\varepsilon}_t$. However, for $\dot{\varepsilon}_t < 10^{-6}\,s^{-1}$ in most materials the linear unloading phase from tension or compression (Fig. 2(b)) becomes curved due to the influence of time-dependent deformation (the material initially relaxes faster than the elastic unloading rate). In this case, the measured parameter ε_p incorporates a creep component (see Chapter 6).

If it is required to perform a low-frequency loop by inserting a constant strain dwell at, say, peak tension (simulating the steady running period of plant) then different considerations apply. Figure 2(d) shows the creep strain ε_{pr} induced as a result of the stress relaxing along PN from σ_T to σ_r, where:

$$\varepsilon_{pr} = \frac{1}{E}(\sigma_T - \sigma_r) \tag{4}$$

Relaxation sometimes occurs in service under conditions of varying strain (elastic follow-up). In the laboratory it occurs down a line PQ dictated by the *machine modulus*, E_m. The induced creep strain in this case is:

$$\varepsilon'_{pr} = \frac{1}{E^*}(\sigma_T - \sigma'_r) \tag{5}$$

where E^* is an effective modulus, in turn defined by:

$$\frac{1}{E^*} = \frac{1}{E} + \frac{1}{E_m} \tag{6}$$

A dwell in tension causes the loop to be biased towards compression so that $\sigma_T/\sigma_C < 1$ [14]. This arises because slow strain rates during relaxation can bring σ_r or σ'_r well below σ_{yT}. The extra strain ε_{pr} or ε'_{pr} must be reversed at a relatively fast strain rate along NR or QR (Fig. 2(d)), requiring a further component of stress after compressive yield in order to close the loop.

When the dwell occurs in the middle of the loop, at σ_{TO} in Fig. 2(b), there is no shift towards compression [14]. Further, if the total strain limits are deliberately offset in continuous cycling there is ample opportunity for the stresses to become equal provided $\varepsilon_t > 2\varepsilon_y$ where $\varepsilon_y = \sigma_y/E$.

2.3. EXPERIMENTATION

The HSF test, which sets out to determine cyclic stress–strain data, naturally divides into two parts: (1) the *evolutionary* stage where hardening or softening or sometimes both set in, and (2) the *saturation* stage where the hysteresis loop becomes stable.

2.3.1. Longitudinal Versus Diametral Control

If diametral extensometry is chosen instead of longitudinal control, one must be certain that oxide growth does not invalidate the strain measurement [15]. The diametral strain is almost always converted into longitudinal strain which is then controlled by computer: the 'unconverted' method has only been used sparingly [16–19]. For dwell tests, arguments involving Poisson's ratio and experimental data suggest, for example, that less relaxation strain occurs with this system [20].

Very little opportunity arises for comparison between the different methods. During testing of $2\frac{1}{4}$Cr–1Mo steel at 427 and 482°C, Ellis *et al.* [21] showed that longitudinal strains computed by the diametrally compensated method on a cylindrical specimen were some 16% greater than values measured directly. Reuchet and Rémy [22] reported that hourglass-shaped specimens of superalloy MAR-M509 tested under diametral control gave stress–strain data lying below those of a cylindrical specimen under longitudinal control. In contrast, no difference was found between cylindrical and hourglass-shaped specimens during hardening of 304 steel at 593°C [23].

2.3.2. Test Techniques

Three basic methods are available for determining cyclic stress–strain data in the saturation period:

(a) *Companion specimens.* Results arise as a by-product of a total endurance determination, measurements being taken at the 'half-life'. Testing separate specimens for the sole purpose of acquiring stress–strain data would be time-consuming and expensive.

(b) *Incremental step test.* As discussed by several authors [24–26] the total strain limits on a single specimen are gradually and symmetrically increased after each half-cycle to a decided maximum, the X–Y plotter being in operation continuously. The superposition of the loops then gives a clearly defined locus of tips as indicated schematically in Fig. 3(a). The increments are then gradually decreased until the starting point is reached and the whole process repeated until the locus is reproducible.

(c) *Multiple step test.* This is similar to the previous method, except that the number of discrete strain levels (steps) and the number of cycles at each level are decided in advance so that the jump in strain between steps is much larger (see Fig. 3(b)). The transition is only made when the stress levels appear to have settled and usually only the last hysteresis loop of each step is plotted (see also Fig. 2(c)).

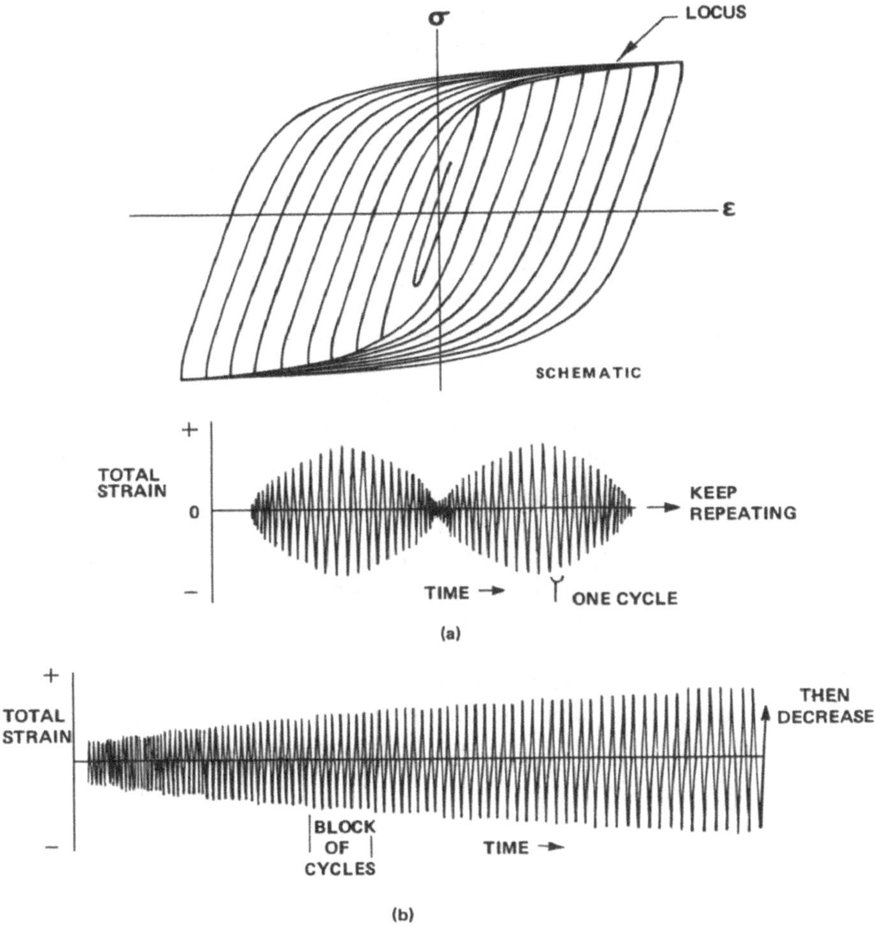

FIG. 3. (a) Incremental step test. Ascending and descending loops are repeated until the locus is stable. (b) In the multiple-step test, several cycles are undergone in each block (see also Fig. 2(c)).

It could be argued that, of all the methods, only the first gives genuine saturation data at each strain level, because (a) it is clear when the evolutionary stage is complete, at least for hardening materials, and (b) there are no complicating history effects (discussed further in Section 2.8). However, there is evidence from the other two methods that materials very quickly adapt to their new levels, and this could prove to be a useful simplification in service assessments.

Methods (b) and (c) thus can produce results quickly and economically, although they do require a computer-controlled testing machine. As a last resort, however, method (c) lends itself to manual adjustments.

2.3.3. Test Parameters

The control limit usually adopted by workers in HSF is the total strain range, ε_t, since this most often simulates the service cycle. However in fundamental studies, for example, energy changes during cyclic hardening or softening, the method is disadvantageous because hysteresis energy per cycle is proportional to the product $\Delta\sigma\varepsilon_p$ and from eqn (1), both these latter parameters are changing. There is thus no simple relation between energy and hardening or softening unless either $\Delta\sigma$ or ε_p is kept constant. Morrow [24] has suggested a stress control test for this purpose but it will be shown in Section 2.5 that this leads to tensile ratchetting. A test at constant plastic strain range is thus preferable and possible with special circuitry [27–29].

In Fig. 4 the upper curve [30] shows the gradual increase in stress range observed during hardening in 316 steel at 625°C at a constant ε_t of 0·006. The plastic strain range was 0·004 initially and at peak hardening had decreased to 0·002. The specimen was solution treated at 1050°C for 1 h and a new test begun at a constant ε_p of 0·002. It is evident from the lower curve of Fig. 4 that the stress range $\Delta\sigma$ was lower but the same final state was obtained, albeit by a different route.

The accompanying Table 1 shows the relative energy changes involved. Whereas in the first test there was a net decrease in energy per cycle

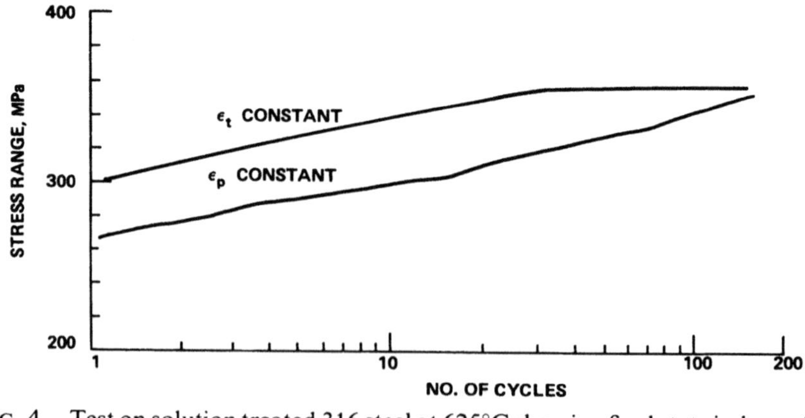

FIG. 4. Test on solution treated 316 steel at 625°C showing final state independent of path of arrival [30].

TABLE 1

RELATIVE ENERGY CHANGE DURING HARDENING OF 316 STEEL AT 625°C

No. of cycles	*Relative energy change ($\propto$ area of hysteresis loop)*	
	ε_t *constant*	ε_p *constant*
2	0·95	0·41
3	0·88	0·42
5	0·83	0·43
11	0·75	0·44
22	0·70	0·46
50	0·66	0·49
100	0·61	0·51
150	0·54	0·54

(increase in $\Delta\sigma$, but offset by a decrease in ε_p), in the second test the energy increased.

2.4. PRESENTATION

There are several ways of plotting saturation cyclic stress–strain data, each of which can cause traps for the unwary. It is important to appreciate that eqns (2) and (3) almost always refer to the *locus* of the hysteresis loop tips, so that if the points A′, B′ and C′ in Fig. 2(c) are superimposed at a new compressive origin O″ as shown in Fig. 5 then this locus does not coincide with the paths adopted by each individual loop. It could be argued that the latter are more relevant for a deforming component, but fortunately the errors in stress and strain values when assessed from any single loop are not large [24] (see also Appendix 2). Indeed, many computer calculations of stress–strain behaviour only require extremity values. Figure 6 illustrates some data obtained by Jaske *et al.* [5] on annealed $2\frac{1}{4}$Cr–1Mo steel at 593°C which recognises this difference in cyclic stress levels. At lower temperatures the scatter was more pronounced.

2.4.1. Stress Range Versus Plastic Strain Range

If a material obeys eqn (3), the ‘stress deviation curve’ [12], a plot of $\log \Delta\sigma$ vs. $\log \varepsilon_p$ gives a straight line, and several examples are given in Fig. 7 [30]. From such plots, values of A and β can be tabulated very

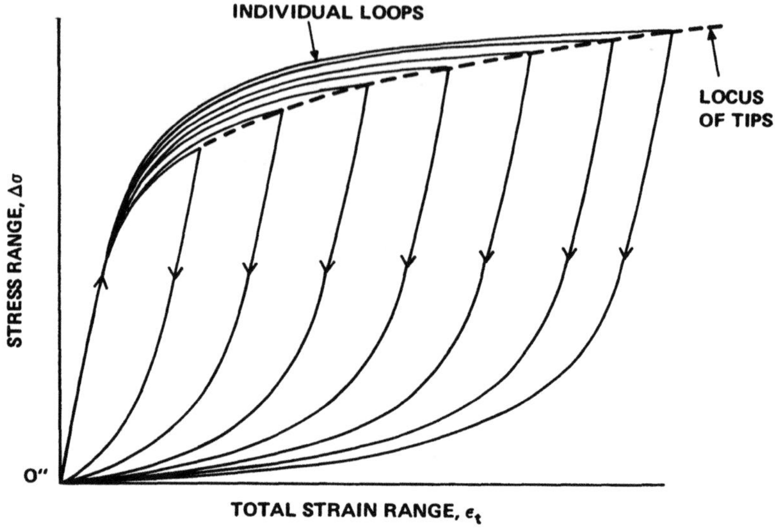

FIG. 5. Illustrating the difference in locus and individual loop shape. Maximum error in A and β occurs for outer loop.

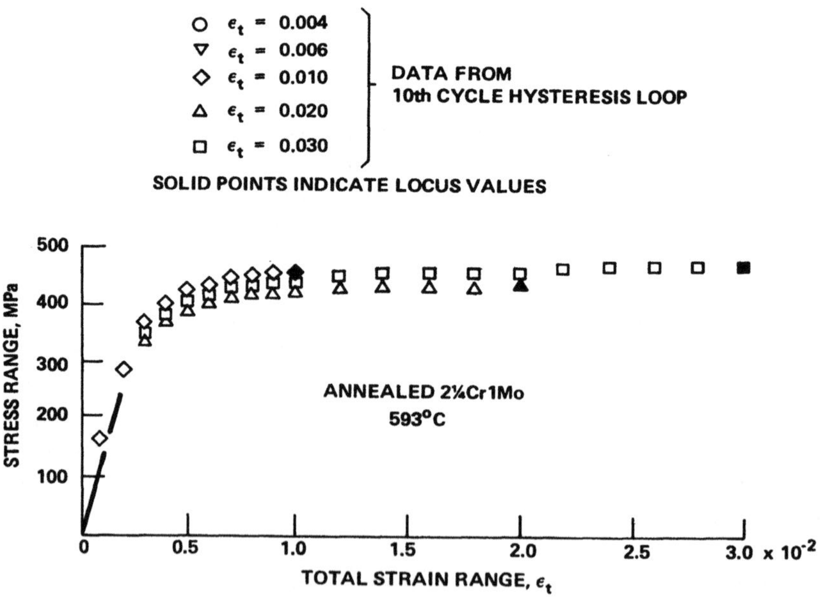

FIG. 6. Showing the difference in locus and individual hysteresis loop values of stress: 10th cycle data, separate specimen tests. (Adapted from Ref. 5 with permission of ASME.)

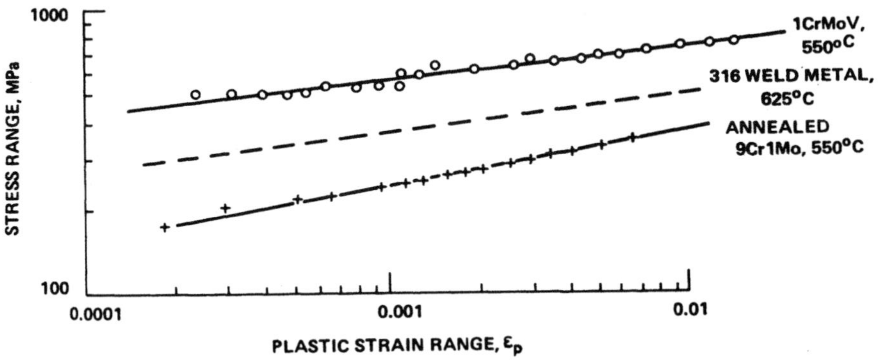

FIG. 7. Data for three high temperature alloys showing agreement with eqn (3) [30].

economically for the temperatures of interest (see Section 2.7). Total strain range ε_t may hence be calculated from eqn (2), provided that the variation of Young's modulus with temperature is known.

Throughout this chapter, *values of A and β will be defined according to the whole range values of eqn (3)* as this is likely to cause the least confusion. Some workers[3, 22, 31–33] prefer to plot $\log \Delta\sigma/2$ vs. $\log \varepsilon_p/2$ (e.g. see Fig. 1(b)), so that to reconcile the data, their quoted strength coefficient must be multiplied by $2^{1-\beta}$. Another variation is to express $\varepsilon_p/2$ as a percentage strain [34] so that the multiplication factor is now $(100)^{\beta}2^{1-\beta}$. In all these possibilities, however, the slope remains constant at β.

Linear plots e.g. $\Delta\sigma$ vs. ε_p [35] or $\Delta\sigma/2$ vs. ε_p [36] are sometimes produced, but are not easy to interpret since the slope of the curve becomes vertical on the stress axis intercept.

2.4.2. Stress Range Versus Total Strain Range

Figure 8 is a linear plot of $\Delta\sigma$ against ε_t for 316 steel and other alloys at several temperatures [11, 18, 28, 34, 37–41]. Equation (2) is also shown in the figure for typical values of A and β. At first sight, the curves appear to be conventional stress–strain plots as for unidirectional deformation. The yield point for a given alloy is therefore easily assumed to be σ_y, whereas of course it is $2\sigma_y$, a parameter commonly used in cyclic analysis.

No particular advantage is offered by plotting on logarithmic coordinates, as the result remains a curve (see eqn (2)). Brinkman *et al.* [42], in yet another variation, presented their data on a high temperature transition joint in terms of $\log \Delta\sigma$ vs. $\log \varepsilon_t/2$.

2.4.3. Linear Plots Involving Stress and Strain Amplitudes

A common method is to plot $\Delta\sigma/2$ against $\varepsilon_t/2$, i.e. halving the coordinates of the method of Fig. 8 [19, 43, 44]. Typical examples for ferritic and austenitic steels at elevated temperature are given in Fig. 9 [9, 45]. A variation is to split the stress range into tension and compression components and plot against total strain range: this is done for two turbine steels at 550°C in Fig. 10 [14, 46]. Such a presentation is very useful for emphasising loop asymmetry or for displaying the stress drop after a dwell

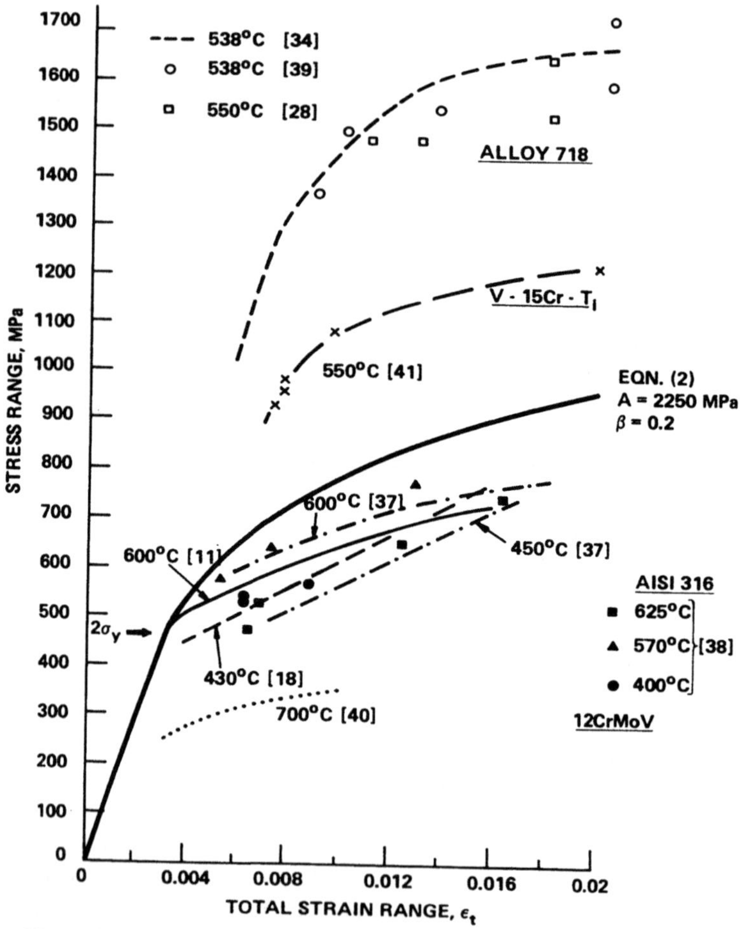

FIG. 8. Plot of stress range versus total strain range for several materials. Note that the 'yield points' are $2\sigma_y$.

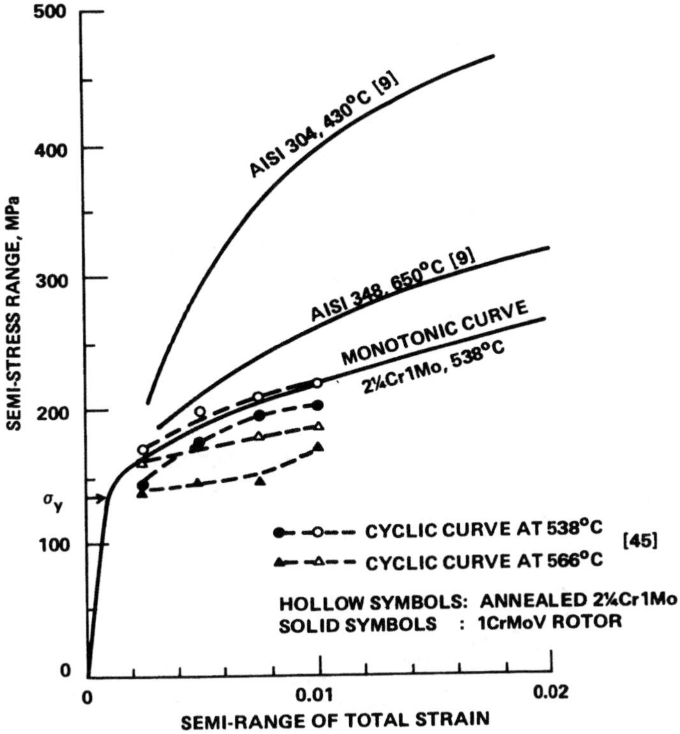

FIG. 9. Typical results plotted in terms of semi-range.

period, but an understandable error is to assume that the slopes OA, OB in Fig. 10 are equal to Young's modulus, whereas they are of course $E/2$.

2.4.4. Plastic Strain Range Versus Total Strain Range

A relatively new method [47] is to plot the logarithm of the total strain range ε_t against $\log \varepsilon_p$ in the manner shown in Fig. 11 [11, 46–49]. The relation between them is found by eliminating $\Delta\sigma$ between eqns (1) and (3):

$$\varepsilon_t = \varepsilon_p + \frac{A}{E}\varepsilon_p^{\beta} \tag{7}$$

Equation (7) is shown plotted in Fig. 11 for the rotor steel at 550°C, for which $A = 1220$ MPa, $E = 1{\cdot}5 \times 10^5$ MPa and $\beta = 0{\cdot}097$ (see also Fig. 7).

Over a limited range of ε_p, eqn (7) may be expressed [47] as

$$\varepsilon_t = B\varepsilon_p^{\gamma} \tag{8}$$

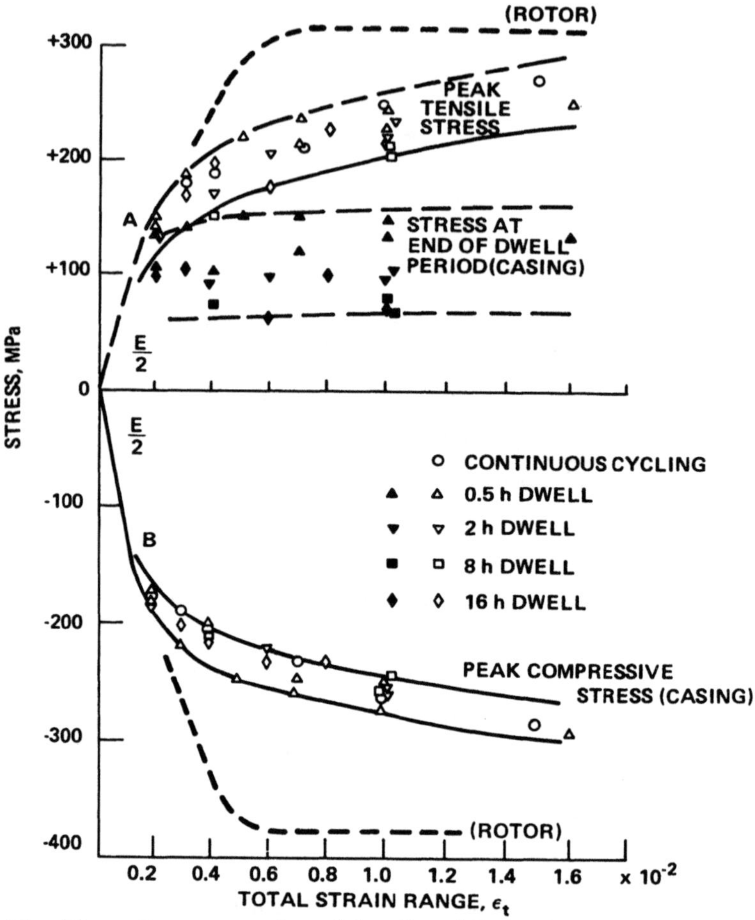

FIG. 10. Alternative presentation of data for a $\frac{1}{2}$Cr–Mo–V casing steel [46] and a 1Cr–Mo–V rotor steel [14] at 550°C.

where B and γ are constants. Equation (8) is shown plotted in Fig. 11 for the same rotor steel, where for plastic strain ranges up to 0·005, $B = 6{\cdot}6 \times 10^{-2}$ and γ, the slope of the line, is 0·375.

Although the stress range is not immediately apparent from either eqns (7) or (8), they have many advantages. Equation (8) is useful in estimating strain concentration factors [47] (see also Section 2.11). White and Horwood have also shown [50] (for mild steel at 350°C and EN58J at 650°C) that total strain versus stress data in reverse bending are in good

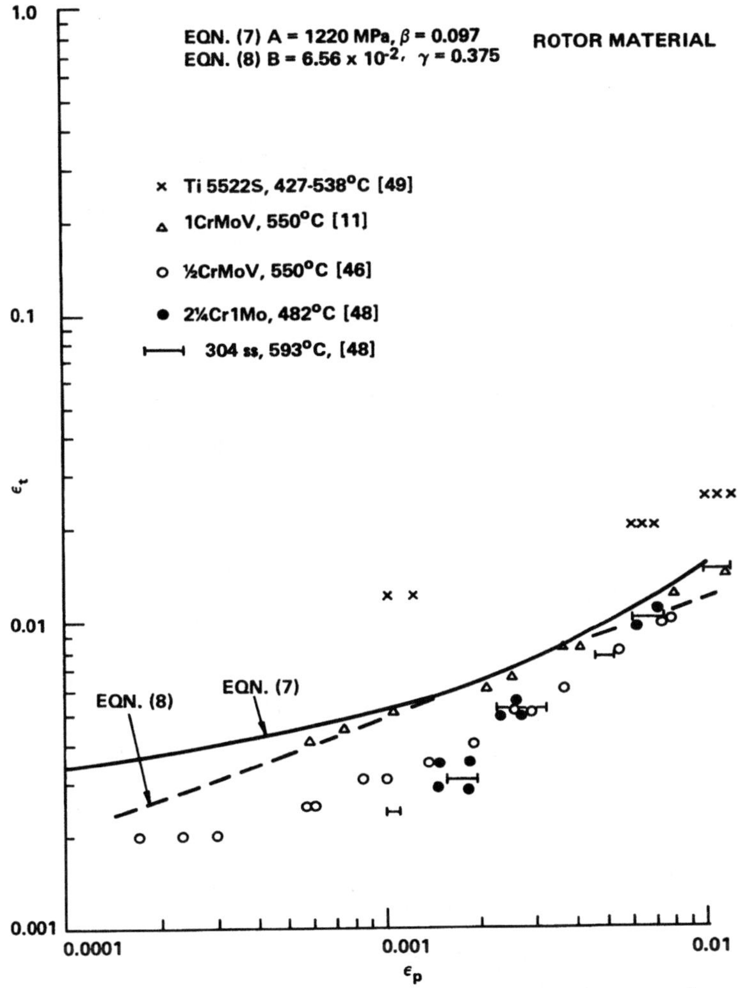

FIG. 11. Relation between total and plastic strain range for several materials.

agreement with push–pull data, so that eqn (8) or the more accurate eqn (7) may be used for estimating surface plastic strain range from measured values of ε_t in a bend test.

2.4.5. Comparison with Monotonic Data

The curve OA in Fig. 2(a) is a monotonic (first loading) curve and the material either hardens or softens with respect to it. In comparing cyclic

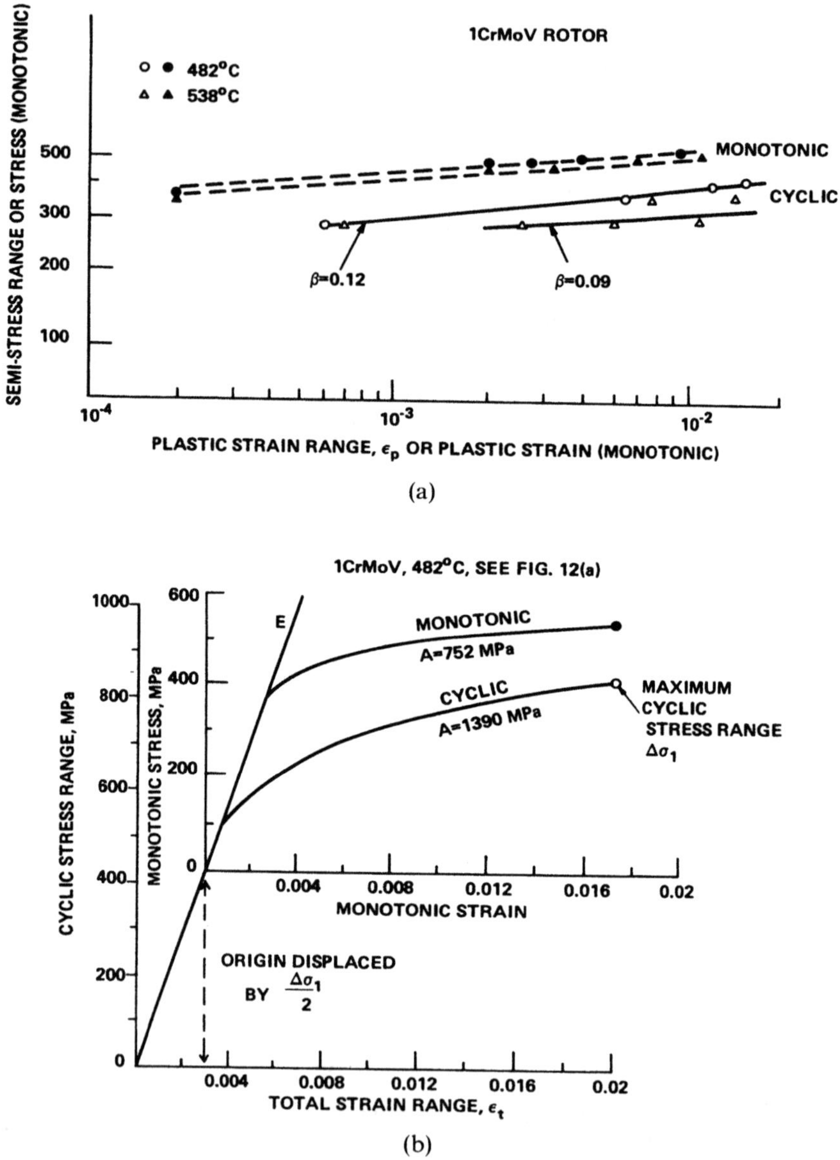

FIG. 12. (a) In order to compare monotonic with cyclic deformation behaviour, plot semi-range of stress against plastic strain range for the cyclic case. Cyclic data from Ref. 45 have been replotted. (b) As for (a); replot data using eqn (2) for cyclic case, and transpose origin for monotonic case.

TABLE 2

CYCLIC AND MONOTONIC LOADING PARAMETERS COMPARED FOR SELECTED ALLOYS

Alloy	$\dot{\varepsilon}_t$ (s^{-1})	*Temperature* (°C)	*A/2 cyclic* (*MPa*)	*A monotonic* (*MPa*)	*β cyclic*	*β monotonic*	*Ref.*
1Cr–Mo–V	$1{\cdot}7 \times 10^{-4}$	482	686	792	0·117	0·083	33
$2\frac{1}{4}$Cr–1Mo, normalised and tempered	$1{\cdot}7 \times 10^{-4}$	538	504	524	0·103	0·050	33
$2\frac{1}{4}$Cr–1Mo, quenched and tempered	$1{\cdot}7 \times 10^{-4}$	482	719	1 088	0·077	0·057	33
AISI 304	4×10^{-3}	538	515	593	0·143	0·293	26
	4×10^{-5}	538	614	599	0·141	0·288	26
	4×10^{-3}	650	433	534	0·120	0·182	26
	4×10^{-5}	650	285	278	0·115	0·183	26
Alloy 800	4×10^{-3}	538	740	207	0·098	0·113	26
	4×10^{-3}	650	597	192	0·087	0·123	26
	4×10^{-3}	760	399	198	0·101	0·100	26

stress–strain with monotonic data, the immediate difficulty is that the former are presented as cusp loci whereas the latter are actual deformation curves. Further, for steady state cycling, the new origin has been displaced to O′ in Fig. 2(c) when considering the tension-going phase alone. Since authors rarely publish actual hysteresis loops, considerable dexterity is required in each of the methods discussed in Sections 2.4.1–2.4.4 to extract the unwanted compressive elastic component, such as O′C′ in Fig. 2(c).

Because of the way Young's modulus is portrayed, monotonic data cannot be plotted in Fig. 10, for example. Sometimes, Masing's theory is invoked [51, 52] (tensile curve is half the locus of tips in terms of both stress and strain, see Fig. 5) and the monotonic curve is thus plotted directly onto the stress amplitude versus strain amplitude diagram, as shown in Fig. 9.

A better system (recognising that a comparison must be made on the basis of actual plastic strain range ε_p) is as follows: plot the cyclic results according to eqn (3) with stress amplitude $\Delta\sigma/2$ replacing $\Delta\sigma$, noting that the cyclic strength coefficient A is now halved for this case only. Also plot the monotonic relation on the same graph. Some results using this method are given in Fig. 12(a). It is seen that softening has taken place [45]. Alternatively, linear axes may be used, as in Fig. 12(b) where stress and strain ranges are employed for the cyclic case, but the monotonic origin has been shifted by the amount shown. This is equivalent to replacing the first term only in eqn (2) by $\Delta\sigma/2E$.

Typical values of cyclic strength coefficients and hardening exponents together with their monotonic counterparts are provided in Table 2 [26, 33].

2.5. CALCULATION

This section contains a detailed examination of hysteresis loop properties and the narrative may be resumed at Section 2.6 without loss of continuity.

2.5.1. Shakedown Behaviour

Even though a material is stable at the outset, that is, neither hardens nor softens, a specified number of cycles will be required before the loop has adjusted to steady alternating plasticity. Such behaviour is termed *shakedown* and the phenomenon arises because of a gradual exploitation of the elastic range in compression (say) after an initial excursion into the tensile (say) plastic range. If $\Delta\sigma < 2\sigma_y$ the system, whether a specimen or a structure, shakes down to complete elasticity.

2.5.1.1. Constant Total Strain Limits

The system is shown in Fig. 13(a). Tangent modulus h and tensile yield stress σ_y are assumed constant, but different from their compressive counterparts h' and σ'_y. First loading takes place into tension from the origin O, and it is assumed that tensile or compressive unloading is immediately elastic.

From the simple geometric paths traced in Fig. 13(a) it may be shown that successive differences δ in peak tension stresses are ever-decreasing and bear a constant ratio to the previous difference, independent of σ_y, σ'_y and ε_t. Thus

$$\delta_{i+1} = \delta_i(hh'/E^2) \tag{9}$$

a similar relation applying to the values δ' in compression. Since δ_{i+1}, δ_{i+2} etc. are in geometric progression with a common ratio <1, a closed loop quickly forms. After an initial loading in tension, the steady state tension stress σ_f can rapidly be determined from the difference δ_2 between the first complete ($1\frac{1}{4}$) and ($2\frac{1}{4}$) cycles. Thus after N cycles:

$$\sigma_f = \sigma_1 - \sum_{1}^{N} \delta_i = \sigma_1 - \delta_2 \left[\frac{1 - \left(\dfrac{hh'}{E^2}\right)^N}{1 - \dfrac{hh'}{E^2}} \right] \tag{10}$$

In eqn (9) for an elastic–perfectly plastic material $h = 0$ and shakedown is immediate after $1\frac{1}{4}$ cycles. Conversely, if the strain limits lie just beyond yield, $h, h' \sim E$ and a relatively large number of cycles is required to close the loop. By similar arguments it is possible to show from eqn (10) that, if the total strain limits of a stable loop are suddenly increased, decreased, or moved in the same direction, the new loop adopted can be calculated after the first new cycle is complete.

2.5.1.2. Constant Stress Limits

The path traced by successive cycles in a load-controlled system is shown in Fig. 13(b). In this case it may be shown by similar arguments that the differences between each peak tensile strain are equal. Putting $h = h'$ for simplicity:

$$\varepsilon_{t_{i-1}} - \varepsilon_{t_i} = [(\sigma - \sigma') + (\sigma'_y - \sigma_y)]\left(\frac{1}{E} - \frac{1}{h}\right) \tag{11}$$

A closed loop results only if this expression equals zero. This is only

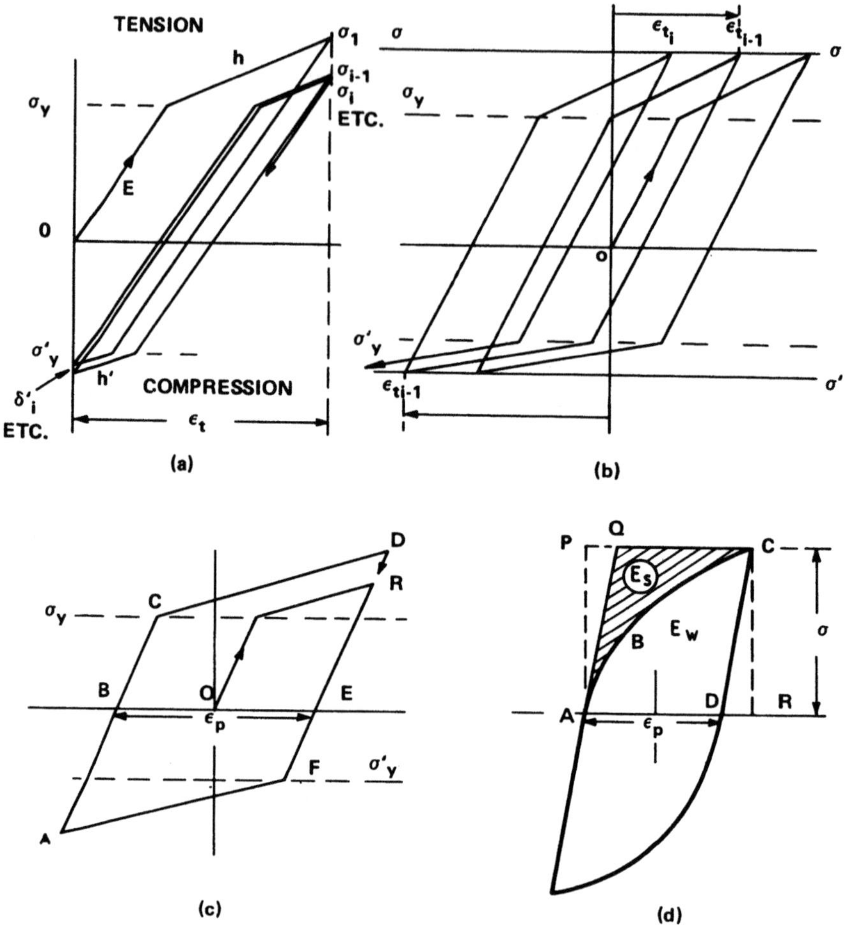

FIG. 13. (a) Shakedown to cyclic plasticity. (b) Ratchetting under load control. (c) Almost immediate shakedown under plastic strain control. (d) Stored energy and plastic work expended.

possible by judicious choice of relative values of the parameters σ_y, h etc. and strains in practice generally accumulate in the tensile sense [53]. Because of this ratchetting, load control is never contemplated in HSF testing unless an overall strain ceiling is specified.

2.5.1.3. Constant Plastic Strain Limits

Shakedown for a constant width of hysteresis loop ε_p (Fig. 13(c)) is almost immediate. The cycle limits are set along the inclined elastic lines ABC and DEF. After a tensile loading to R the system unloads elastically then deforms in compression to A (Fig. 13(c)). Elastic unloading in compression is along AB and the next tensile loading rejoins the inclined limit at D. Provided therefore that h, h', σ_y and σ'_y remain constant, only this *one* cycle is required to close the loop.

2.5.2. Energy Changes Around Loop

Consider the tension half-cycle only of a symmetrical and stable hysteresis loop (Fig. 13(d)). Upon deformation along ABC the total work done per unit volume is the area under the curve, viz. ABCR. Of this, however, CRD is the elastic energy per unit volume $= \sigma^2/2E$. This is recoverable on unloading to D and is represented by area APQ above the curve. The argument also applies to compressive loading and because no elastic work is lost in a cycle, this energy term can be ignored for the present.

In Fig. 13(d) the total plastic work expended along ABC is thus E_w and is given by $\int_0^{\varepsilon_p} \sigma \, d\varepsilon_p$. Taking the origin as A in Fig. 13(d) and putting $\sigma = \Delta\sigma/2$ it is found from eqn (3) that the plastic work done per half-cycle is:

$$E_w = \sigma\varepsilon_p/(1+\beta) \tag{12}$$

Now the shaded area AQC in Fig. 13(d) is also a work term given by $E_s = \int_0^{\sigma} \varepsilon_p \, d\sigma$ which from eqn (3) reduces to:

$$E_s = \beta\sigma\varepsilon_p/(1+\beta) \tag{13}$$

Equation (13) may be identified as the stored energy of deformation per half-cycle [54] which remains in the specimen upon unloading to point D. The heat liberated per half-cycle, dQ, is thus given by $E_w - E_s$, i.e. from eqns (12) and (13),

$$dQ = \sigma\varepsilon_p \frac{1-\beta}{1+\beta} \tag{14}$$

In a *complete* cycle the heat dissipated is thus given by eqn (14), with $\Delta\sigma$ replacing σ. This form was derived by Halford [55] who denoted it the 'plastic strain hysteresis energy per cycle'. Similarly, the total plastic work done per cycle is given by eqn (12) where σ now becomes $\Delta\sigma$. As regards the stored energy, however, only the half-cycle value is conserved in each closed loop, as will be discussed in Section 2.5.5.

When $\beta = 0$ we have an elastic–perfectly plastic solid (AQC in Fig. 13(d)) and all the work of plastic deformation is converted into heat, i.e. from eqns

(12) and (14) $E_w = dQ = \sigma\varepsilon_p$, while from eqn (13), $E_s = 0$. For all positive values of β, therefore, we find from eqns (12) and (13)

$$\beta = E_s/E_w \tag{15}$$

Some idea of the magnitude of E_s for two ferritic steels is provided in Table 3(a), these results being taken from the relevant areas of stabilised hysteresis loops [30]. Table 3(b) gives similar data for two other steels in the evolutionary stage [30]. The measured values of E_s for the austenitic steel are compared with eqn (13). Agreement is seen to be reasonable. It is also of interest to note that eqn (15) gives approximate values of β for Réné 77 at

TABLE 3(a)
STORED ENERGY VALUES DERIVED FROM STABLE HYSTERESIS LOOPS OF TWO STEELS AT 550°C

1Cr–Mo–V		*½Cr–Mo–V*	
Tension stress (MPa)	*Stored energy ($J\ mm^{-3}$)*	*Tension stress (MPa)*	*Stored energy ($J\ mm^{-3}$)*
300	9.5×10^{-5}	250	1.5×10^{-4}
250	3.6×10^{-5}	225	9.5×10^{-5}
200	1.4×10^{-5}	200	6.5×10^{-5}
150	5.9×10^{-6}	175	3.2×10^{-5}
100	2.1×10^{-6}	150	2.4×10^{-5}
		125	1.3×10^{-5}
		100	8.2×10^{-6}
		50	2.0×10^{-6}

TABLE 3(b)
STORED ENERGY VALUES DURING EVOLUTION

S.T. 316 Steel, 625°C; $\varepsilon_p = 0.01$, $\beta = 0.25$, hardening				*N&T 9Cr–1Mo Steel, 550°C $\varepsilon_p = 0.01$, $\beta = 0.19$, softening*		
Cycle no.	*A (MPa)*	*E_s, eqns (13) and (17) ($J\ mm^{-3}$)*	*E_s, from loop ($J\ mm^{-3}$)*	*Cycle no.*	*A (MPa)*	*E_s, eqns (13) and (17) ($J\ mm^{-3}$)*
1	1 186	3.75×10^{-4}	3.48×10^{-4}	1	1 842	6.12×10^{-4}
7	1 450	4.59×10^{-4}	4.47×10^{-4}	34	1 593	5.29×10^{-4}
20	1 929	6.10×10^{-4}	5.16×10^{-4}	53	1 431	4.75×10^{-4}
254	2 172	6.90×10^{-4}	6.14×10^{-4}	Stabilised	1 282	4.26×10^{-4}

927°C from the work of Antolovich and Rosa [35], who computed the plastic work from eqn (12) but derived E_s by calculating the stored energy of dislocations.

2.5.3. Alternative Expressions

Combining eqns (3) and (12) an alternative form for the work done per half-cycle is

$$E_w = \frac{A}{2} \frac{\varepsilon_p^{1+\beta}}{1+\beta} \tag{16}$$

Similarly, the corresponding form for stored energy, eqn (13), is

$$E_s = \frac{A}{2} \frac{\beta \varepsilon_p^{1+\beta}}{1+\beta} \tag{17}$$

Equation (11) suggests that alternative expressions may be derived for hysteresis loop parameters using the simplified model of Fig. 2(b), i.e. in terms of h, σ_y, E and σ. The link between this model and eqn (3) is via the plastic strain term:

$$\varepsilon_p = (\sigma - \sigma_y)\left(\frac{1}{h} - \frac{1}{E}\right) \tag{18}$$

Corresponding expressions are summarised in Table 4 and, in the following, use will be made of both the simplified and Ramberg–Osgood forms.

2.5.4. External Changes with Strain

Equations (16) and (17) or their equivalents in Table 4 can throw further light on stress–strain behaviour, as the following examples show.

2.5.4.1. Variation of Stored Energy/Cycle with Plastic Strain

Suppose a stable hysteresis loop is achieved at constant ε_p, i.e. from eqn (17), A is constant. Then

$$\frac{1}{E_s}\frac{dE_s}{d\beta} = \frac{1}{\beta} - \frac{1}{1+\beta} + \ln \varepsilon_p$$

and since $dE_s/d\beta = 0$ at a stationary point we find

$$\beta_{max} = \frac{1}{2}\left[\left(1 - \frac{4}{\ln \varepsilon_p}\right)^{1/2} - 1\right] \tag{19}$$

TABLE 4

ALTERNATIVE EXPRESSIONS FOR PARAMETERS

Parameter	Definition			Comments
β	$\dfrac{\sigma-\sigma_y}{\sigma+\sigma_y}$	$\varepsilon_p \Big/ (\sigma+\sigma_y)\left(\dfrac{1}{h}+\dfrac{1}{E}\right)$	$h\left(\dfrac{\varepsilon_t}{\sigma+\sigma_y}-\dfrac{1}{E}\right)$	
E_w, eqns (12) and (16)	$\tfrac{1}{2}\varepsilon_p(\sigma+\sigma_y)$	$\tfrac{1}{2}(\sigma^2-\sigma_y^2)\left(\dfrac{1}{h}-\dfrac{1}{E}\right)$	$\dfrac{1}{2}\left(\varepsilon_t-\dfrac{2\sigma}{E}\right)(\sigma+\sigma_y)$	
E_s, eqns (13) and (17)	$\tfrac{1}{2}\varepsilon_p(\sigma-\sigma_y)$	$\tfrac{1}{2}(\sigma-\sigma_y)^2\left(\dfrac{1}{h}-\dfrac{1}{E}\right)$	$\tfrac{1}{2}\varepsilon_p^2\left(\dfrac{hE}{E-h}\right)$	
E_{sec}	$\dfrac{\Delta\sigma}{\varepsilon_t}$	$E\Big/\left(1+\dfrac{E\,\Delta\sigma^{(1-\beta)/\beta}}{A^{1/\beta}}\right)$		Decreases as ε_t increases; derived from eqn (2)
E_{tan}	$\dfrac{d\,\Delta\sigma}{d\varepsilon_t}=h$	$E\Big/\left(1+\dfrac{E\,\Delta\sigma^{(1-\beta)/\beta}}{\beta A^{1/\beta}}\right)$		Decreases as ε_t increases; derived from eqn (2)

TABLE 5
VALUES OF β FOR MAXIMUM STORED ENERGY PER HALF CYCLE

ε_p	β	ε_p	β
0·000 1	0·098	0·008	0·176
0·000 3	0·111	0·01	0·183
0·000 5	0·118	0·02	0·211
0·001	0·128	0·03	0·231
0·002	0·141	0·04	0·249
0·004	0·157	0·05	0·264
0·006	0·167	0·10	0·327

Values of β_{max} for several values of plastic strain calculated from eqn (19) are summarised in Table 5. It is noteworthy that their magnitudes are typical of those determined empirically (Section 2.7), suggesting that a loop seeks to maximise the stored energy in each half-cycle.

2.5.4.2. *Variation of Energies at Constant Total Strain Range*

Assume that hardening or softening sets in equally in tension and compression. Because the loop is always very nearly closed, the rate of energy change with peak stress change is given (see Table 4) by:

$$\frac{dE_w}{d\sigma} = \sigma\frac{d\varepsilon_p}{d\sigma} + \varepsilon_p + \sigma_y\frac{d\varepsilon_p}{d\sigma} + \varepsilon_p\frac{d\sigma_y}{d\sigma} \tag{20}$$

It is reasonable to suppose that as σ changes during evolution, so does the yield stress σ_y, and we may put $d\sigma_y/d\sigma = k$ where $k < 1$. Also, since $\varepsilon_p = \varepsilon_t - 2\sigma/E$ we have, since ε_t is constant, $d\varepsilon_p/d\sigma = -2/E$. There is thus a stationary value of E_w when

$$\varepsilon_p = \frac{2(\sigma + \sigma_y)}{E(1 + k)} \tag{21}$$

The term $d^2E_w/d\sigma^2$ is found to be negative, so from Table 4 the maximum work done per half-cycle occurs when

$$E_{w\,max} = \frac{(\sigma + \sigma_y)^2}{E(1 + k)} \tag{22}$$

Similar considerations show that the maximum stored energy is given by:

$$E_{s\,max} = \frac{\sigma^2 - \sigma_y^2}{E(1 - k)} \tag{23}$$

2.5.4.3. Variation of Plastic Strain Rate Around Loop

Most continuous cycling fatigue testing is performed with triangular waveshape between fixed total strain limits so that total strain rate $\dot{\varepsilon}_t$ is constant. It follows from eqn (1) that

$$\dot{\varepsilon}_t = \dot{\varepsilon}_p + \frac{\Delta\dot{\sigma}}{E}$$

i.e. the *plastic* strain rate $\dot{\varepsilon}_p$ must vary because the rate of change of stress $\Delta\dot{\sigma}$ is not constant during elasto-plastic deformation. By differentiation of eqn (3)

$$\Delta\dot{\sigma} = \frac{A\beta}{\varepsilon_p^{1-\beta}}\dot{\varepsilon}_p$$

and so the plastic strain rate is thus given by

$$\dot{\varepsilon}_t = \dot{\varepsilon}_p\left(\frac{A\beta}{E\varepsilon_p^{1-\beta}} + 1\right) \qquad (24)$$

For a typical material, $A = 1500$ MPa, $\beta = 0{\cdot}2$ and $E = 1{\cdot}5 \times 10^5$ MPa. The variation of plastic strain rate as the plastic strain increases during the cycle is shown in Fig. 14, assuming that $\dot{\varepsilon}_t = 10^{-3}\,\mathrm{s}^{-1}$. It is seen that $\dot{\varepsilon}_p$ increases by an order of magnitude, only approaching $\dot{\varepsilon}_t$ at the upper limit when plastic strain is large. Thus, when comparisons are made of endurances or

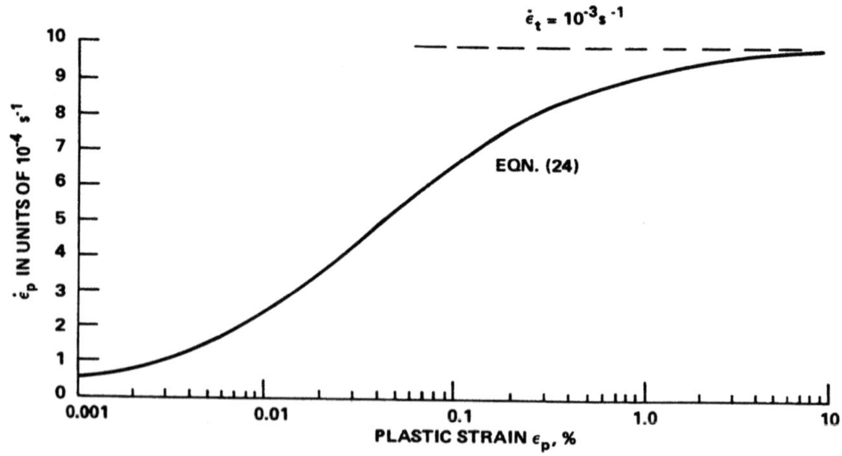

FIG. 14. Variation of plastic strain rate during a cycle in which total strain rate is constant.

microstructures after high temperature tests in the region 10^{-3}–10^{-6} s^{-1} total strain rate, it should be realised that a considerable overlap in plastic strain rates can occur.

2.5.5. Internal Stress Changes Around Loop

2.5.5.1. Definitions

A rather subtle phenomenon, which is not obvious from the steady state loop of Fig. 2(b), is the *Bauschinger effect* which is illustrated in Fig. 15(a) (r.h.s.). On a first tensile loading, the material yields at σ_y. Upon unloading into compression the material yields at $\sigma'_y < \sigma_y$, but if the material were instead unloaded from tension and then loaded in the same direction the yield point would be $> \sigma_y$. Such effects are best interpreted in terms of a *back-stress* σ_b [56, 57], which either hinders or helps yield according to the straining sense. Wilson [58] found that σ_b could be reduced to zero by applying a reverse strain of between one-third and half of the previously applied forward strain so that the initial yield conditions could be re-created (see Fig. 15(a)).

In the steady state loop of Fig. 15(a) (l.h.s.) therefore, the back-stress changes sign during each half-cycle of the same loading sense, reaching its maximum value at peak strain, equal but opposite in sign to its value at yield. It is opposed by a *friction stress* σ_f which acts independently of straining direction. Thus, on deforming along O′A in Fig. 15(a), we have at yield [56, 57]:

$$\sigma_y = \sigma_f - \sigma_b \tag{25}$$

while at maximum strain:

$$\sigma = \sigma_f + \sigma_b \tag{26}$$

i.e. the back-stress first assists then opposes the flow. From these two equations:

$$\sigma_b = \tfrac{1}{2}(\sigma - \sigma_y) \tag{27}$$

and

$$\sigma_f = \tfrac{1}{2}(\sigma + \sigma_y) \tag{28}$$

Hence the Bauschinger effect: on the first loading, $\sigma_b = 0$ and $\sigma_y = \sigma_f$, subsequently reduced on the next cycle (eqn (25)). If $\sigma_b > \sigma_f$ yield occurs *before* zero stress is reached on unloading, as has been observed, for example, in 304 steel at 565°C [33].

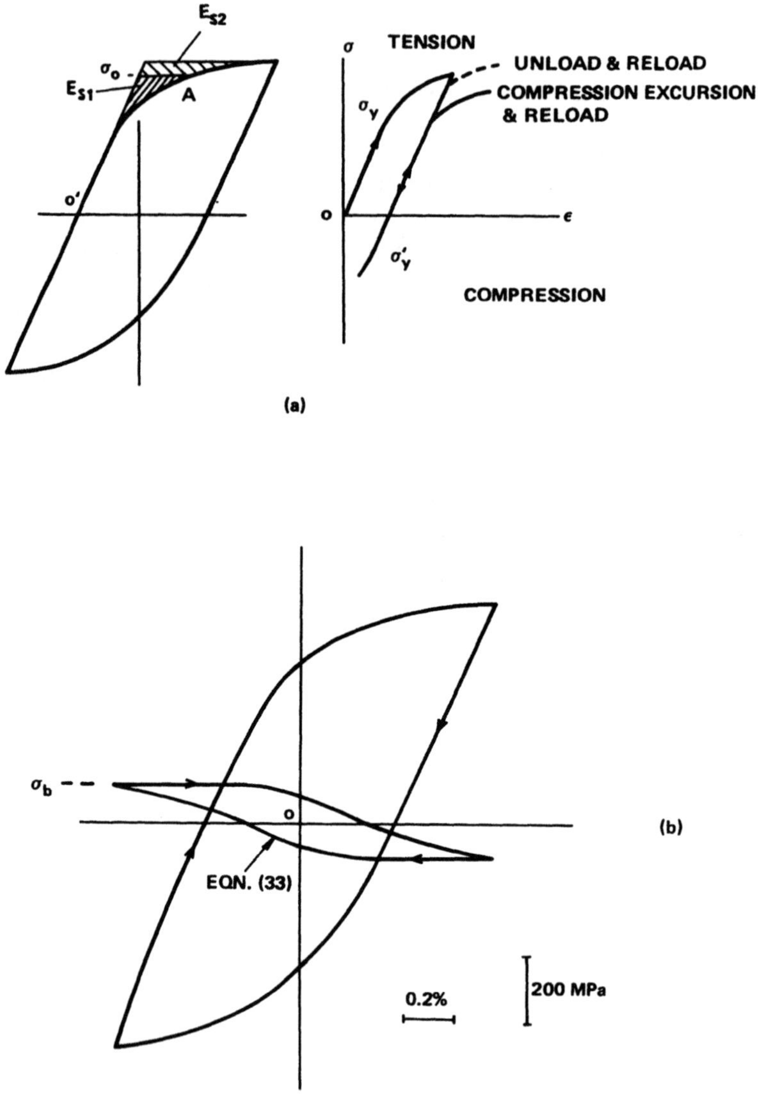

FIG. 15. (a) The Bauschinger effect and associated energies. (b) Proposed internal loop adopted by the back-stress.

Dividing eqns (27) and (28) we have, from Table 4:

$$\frac{\sigma_b}{\sigma_f} = \frac{\sigma - \sigma_y}{\sigma + \sigma_y} = \beta \tag{29}$$

i.e. the cyclic hardening exponent may also be identified as the ratio of the maximum back-stress to the friction stress.

Equations (27) and (28) may be combined with the expressions in Table 4 to give $\sigma_b = E_s/\varepsilon_p$ and $\sigma_f = E_w/\varepsilon_p$ so that from eqns (12) and (13) alternative forms for back- and friction stress respectively are:

$$\sigma_b = \sigma\beta/(1 + \beta) \tag{30}$$

and

$$\sigma_f = \sigma/(1 + \beta) \tag{31}$$

2.5.5.2. An 'Internal' Hysteresis Loop

In the course of a hysteresis cycle, the back-stress reduces to zero between σ_y and σ in the tensile and compressive quadrants. Because of this change, we therefore expect the back-stress itself to execute a 'latent' hysteresis loop.

On a stored energy argument, the external stress σ_0 at which $\sigma_b = 0$ occurs when area E_{s1} is equal to area E_{s2}, i.e. $E_{s1} = \frac{1}{2}E_s$ in Fig. 15(a). This arises because the previous build-up of stored energy with compressive plastic strain, given by eqn (17), must be shed as the new half-cycle begins to accumulate its own stored energy. If this were not so there would be an enormous energy reservoir after relatively few cycles.

To calculate this variation of the back-stress σ'_b, corresponding to any intermediate external stress σ' in the loop, we first note that eqn (17) may be expressed in terms of stress, using eqn (3):

$$E_s = \frac{\beta\sigma^{(1+\beta)/\beta}}{(1 + \beta)(A/2)^{1/\beta}} \tag{32}$$

A general relation from eqn (32) obeying the criterion $E_{s1} = \frac{1}{2}E_s$ is $\sigma' = \sigma(\sigma'_b/2\sigma_b)^{\beta/(1+\beta)}$, which may be rearranged as

$$\Delta\sigma'_b = 2\sigma_b(\sigma'/\sigma)^{(1+\beta)/\beta} \tag{33}$$

The range $\Delta\sigma'_b$ has been introduced since the origin is taken at the maximum level σ_b, as is now shown.

Figure 15(b) is taken from an actual hysteresis loop on quenched and tempered $2\frac{1}{4}$Cr–Mo steel at 482°C [33]. The value of β deduced from this loop is 0·207. The peak back-stress σ_b calculated from eqn (30) is 105 MPa. The proposed internal hysteresis loop has the following properties:

(a) The back-stress is constant during the unloading phases.
(b) It assists tension or compression yield but thereafter decays according to eqn (33), crossing the zero-stress axis.
(c) It builds up to its original value but acting in the opposite sense, again according to eqn (33).
(d) The area of the loop (energy dissipated per unit volume per cycle) is $2E_s$.

Other forms of eqn (33) may be derived either in terms of plastic strain or using the bilinear hysteresis loop (see Table 4 and Fig. 13(a)).

Stored energy is thus destroyed twice each complete cycle, but a quantity given by eqn (17), for example, is retained if the cycle is stopped at zero stress. Such a 'storage–release' mechanism has been detected by calorimetric experiments [59]. Stored energy and back-stress (which is elastic in nature) are thus alternative descriptions of the same microstructural event, as indeed are dissipated energy and friction stress.

2.6. EVOLUTION

2.6.1. Alloy Behaviour

High temperature alloys are either classified as hardening, softening or sometimes as stable if no obvious change in initial stress range is observed at the beginning of a HSF test. Cyclic response is usually plotted simply as the peak load range versus time (i.e. cycle) envelope on a chart recorder, and a family of curves constructed for each level of strain. Evolutionary effects are much more pronounced at higher strain levels and behaviour is strongly influenced by temperature, prior heat treatment, strain rate and time of exposure (see also Section 2.8).

Beginning with ferritic alloys, low-carbon steels generally harden [32], while $\frac{1}{2}$Cr–Mo–V steel with a ferrite–pearlite microstructure is stable [30] though in really long-term tests it may soften [60]. Bainitic 1Cr–Mo–V steel, as used in rotor forgings, shows pronounced softening independent of temperature and strain and is more pronounced at low frequency [10]. Normalised and tempered $2\frac{1}{4}$Cr–1Mo, 9Cr–1Mo and 12Cr–Mo–V steels also soften [33, 61–63], the latter two having a martensitic structure. In contrast, isothermally annealed 9Cr–1Mo steel shows fairly stable behaviour [30].

Solution-treated austenitic steels in the series AISI 304, 316 etc. show pronounced hardening in the range 400–600°C [9, 18, 37], as do the more complex alloys, Incoloy 800 [9] and A286 [64] for example. Superalloy behaviour is extremely complicated as this class of alloys can display metallurgical instability during fatigue (precipitation, over-ageing etc.). Udimet 500 hardens at 730°C, but at 815°C hardens or softens according to testing frequency [65]. At 760°C, Udimet 700 softens at low plastic strain, but at higher values first hardens then softens [66].

In duplex structures, such as a weld, it is of interest to note that, in a typical fatigue test of a 304/308 weldment at 593°C, the base metal hardened while at the same time the weld metal softened [67].

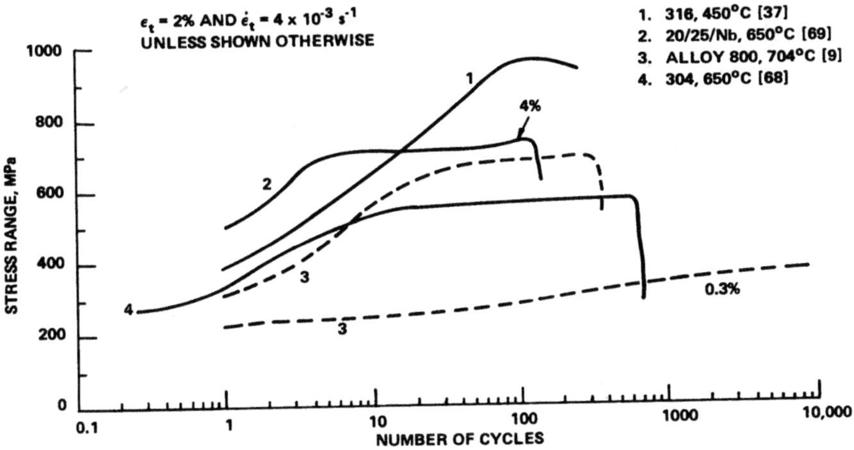

FIG. 16. Examples of cyclic hardening.

Figure 16 gives some [9, 37, 68, 69] of the many examples of hardening which have been published, while Fig. 17 does likewise for softening behaviour [33, 62, 70, 71]. In Table 6 an attempt has been made to summarise the overall stress–strain response of some common alloys at elevated temperature [3, 32, 34, 44, 53, 61, 64, 73–76].

2.6.2. Hardening/Softening Relations

There is unfortunately no universal parameter for expressing hardening or softening behaviour with respect to strain and cycles. One possible candidate from Fig. 2(b) is the secant modulus, $\Delta\sigma/\varepsilon_t$ (see Table 4), which clearly increases during hardening (for tests between total or plastic strain

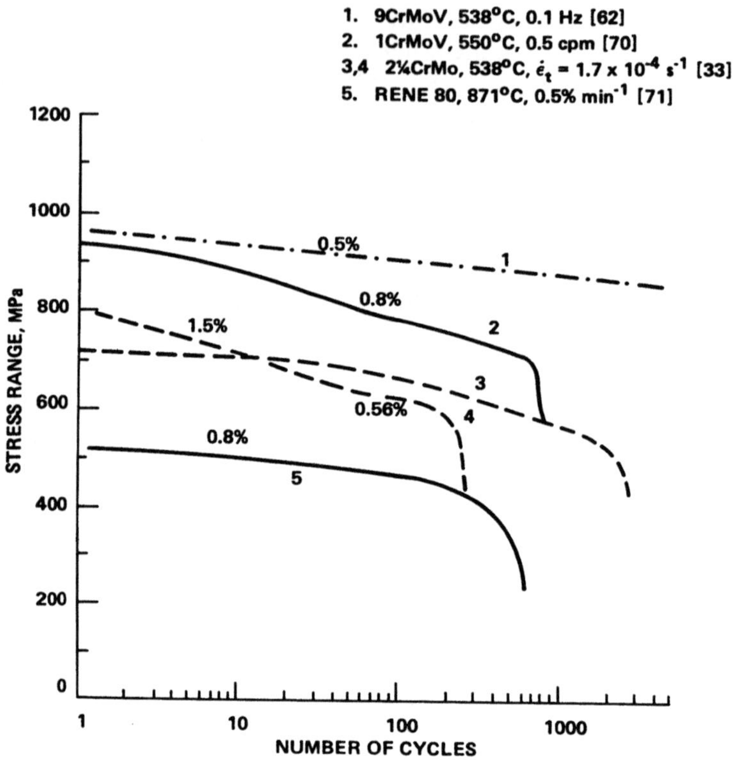

FIG. 17. Examples of cyclic softening.

limits) and conversely for softening. However, the initial values of E_{sec} depend on strain level. In all the following examples where investigators have suggested relations for the variation of $\Delta\sigma$ with cycles, N, it will be noted that the saturation stress $\Delta\sigma_{sat}$ must be known in advance.

(1) A simple expression for the degree of hardening, H, is given by the ratio (e.g., see Ref. 74):

$$H = \frac{\Delta\sigma_{sat} - \Delta\sigma_{init}}{\Delta\sigma_{init}} \tag{34a}$$

where $\Delta\sigma_{init}$ refers to the first full-cycle value, while $\Delta\sigma_{sat}$ is the saturation value. Similarly for the degree of softening, S:

$$S = \frac{\Delta\sigma_{init} - \Delta\sigma_{sat}}{\Delta\sigma_{init}} \tag{34b}$$

TABLE 6
HARDENING (H) OR SOFTENING (S) BEHAVIOUR OF SOME COMMON ALLOYS

Material	*Temperature (°C)*	*H*	*S*	*Comments*	*Ref.*
$2\frac{1}{4}$Cr–Mo, annealed	538	√		Pronounced above 2% ε_t	53
1Cr–Mo–V, rotor	538		√	Up to 25%	72
9Cr–1Mo, modified	538		√	Some *H* in early stages	61
AISI 1010	650	—	—	Stable	32
304/308, weld	593	√		Typically 8%	73
AISI 347	538 760	√ √		~25% ~13%	44
AISI 316	600	√		up to 138%	74
SUH 660 (A286)	600		√	Some *H* above 1·5% ε_t	64
IN 617	760	√		15–40%	75
Inconel 718	650		√	Slight	34
Hastelloy X	538 760	√ √		75% at 3% ε_t 12% at 3% ε_t	44
OFHC Cu	300 650	√		~100% ~25%	3
Nb–1Zr	700	√		25% at 1·5% ε_p	76

Typical values of *H* for solution-treated 316 steel at 600°C were 130%, while *S* for the cold-worked state was ~16% [74]. Variations of eqns (34a) and (34b) are to use the $\Delta\sigma_{sat}$ term in the denominator [75], or simply to quote the term $\Delta\sigma_{sat}/\Delta\sigma_{init}$ [77]. If materials first harden then soften, for example, the peak state must be defined separately from the end state.

(2) In an earlier study on 1Cr–Mo–V rotor steel, Krempl and Walker [78] suggested the following relation for softening:

$$\sigma/\sigma_{init} = 1 - K \log N \tag{35}$$

where *K* depends on total strain ε_t. It was found that a high-ductility cast softened more than a low-ductility version, and that eqn (35) applied also to the variation in relaxed stress as cycles proceeded in a dwell time test.

(3) A similar relation for hardening of 304 steel at 593°C was proposed by Swindeman *et al.* [23]:

$$\frac{\sigma - \sigma_{\text{sat}}}{\sigma_{\text{init}} - \sigma_{\text{sat}}} = \exp(-\sqrt{kN}) \tag{36}$$

where k relates the rate of approach to stabilisation. These authors further noted that hourglass and cylindrical specimens gave the same hardening behaviour, that dwell periods up to 24 h did not affect the *rate* of hardening given by eqn (36) (though stress levels were lower generally), and that some *secondary hardening* was present. This set in after an apparent period of stability (see Chapter 4).

(4) In their tests on 316 steel at 600°C, Goodall *et al.* [79] correlated hardening behaviour with accumulated plastic strain ε_p^* (i.e. values of ε_p per half-cycle added without regard to sign) the result being:

$$\frac{\sigma - \sigma_{\text{init}}}{\sigma_{\text{sat}} - \sigma_{\text{init}}} = 1 - \exp(-b\varepsilon_p^*) \tag{37}$$

The data from several strain ranges could be described by putting the constant $b = 3{\cdot}85$. It was further observed that hardening was more rapid (in terms of cycles) the longer the dwell time, so that, for example, a specimen which took 500 cycles to saturate with a 5 s dwell in tension would do so in about 25 cycles with a 1000 min dwell. A similar effect occurred with 304 steel at 650°C [80], but entirely the opposite has been reported for 316L steel at 600°C [81].

(5) It follows from eqn (1) that when hardening or softening occurs at constant total strain then $d\varepsilon_p = -d\,\Delta\sigma/E$, so that changes in plastic strain may be used instead. This method was used by Kuwabara and Nitta [70] for a softening steel at 550°C, expressing their results as a softening parameter, ε_r:

$$\varepsilon_r = \ln(\Delta\sigma_{\text{init}}/\Delta\sigma) \tag{38}$$

(6) For experiments performed at constant plastic strain, then a replotting in Fig. 18 of the author's previous results on 316 steel [82] (see also Fig. 4) on a logarithmic basis shows the following hardening relation:

$$\Delta\sigma = CN^{\lambda} \tag{39}$$

where C and λ are constant for given plastic strain. Clearly, this relation is only valid up to a specified maximum $\Delta\sigma_{\text{sat}}$. These results were obtained by

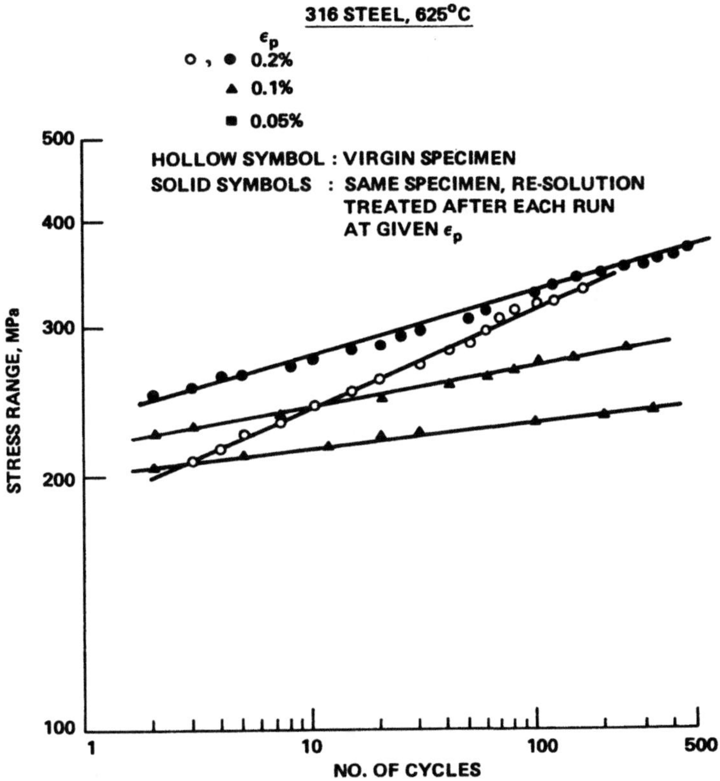

FIG. 18. Cycle by cycle hardening on same specimen using intermediate re-solution treatments of 1000°C. Note that virgin behaviour is not reproduced.

using the same specimen in a cycle–anneal technique, and show quite distinctly how the hardening for virgin material was not reproduced (see also Frontispiece).

Equations (34)–(39) have thus indicated several ways of describing evolutionary behaviour, which could be important in assessing the total response of a structure (see Section 2.10). Changes in Vickers hardness, H_v, or cyclic yield stress, σ_y, can also be used as approximate measures of hardening or softening (see Sections 2.8.9 and 2.8.10 respectively).

If details of the evolutionary path are not required, perhaps the most useful approach is to plot, for each value of strain, the parameters H or S as given by eqn (33) or its variants. Some examples are given in Fig. 19 [83, 84].

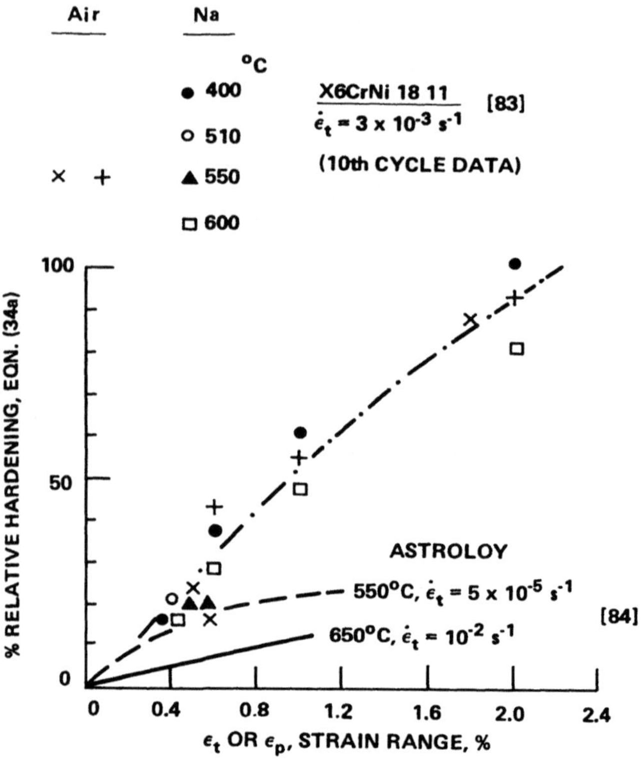

FIG. 19. Indicating how hardening increases with total strain (X6CrNi1811) and plastic strain (Astroloy).

2.6.3. Variation of Cyclic Hardening Exponent

The power law relation, eqn (3), is strictly applicable in the steady state, but if it is assumed to be valid at all stages of evolution then, differentiating logarithmically, we have:

$$\frac{d\,\Delta\sigma}{dN} = \varepsilon_p^{\beta}\frac{dA}{dN} + A\beta\varepsilon_p^{\beta-1}\frac{d\varepsilon_p}{dN} + A\varepsilon_p^{\beta}\ln\varepsilon_p\frac{d\beta}{dN} \tag{40}$$

so that overall behaviour of stress range with respect to cycles is governed by the properties of the quantities dA/dN etc. If it is decided to test at constant ε_p, eqn (40) is immediately simplified, since $d\varepsilon_p/dN = 0$.

Bressers *et al.* [84] tested the nickel-base alloy Astroloy at a variety of constant plastic strain ranges, strain rates and temperatures, and determined intermediate cyclic stress–strain relations during hardening.

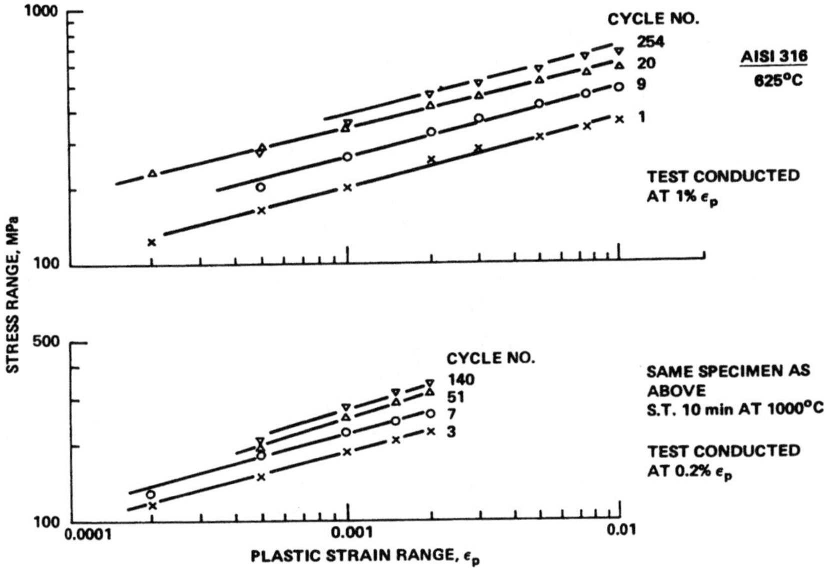

(a)

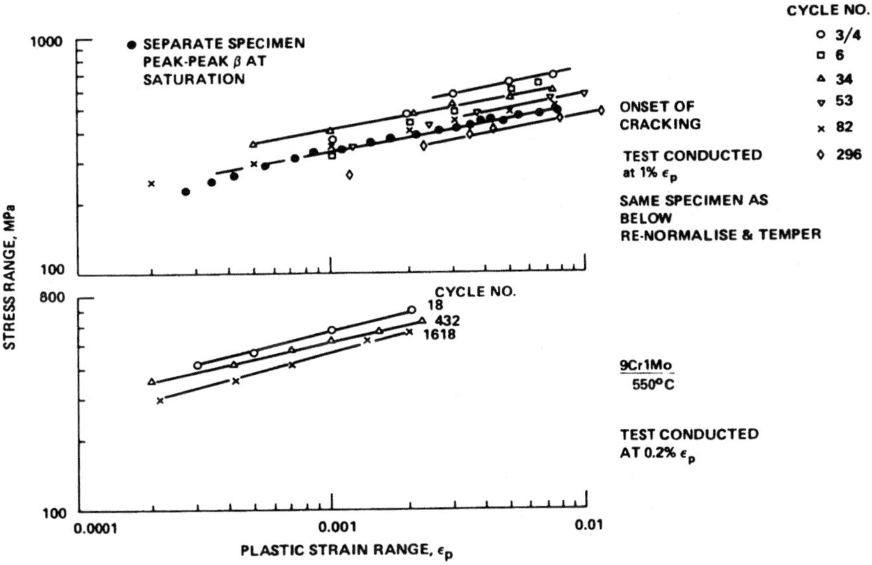

(b)

FIG. 20. (a) Variation of cyclic hardening exponent β with cycles for a hardening steel. (b) Variation of cyclic hardening exponent β with cycles for a softening steel.

TABLE 7
VARIATION OF CYCLIC HARDENING EXPONENT AND STRENGTH COEFFICIENT DURING HARDENING: SEPARATE SPECIMEN TESTS ON ASTROLOY
(Taken from Ref. 84, courtesy of Dr J. Bressers.)

Temperature (°C)	$\dot{\varepsilon}_t$ (s^{-1})	*1st cycle*		*Peak tension*		*Half-life*	
		β	A (MPa)	β	A (MPa)	β	A (MPa)
400	1×10^{-2}	0·079	1 713	0·090	2 085	0·095	2 143
	5×10^{-5}	0·042	1 424	0·049	1 726	0·048	1 695
550	1×10^{-2}	0·071	1 663	0·107	2 258	0·104	2 194
	5×10^{-5}	0·038	1 372	0·110	2 497	0·099	2 285
650	1×10^{-2}	0·052	1 493	0·099	2 147	0·182	3 354
	5×10^{-5}	0·054	1 538	0·070	1 819	0·097	2 115
730	1×10^{-2}	0·071	1 660	0·099	2 067	0·211	3 833
	5×10^{-5}	0·049	1 418	0·049	1 418	0·049	1 418

Table 7, adapted from their data, shows that values of β and A indeed increase with cycles.

In the author's laboratory, similar tests at two constant ε_p levels were conducted on a hardening steel (316) at 625°C and a softening steel (9Cr–1Mo) at 550°C [30]. The values of β and A were deduced from the shape of the individual loops, as discussed previously. The results are given in Fig. 20, where it is seen that β very rapidly adjusted to a fixed value which was maintained nearly constant throughout each test (i.e. $d\beta/dN = 0$ in eqn (40)). Hardening or softening behaviour in these two cases is thus simply described by the variation of A with N. Similar results for a 9Cr–1Mo variant at 538°C have also been reported [62].

2.7. SATURATION

In this section we consider a range of high temperature alloys and summarise some values of A and β according to eqn (3). Cyclic stress–strain behaviour is hence defined in eqn (2), provided also that one knows the variation of E with temperature (see Section 2.8.1).

For some alloys, notably a low-carbon steel at 427 and 538°C [32], a $\frac{1}{2}$Cr–Mo–V steel at 550°C [47] and a 20Cr–25Ni–Nb alloy at 750°C [85], the value of β increases below a plastic strain of 0·007, although it must be noted that very few investigators have explored much below this level at high temperature. In Alloy 800, β *decreases* at low plastic strains [86], the effect being attributed to γ' formation. For 321 steel in the range 180–760°C it has been reported [87] that β *increases* above 0·5% strain. Generally, the effect of dwell time is to lower the peak–peak stress (hence A value) compared with continuous cycling, but some exceptions have been observed, e.g. in a 316 steel at 750°C [88]; see also the previous section on evolutionary behaviour. Effects on the parameters A and β due to other causes are examined in the next section.

Values of A and β are given in Table 8(a) for low-alloy ferritic steels [21, 32, 45, 47, 72, 89–94], Table 8(b) for the higher-alloyed ferritics [30, 62, 63, 95], Table 8(c) for the austenitics [26, 30, 42, 73, 85, 96–101], Table 8(d) for some superalloys [22, 34, 36, 64, 71, 75, 102–107] and, finally, in Table 8(e) for other alloys [3, 108–112]. Each table records also the strain rate (or frequency ν; see Appendix 1) of testing, the temperature, and, perhaps most important, the limits of ε_p between which the values of A and β are valid. The parameter A is defined as the theoretical strength at a plastic strain of unity, so the higher the value of A, the stronger the material at that hypothetical strain. However, values of β also vary greatly amongst the alloys (Tables 8(a)–(e)), and so the relative strengths at more realistic strain ranges must be calculated, either from eqns (2) or (3), as crossovers may occur. It has only been possible to include a selection of data in the tables, and the original references should be consulted for further values.

Very generally, low values of β tend to be favoured by the strain-softening materials (normalised and tempered 1Cr–Mo–V, 2¼Cr–1Mo which are bainitic, normalised and tempered 9Cr–1Mo, 12Cr–1Mo which are martensitic) and some superalloys. High values of β occur with the hardening austenitics. Well annealed ferritic steels which show approximately stable behaviour have β values lying between 0·15 and 0·2.

2.8. OBSERVATION

2.8.1. Effect of Temperature

It is well known that the value of Young's modulus E decreases with increase in temperature, and some examples for typical materials are provided in Fig. 21(a) [5, 39, 113, 114]. The cyclic hardening coefficient A generally also decreases as the temperature is raised, and Fig. 21(b) gives several examples [3, 5, 26, 113]. Figure 21(c) illustrates that there is much variation in the values of β with temperature. The parameters A and β are

TABLE 8
SATURATION STRESS–STRAIN DATA
(a) Low-alloy Ferritic Steels

Material	$\dot{\varepsilon}_t$ *or* v	*Temperature* (°*C*)	*A* (*MPa*)	β	r^2 (*fit*)	ε_p *range* (%)	*Ref.*
AISI 1010	$4 \times 10^{-3}\,s^{-1}$	538	915	0·228		0·02–0·94	32
	$4 \times 10^{-3}\,s^{-1}$	538	389	0·047		0·94–2·0	
0·2C steel	0·4% min^{-1}	500	641	0·109		0·2–2·5	89
1Cr–$\frac{1}{2}$Mo	0·4% min^{-1}	600	709	0·137		0·2–2·0	90
1$\frac{1}{4}$Cr–$\frac{1}{2}$Mo	0·4% min^{-1}	500	909	0·087		0·2–1·6	91
$\frac{1}{2}$Cr–Mo–V	$\sim 10^{-4}\,s^{-1}$	550	3 660	0·307	0·874	0·007–0·27	47
1Cr–Mo–V	1 cycle min^{-1}	538	1 533	0·154		0·07–4·5	72, 92
1Cr–Mo–V, rotor	$1{\cdot}6 \times 10^{-4}\,s^{-1}$	538	1 008	0·09		0·06–1·4	45
2$\frac{1}{4}$Cr–1Mo, N&T	$10^{-4}\,s^{-1}$	500	887	0·075	0·93	0·07–1·6	93

$2\frac{1}{4}$Cr–1Mo, annealed	$4 \times 10^{-3}\,s^{-1}$	593	979	0·173	0·93	0·14–3·0	21
$2\frac{1}{4}$Cr–1Mo, 130 000 h exposure	0·1 cycle min^{-1}	538	912	0·205	0·97	0·6–1·25	94
		(b) Higher-alloyed Ferritic Steels					
9Cr–1Mo, N&T	$\sim 10^{-4}\,s^{-1}$	550	2 417	0·260	0·991	0·014–0·66	30
9Cr–1 Mo, annealed	$\sim 10^{-4}\,s^{-1}$	550	942	0·197	0·997	0·018–0·65	30
9Cr–1Mo, N&T, dwell	$4 \times 10^{-2}\,min^{-1}$	525	1 113	0·124	0·814	0·09–1·60	95
9Cr–1Mo–V–Nb	0·1 Hz	538	676	0·11		0·16–1·0	62
12Cr–Mo–V, N&T	$\sim 10^{-4}\,s^{-1}$	550	2 476	0·273	0·982	0·02–0·69	30
12Cr–Mo–1W–0·3V	40% min^{-1}	600	1 357	0·10		0·6–3·0	63
12Cr	40% min^{-1}	600	559	0·10		0·6–3·0	63

(continued)

TABLE 8—*contd.*
(*c*) *Austenitic Steels*

Material	ε_t *or* v	*Temperature* (°*C*)	*A* (*MPa*)	β	r^2 (*fit*)	ε_p *range* (%)	*Ref.*
AISI 304	0·4% min^{-1}	450	3 614	0·340		0·4–1·8	96
	0·4% min^{-1}	600	833	0·073			
	0·4% min^{-1}	700	650	0·146			
	0·4% min^{-1}	800	278	0·059			
AISI 304, incremental step	$4 \times 10^{-3} s^{-1}$	650	867	0·12		0–1%	26
AISI 316	0·4% min^{-1}	450	4 354	0·338		0·2–1·8	96
	0·4% min^{-1}	600	1 375	0·134			
	0·4% min^{-1}	700	1 222	0·202			
(2nd cast)	0·4% min^{-1}	800	823	0·142			
316, aged	$4 \times 10^{-3} s^{-1}$	593	2 572	0·303	0·941	0·2–1·5	97
316L	3 cycles min^{-1}	650	2 243	0·341	0·989	0·4–6·0	98

AISI 321	0·4% min^{-1}	450	4 199	0·395		0·2–2·0	99
	0·4% min^{-1}	600	1 256	0·171			
	0·4% min^{-1}	700	833	0·142			
AISI 347	0·4% min^{-1}	450	3 345	0·336		0·2–2·0	100
	0·4% min^{-1}	600	1 354	0·163			
	0·4% min^{-1}	700	1 148	0·185			
	0·4% min^{-1}	800	491	0·143			
20/25/Nb	$2 \times 10^{-6} s^{-1}$	750	293	0·029		0·07–10·0	85
16/8/2 ASA, weld	$4 \times 10^{-3} s^{-1}$	593	1 218	0·175		0·15–1·4	101
ERNiCr–3, HAZ	$4 \times 10^{-3} s^{-1}$	538	741	0·154		0·4–3·0	42
304/308, weldment (transverse)	$4 \times 10^{-3} s^{-1}$	482	1 330	0·142	0·671	0·09–1·3	73
316, weld metal (parallel)	$\sim 10^{-4} s^{-1}$	625	933	0·170	0·980	0·02–0·56	30

(continued)

TABLE 8—*contd.*
(*d*) *Superalloys*

Material	$\dot{\varepsilon}_t$ *or v*	*Temperature* (°*C*)	*A* (*MPa*)	β	r^2 (*fit*)	ε_p *range* (%)	*Ref.*
Alloy 800, incremental step	$10^{-3}\,s^{-1}$	500	1 437	0·085		0·01–1·2	102
		600	1 372	0·088		0·01–1·2	
		700	1 038	0·092		0·01–1·2	
		800	596	0·080		0·04–1·3	
Alloy 800	$4 \times 10^{-3}\,s^{-1}$	538	1 853	0·185	0·931	0·05–3·6	107
Alloy 800H	$4 \times 10^{-3}\,s^{-1}$	500	2 778	0·231	0·839	0·13–1·28	103
Hastelloy X	$2 \times 10^{-3}\,s^{-1}$	760	1 772	0·215		0·10–0·6	104
SUH 660 (A286)	0·4% min^{-1}	450	3 212	0·123		0·1–3·0	64
		600	4 390	0·229			
		700	3 873	0·288			
IN 738	$10^{-2}\,s^{-1}$	850	5 288	0·226	0·861	0·04–0·32	105
Inconel 718	$4 \times 10^{-3}\,s^{-1}$	538	2 212	0·056		0·015–4·0	34
		650	2 896	0·138		0·04–4·0	
Waspalloy	3 cycles min^{-1}	650	4 332	0·163	0·988	0·2–2·0	106

Rene 80	0·5% min^{-1}	871	1 666	0·264		0·01–0·25	71
		982	817	0·139		0·01–0·25	
IN 617	0·1 Hz	760	1 590	0·140		0·01–0·8	75
		871	1 142	0·300		0·02–0·6	
MAR-M509	$7 \times 10^{-4} s^{-1}$	700	1 617	0·139		0·1–1·0	22
Waspalloy	10^{-2}, $5 \times 10^{-5} s^{-1}$	650	2 255	0·15		0·28–1·8	36
		(e) Other Alloys					
OFHC Cu	$4 \times 10^{-4} s^{-1}$	300	669	0·20		0·3–4·0	3
		600	148	0·13			
Ferrovac E, iron	$4 \times 10^{-4} s^{-1}$	370	1 527	0·23		0·8–10·0	3, 108
		540	479	0·13			
Zircalloy, weld	1–3 cycles min^{-1}	350	1 262	0·15		0·2–6·0	109
W	$4 \times 10^{-3} s^{-1}$	815	1 007	0·091	0·760	0·4–4·0	110
W (recrystallised)	$4 \times 10^{-3} s^{-1}$	815	1 702	0·291	0·787	0·18–0·95	
Al (pure)	$5 \times 10^{-4} s^{-1}$	205	26	0·132	0·984	0·04–0·30	111
		258	30	0·229	0·982	0·17–0·30	
Ti alloy (air)	0·25, 0·4 Hz	525	2 461	0·163	0·666	0·2–1·4	112
Ti alloy (vacuum)		525	4 413	0·311	0·882	0·45–1·5	

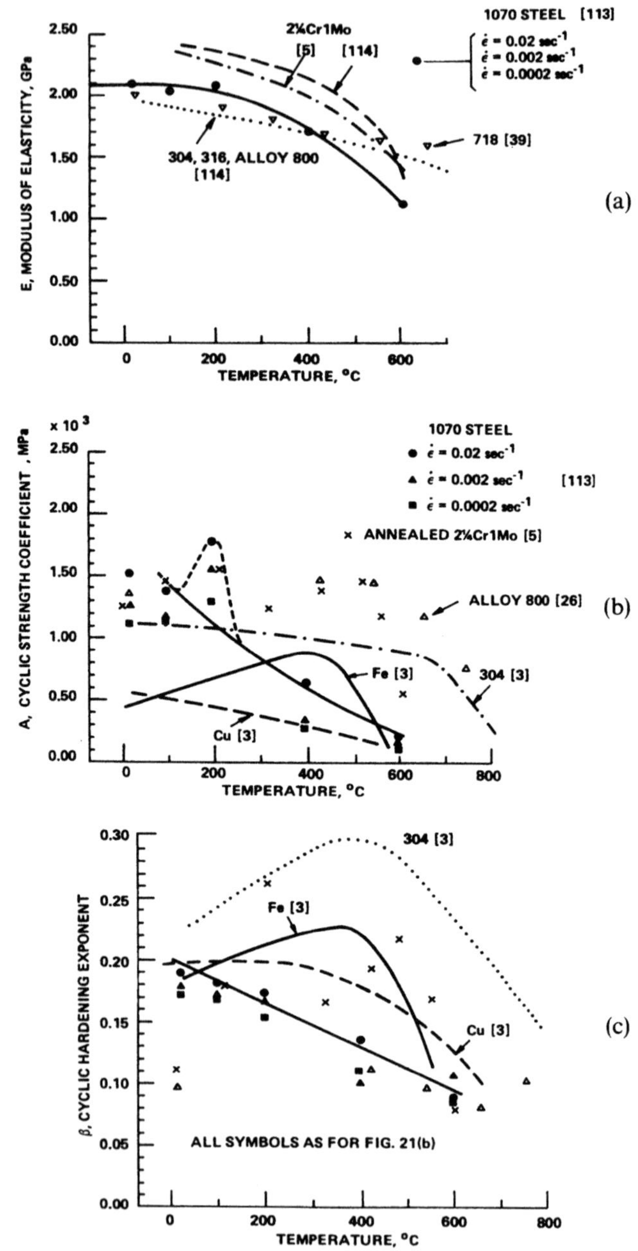

FIG. 21. Effect of temperature on (a) the modulus of elasticity, (b) the cyclic strength coefficient, and (c) the cyclic hardening exponent.

thus best combined via eqn (3) into the stress range to give 0·2% plastic strain (analogous to the 0·2% proof stress): this will generally be found to decrease as the temperature is raised.

Exceptional behaviour is provided by the low-carbon steels, which show a peak in cyclic strength in the range 200–300°C owing to the effects of strain ageing [32, 43, 90, 113, 115–117]. Similarly, the 300 series of austenitic steels show peak evolutionary hardening in the range 400–600°C [37], whereas in a 20Cr–25Ni–Nb steel, for example, peak hardening occurs at room temperature [69].

2.8.2. Effect of Strain Rate

By analogy with monotonic behaviour, it is generally found that as the cyclic strain rate is lowered, stress levels fall everywhere, i.e. the values of A and β both decrease. Wareing *et al.* [85] showed that as the plastic strain rate was reduced from 5×10^{-3} to 2×10^{-6} s^{-1} in a 20Cr–25Ni–Nb alloy at 750°C, the value of A fell from 1008 to 293 MPa, while the corresponding value of β dropped from 0·17 to 0·03. Other examples may be found in Tables 8(a) and 8(c). In contrast, the value of β increased from 0·13 to 0·21 for decreasing strain rate in IN 738 alloy at 850°C [118].

Numerous examples can be found in the literature of materials which demonstrate a negative strain rate effect, that is, the cyclic strength *increases* when the imposed strain rate is reduced (dynamic strain ageing). The critical temperature range for the phenomenon is 200–400°C for the low-carbon steels [108] and the low-alloy steels [5, 53, 94]. However, for ~0·1%C–2Mo steel, the effect was observed in the range 550–650°C [116], while in a 0·1%C–1·5%Mn steel the effect could not be found at any temperature [119].

A critical temperature range of 400–600°C applies for austenitic steels [3, 26, 120] and 730°C in Udimet 500 (for example) [65]. Serrations sometimes appear in the hysteresis loop itself during dynamic strain ageing, as has been reported for 304 steel between 500 and 700°C [121] and Alloy 800H at 550°C [122].

This negative strain rate sensitivity can be quite large. For example, Abdel Raouf *et al.* [108] found about 40% increase in cyclic stress range in Ferrovac E iron at 370°C as the strain rate $\dot{\varepsilon}_t$ was reduced from 2×10^{-1} to 4×10^{-4} s^{-1}.

2.8.3. Irradiation Effect

Studies have mostly been confined to the austenitic steels at fluences of up to 2×10^{26} N m^{-2} (energy > 0·1 MeV). In the following, irradiation temperature approximately corresponds to testing temperature. Under ambient conditions, solution-treated 304 steel demonstrated a marked increase in initial cyclic strength and considerable cyclic softening [123]. At 700°C there was very little effect upon saturation properties [124], while at 500 and 400°C respectively the saturation stress levels at 0·2% plastic strain were about 20% and 60% above their unirradiated counterparts [125].

In solution-treated 316 steel the irradiated cyclic strength at 700°C was some 10% below that of control specimens [124], whereas at 593°C it appeared to be stronger [97]. At 550°C, 20% cold-worked 316 steel had a steady-state stress range about 25% below the unirradiated control [126], although this situation was reversed at 430°C [127]. After irradiation, Incoloy 800 was about 16% stronger in the cyclic state when tested at 700°C [124]. For 304/308 weldments tested at 482°C after irradiation, there was a 17–25% increase in cyclic strength [73].

2.8.4. Environment

A recent comparison [128] of saturation behaviour of 304 steel in vacuum [129] and air [130] at 593°C has shown slight differences easily attributable to normal specimen-to-specimen scatter. Similarly, cyclic softening of IN 738LC alloy for several strain levels at 850°C [131] shows very good concord by the 10th cycle, whether in air or vacuum.

Other workers, however, report differently [128]. Furuya *et al.* [132] compared their results on 316 steel at 2% plastic strain in the range 25–850°C under vacuum with some earlier air data of Kanazawa and Yoshida [37], and found that the vacuum stress ranges were consistently lower. The maximum divergence of 300 MPa occurred at ~450°C. The observation is supported by Asada and Mitsuhashi [17] on an independent supply of 304 steel tested at 650°C. From Coffin's [133] data on A286 at 595°C it can be shown that the alloy appeared 50% stronger at 0·5% strain in air than it was in vacuum. Hoffman *et al.* [112] showed that a Ti alloy at 525°C was some 40% stronger in air at 0·2% plastic strain range than in vacuum. They suggested an oxygen diffusion mechanism to be responsible in air. It is also possible in many of these cases that selective evaporation of certain elements took place in vacuum.

Turning to other environments, Smith *et al.* [134] noted that the value of β in 316 steel increased from 0·31 to 0·41 after exposure to flowing sodium at 600°C for 1500 h. This was attributed to surface carburisation (to a depth of 0·1 mm), the material overall being *weaker* than the annealed control specimens. Later work at 550 and 600°C [135] also showed that the Na-exposed steel did not harden so rapidly with cycles. The same effect was found with 304 steel at 600 and 700°C, but not at 550°C where it became stronger. During hardening of X6CrNi1811 at 550°C, no difference was detected between air and sodium tests [83]. Finally, in testing both 347 steel and Hastelloy X in hydrogen at 538–871°C, Jaske and Rice observed [44] that hardening was marginally slower than in air.

2.8.5. Effect of Prior Exposure

The most marked changes in evolutionary and saturation behaviour in the short term are conferred by simple heat treatment (e.g. normalising and tempering the ferritic steels), by quite elaborate heat treatments in the superalloys which bring about complex precipitation reactions promoting high temperature strength, or by cold working. Prior ageing of austenitic steels, whether by periods of up to a year at the service temperature [136] or at an artificially high temperature [137], reduces the ability to harden and so saturation strengths lie below those of solution-treated material.

The most valuable materials of all to test are those which have had long periods of service exposure, the ageing characteristics of which may indeed have been influenced by actual service cycles. Some examples for ferritic and austenitic steels are given in Fig. 22 [30, 94, 137, 138]. Unfortunately, samples of the materials in their initial state are not always available.

2.8.6. History and Sequential Effects

The observation that a material quickly adapts to its new cyclic state (after a change in strain range) which is independent of previous history, e.g. Goodall *et al.*'s work on 316 steel at 600°C [79], is at best a very good approximation. In earlier studies at room temperature, Landgraf *et al.* [25] recommended that in multiple-step tests, peak stress–strain data obtained in the descending mode were preferable to those obtained while the strain limits were increasing. At elevated temperature, Bernard-Connolly *et al.* [139] determined two 'standard' hardening curves ($\Delta\sigma$ vs. N) for 304 steel at 650°C, using separate specimens at 1·6% and 0·5% total strain ranges respectively. They then attempted to reproduce these curves using another specimen in a multiple-step test using alternate 'blocks' of cycles at these strain levels. It was found that the higher hardening curve could be reproduced accurately but that the lower curve could not, levels lying consistently above despite repeated attempts at softening during the appropriate block. The net effect was to lower the equilibrium value of β, showing that in such tests, low-strain parameters can be conditioned by previous exposure to high strain. However, for a 0·2%C steel at 538°C, no such problem was encountered [140].

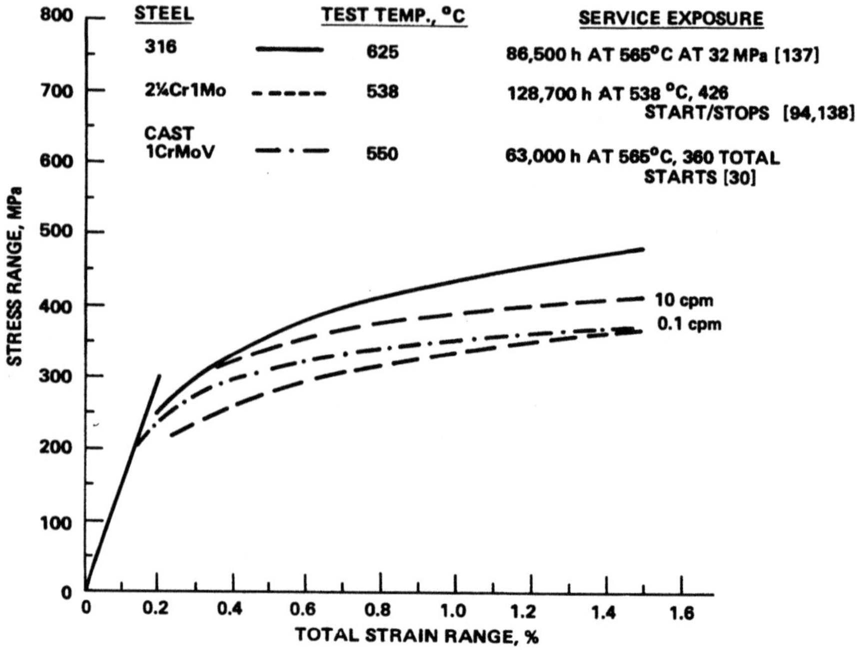

FIG. 22. Cyclic stress–strain curves for service-exposed materials.

A related series of tests was performed by Kanazawa and Yoshida [141], in which the strain *rate* was altered in a stepwise manner on a 17Cr–20Ni–2Mo steel at 450°C and 700°C. These authors experienced no difficulty in reproducing the original $\Delta\sigma$ vs. N curves, even at the lower temperature where a negative strain rate effect was exhibited. Similarly, in temperature cycling tests, Wells and Sullivan [142] found that hardening of Udimet 700 at 315°C was followed by softening at 760°C, the effect being reversible.

During an incremental step test on Alloy 800 at temperatures in the range 427–760°C, and 304 steel in the range 427–650°C, Jaske *et al.* [26] also found that stress ranges at lower strains were also higher than for separate specimen tests because of the memory effect. The same applied to block loading tests, and the authors concluded that single specimen tests were better for reproducing the service situation, where the magnitude of cycles also varies.

These experiments demonstrate that, for some materials at least, the higher degree of hardening at high strain levels is not easily given up during low-strain excursions. Recently, Polak *et al.* have shown [143], for $2\frac{1}{4}$Cr–1Mo steel which softens at 550°C, that random block loading lowers the cyclic curve with respect to that obtained from separate specimen tests.

2.8.7. Thermomechanical Tests

This type of test is undertaken as more accurately representing service conditions than an isothermal test, and is fully discussed in Chapter 3. Because the strength coefficient A and cyclic hardening coefficient β vary with temperature, stress–strain properties would be expected to be represented by some average temperature in the cycle, though this has not been rigorously explored. In studying cyclic softening of $2\frac{1}{4}$Cr–1Mo steel at 538°C, Kuwabara *et al.* [94] showed that an 'in-phase' test in the range 300–538°C consistently gave a greater stress range under the same (total strain) conditions, while the 'out-of-phase' test gave a bigger range still. For a 1Cr–Mo–V rotor steel, however, isothermal tests at 550°C gave the largest stress range [70]. Jaske [43] reported much more hardening for AISI 1010 steel in the range 100–316°C than for isothermal tests at peak temperatures.

Although overall behaviour may be expressed in terms of eqn (3), thermomechanical loops are frequently offset. Typical peak tension:peak compression stress ratios are 0·80 and 1·12 for in-phase and out-of-phase cycling respectively, for 304 steel cycled between 200 and 550°C [144].

2.8.8. Anomalous Behaviour

Hysteresis loops are not always as regular as Fig. 2 etc. might suggest. In their experiments on annealed aluminium, Tsou and Quesnel [111] observed, alongside a sharp fall in $\Delta\sigma$ as the test temperature was raised towards 400°C, a marked discontinuity in the plastic parts of the loop. This shape change was not reflected in the cyclic stress–strain curve (the maximum strain range of the tests was 0·3%). On the other hand, in experiments on a magnesium alloy over the same temperature range, Wareing *et al.* [85] reported a marked kink in the cyclic stress–strain plot, so that below a plastic strain range of 1% the exponent β was negative! At ambient temperature, mechanical twinning was held responsible, whereas at elevated temperature the phenomenon was attributed to grain growth [145]. The hysteresis loop was distorted, its shape reflecting the cyclic stress–strain plot.

During high-strain fatigue of OFHC Cu, the hysteresis loop was symmetrical at all temperatures above ambient except for 650°C, where the compression:tension stress ratio was 1·7 [3]. The effect was attributed to 'diamond grain' formation which occurred at a strain rate of $4 \times 10^{-4}\ s^{-1}$.

TABLE 9

HARDNESS CHANGES OBSERVED AFTER FATIGUE

Material	*Temperature* (°*C*)	*Hardness*, H_v		*Comments*	*Ref.*
		Initial	*Final*		
316, solution treated	600	169 (average)	208 (average)	Test duration 30 h	74
316, cold worked	600	387 (average)	310 (average)	Stress-free exposure for $1{\cdot}2 \times 10^4$ h shows 14% variation maximum	74
9Cr–1Mo, N&T	550	220	180	Average 1000 h fatigue testing time, ~10 000 cycles	
9Cr–1Mo, annealed	550	155	165	Stress-free equivalent 100 h at 750°C	30
Martensitic 12Cr	400–600	238	210–180	0·6%, 2·0% strain	63
$2\frac{1}{4}$Cr–1Mo	593	230 163	200 167	Bainite } single specimen Ferrite }	146
1Cr–Mo–V	538	—	—	$5H_v$ drop at 0·4% ε_t increasing to $35H_v$ at 1·0% ε_t, then constant at $35H_v$; unaffected by dwell	78
1Cr–Mo–V	550	—	—	$8H_v$ drop at 0·06% ε_p increasing to $35H_v$ drop at 1·0% ε_p; less for thermomechanical tests (see text)	70
$2\frac{1}{4}$Cr–1Mo pipe, 128 700 h exposure, 426 shutdowns	538	133	145	No correlation with ε_t (post-service fatigue tests)	138
Cr–Mo–V rotor, 140 000 h exposure, 230 starts	566	260	250	Measured at groove root; beyond 2 mm from root H_v remains at original value (measured after service)	147

2.8.9. Hardness Changes

As might be surmised, the notable cyclic hardening or softening exhibited by many alloys (see, for example, Figs 16 and 17), is also manifested by changes in Vickers hardness, H_v. Several investigators have measured such changes before and after fatigue testing, and their results are summarised in Table 9 [30, 63, 70, 74, 78, 138, 146, 147]. Only in one instance [70] have changes in H_v been related quantitatively to the intervening stages in a test, in this case for softening of a rotor steel at 550°C, by a modified form of eqn (38):

$$\Delta H_v = -\text{constant} \times \ln (\Delta\sigma_{\text{init}}/\Delta\sigma) \tag{41}$$

where the constant equalled 82 for anisothermal testing and 106 for isothermal testing. Table 9 shows that most results have been obtained on softening alloys. An interesting feature reported by Teranishi and McEvily [146] was that in a single specimen of $2\frac{1}{4}$Cr–1Mo the ferritic regions hardened, while the bainitic regions softened from their respective initial conditions after fatigue at 593°C (see Table 9).

A recent examination of some very long-term fatigue tests (up to 25 000 h) of $\frac{1}{2}$Cr–Mo–V steel at 500 and 550°C has shown [148] that, in terms of time, hardness reduced from 160H_v to 140H_v, but the data fit was not improved by plotting hardness against cycles to failure.

2.8.10. Cyclic Yield Stress Changes

The individual stable loops indicated in Fig. 5 suggest that cyclic yield stress increases with increasing strain, and that there is also a change in general deformation behaviour. It is notable that very few authors have used changes in σ_y as an indication of evolutionary behaviour, possibly because its determination is rather subjective; however, Kanazawa *et al.* [63] used the drop in 0·2% proof stress in determining the softening of a 12Cr martensitic stainless steel at temperatures in the range 400–600°C, and demonstrated that a change from 370 to 170 MPa occurred in 200 cycles at a plastic strain of 2·7%, but the process took 6660 cycles when ε_p was 0·3%. Correspondingly fewer cycles were required at intermediate strains, but these authors' results show no obvious correlation of hardness with cumulative plastic strain.

A selection of results involving the cyclic yield stress is given in Table 10 [26, 33, 53, 63, 67, 101, 149]. The 0·2% offset cyclic yield stress seems rather a large criterion to adopt in view of the much smaller plastic strains which occur in service.

TABLE 10
YIELD STRESS CHANGES AFTER FATIGUE

Material	*Temperature* (°*C*)	*Parameter*	*Values/comments*	*Ref.*
$2\frac{1}{4}$Cr–Mo, annealed	538	Ratio cyclic to monotonic yield strength	1·05	33
$2\frac{1}{4}$Cr–1Mo, N&T	482		0·72	
304, solution treated	568		1·99	
$2\frac{1}{4}$Cr–1Mo, annealed	427 539	0·2% offset (cyclic)	220:331 MPa 1st value monotonic, 193:262 MPa 2nd value cyclic	53
Alloy 800	650	Ratio cyclic to monotonic yield strength	4·38 3·0% max. ε_t	26
	760		2·60 $4 \times 10^{-3}\,s^{-1}$	
304	650		2·28 $4 \times 10^{-3}\,s^{-1}$ } 1% ε_t	
			1·86 $4 \times 10^{-5}\,s^{-1}$ } 1% ε_t	
304/308, weld	593 593	As above	2·03 base metal 0·93 weld (parallel)	67
316	593	0·2% offset (cyclic)	258 MPa	149
316, irradiated	593		340 MPa	
16-8-2, weld metal	593	As above	214 MPa monotonic 232 MPa cyclic (comparison specimens) 270 MPa cyclic (after spectrum straining)	101
12Cr steel, martensitic	600	As above	Decreased from 370 MPa to 170 MPa during softening; cycles required depends on strain (see text)	63

2.9. INTERPRETATION

The seat of all stress–strain changes during high temperature fatigue lies in the microstructural response to alternating deformation, and a detailed discussion is reserved for Chapter 4. Very broadly, hardening phenomena involve the *introduction* of an entity to the matrix; for example, a discrete dislocation cell size followed by precipitation processes while softening comes about by a *modification* to an existing microstructure, for instance precipitate shearing or over-ageing by accelerated diffusion, or dislocation redistribution in a cold-worked material. It is clear that many microstructures which are regarded as relatively stable during long-term creep do not remain so under reversed-loading conditions, because activation processes are accelerated many times over (in terms of time).

There now follows a brief list of the major characteristics of fatigue-induced changes. Some attention will be given, however, to possible causes of the back-stress discussed in Section 2.5.5:

(a) Above about 400°C in ferritic [117, 150] and austenitic [68, 98, 151–154] steels, a distinct cell structure evolves. However, the plastic strain range of testing also has a strong influence.

(b) In the austenitic steels particularly, there is an inverse relation between cell size and saturation stress range [151, 155, 156]. Rapid adjustment from one level to another in multiple-step tests is thus best explained by an immediate alteration in cell size.

(c) A decrease in cell size has been detected during evolutionary hardening [68].

(d) Prolonged dwells in austenitic steels [157] or higher temperature testing ($\sim$816°C) [151] enhances the subgrain structure.

(e) Fatigue accelerates fine carbide precipitation, particularly at cell boundaries in 316 steel at 625°C [137, 154], in 304 steel at 593°C [23], in a Co-base alloy at 700°C [22], in an Ni-base alloy at 760°C [75] and in Alloy 800H [122]. On the other hand, some workers report no change in microstructure after fatigue [158]. The susceptibility to change depends on plastic strain range, temperature and cycles undergone, and intrinsic alloy stability.

(f) Sometimes this precipitation causes secondary hardening, i.e. hardening resumes after a period of apparent stability [4, 22, 23, 108, 115, 116].

(g) Growth of carbides on a dislocation substructure may also occur in softening ferritic steels, e.g. 1Cr–Mo–V at 565°C [159] and

9Cr–1Mo at 538°C [62]. There is also evidence that the original lath structure can disappear, being replaced by equiaxed sub-grains [61–63].

(h) In superalloy systems, alloys with ordered precipitates γ' and θ'' harden then soften after a critical accumulated plastic strain [160]. Slip lines on the surface are not detected until peak hardening and early softening in wrought Ni-base alloys is complete [161]. Behaviour resulting from mechanical disordering of precipitates (shear) can be explained from the slip system involved [6, 28].

(i) Outside the classical dynamic strain-ageing range, $2\frac{1}{2}$Cr–1Mo is strengthened by 'interaction solid solution hardening' between 427 and 600°C [4, 162].

(j) Over-ageing by prior stress-free exposure at artificially high temperature cannot model over-ageing during fatigue. Accelerated coarsening during fatigue of superalloys [71, 163] can be explained by vacancy generation, leading to enhanced diffusion rates [164].

(k) Although precipitates may be returned into solid solution by re-annealing, fatigue 'damage' in the form of a cell structure cannot be removed (see Frontispiece). Consequently, first-cycle behaviour can never be exactly reproduced.

Challenger and Vining have recently [165] offered an explanation of the back-stress effect deduced from fatigue of $2\frac{1}{4}$Cr–1Mo steel at 482°C. During continuous cycling, the yield stress occurred 35 MPa *before* zero load was reached during the unloading phase. When a tension or compression dwell was introduced, σ_y occurred 25 MPa *after* zero load. Use of eqn (27) showed that the back-stress had decreased from 72·5 MPa in continuous cycling to 41·5 MPa after a dwell. In continuous cycling, transmission electron microscopy showed uniform dislocations and loops, whereas for asymmetric (hold) cycling there were high dislocation density, tangles, but no loops.

The reduction in σ_b during dwell (at constant ε_t when the peak stress relaxes) was interpreted as thermally activated recovery of pile-ups, the dislocations becoming entangled with carbides and therefore unable to assist reverse deformation (as would normally occur with dislocation shuttling between cell walls during continuous cycling). In order to complete the hysteresis loop, therefore, new dislocations must be created, thus raising the yield point.

2.10. REPRESENTATION

When analysing the behaviour of engineering structures, simplifications are frequently made to the cyclic stress–strain curve (as measured in the laboratory), the approach requiring constitutive equations and elastic–plastic models. One variation takes a '10th cycle' curve, recognised as being a compromise between monotonic and saturation data. This in turn is then assumed to be history-independent [166, 167] and is further approximated by a bilinear representation in which separate constants define the elastic–plastic portion of the curve [167] (see also Fig. 13(a)). For example, in the case of 316 steel at 600°C, yield is assumed to occur at the departure from elasticity, and the elastic–plastic part of the curve is given by $k = 0{\cdot}05$, where k is the ratio of 'plastic modulus' to Young's modulus [168]. If monotonic description is preferred, yield is taken at 0·2% proof stress and k is reduced to 0·016.

Other models employ a hardening rule which specifies the change in yield stress in the course of each subsequent cycle. The two most common are

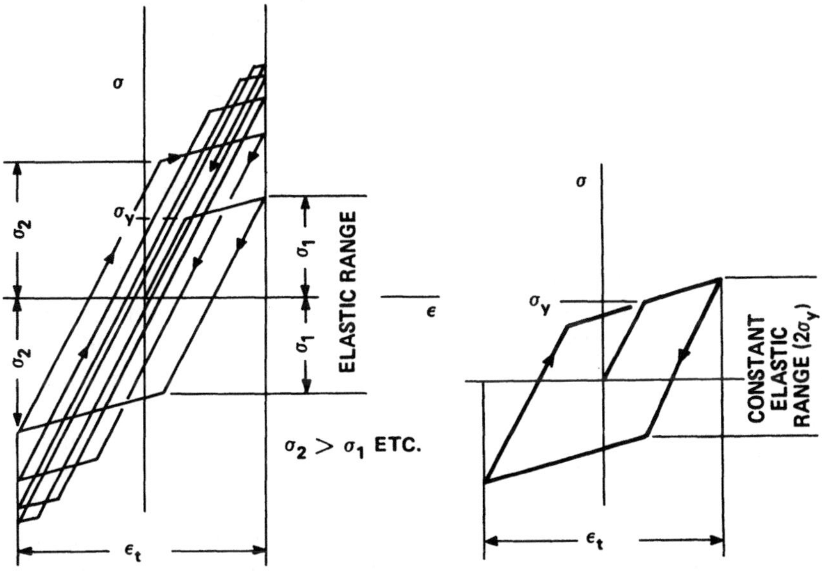

FIG. 23. Two types of hardening. (After Ref. 169, reproduced with permission of ORNL (US Department of Energy).

isotropic hardening and *kinematic hardening*, which are shown in Fig. 23, using the same bilinear convention as Fig. 13 [169]. Isotropic hardening (Fig. 23(a)) between constant limits of ε_t, assumes that yielding on each reversal occurs at a stress level equal to the peak level (at the loop tip) of the previous half-cycle. On the other hand, kinematic hardening assumes that yield always occurs $2\sigma_y$ away from the previous half-cycle tip, i.e. the elastic loading/unloading range is $2\sigma_y$, where σ_y is the initial (first half-cycle) yield stress (see Fig. 23(b)).

In isotropic hardening, therefore, the elastic stress range is forever increasing and, in the limit, deformation is entirely elastic (unlike the model of Fig. 13(a) which permits shakedown to cyclic plasticity). Kinematic hardening, however, most closely resembles the constant plastic strain model (Fig. 13(c)), except that the $2\sigma_y$ unloading criterion incorporates a 'Bauschinger effect' because compressive yield is displaced (see Section 2.5.5).

It is considered [169] that kinematic hardening most closely portrays steady-state behaviour, at least for 304 steel. Hardening predicted by the isotropic model is too severe for this steel. There appears to be no analogous model describing the important cyclic softening class of steels.

2.11. APPLICATION

In the following, we consider the practical application of saturation cyclic stress–strain data.

2.11.1. Stress and Strain Concentration

In conventional plant (rotors, steam turbines etc.) saturation data find most application at a stress concentrator such as a groove in a shaft. The real situation may involve a thermal transient, in turn requiring a costly finite element analysis or a biaxial correction for conditions at the notch root [11]. However, much insight can be gained by considering the uniaxial case and the problem reduces to the following enquiry: Given the stress–strain state remote from the groove, how much further along the stress–strain curve do the yielded conditions at the root of the groove lie? (It may be noted that the remote behaviour may be totally elastic or yield may have occurred.) In the following, we compare three alternative interpretations of each of three simple methods. The parameter of principal interest is the *strain concentration factor* K_ε, although similar expressions may be obtained for the *stress concentration factor* K_σ.

2.11.1.1. Area Construction

The Neuber relation [170], probably the most often used in engineering assessment, is expressed as:

$$K_\sigma K_\varepsilon = K_t^2 \tag{42}$$

where K_t is the *elastic stress concentration factor*. At this stage no distinction is made between full-range and semi-range parameters, so that K_σ is defined as σ_{max}/σ and K_ε is defined as $\varepsilon_{t\ max}/\varepsilon_t$, where the numerator denotes values at the notch root and ε_t means total strain. Equation (42) may thus be interpreted geometrically as follows. In Fig. 24(a), construct OF = K_tOC so that ΔOFB/ΔOAC = K_t^2. The strain concentration factor is thus OE/OC, where ΔODE/ΔOAC = K_t^2, and D represents the enhanced stress–strain state at the root.

The equivalent energy method [171, 172], on the other hand, utilises the actual areas W_1, W_2 under the remote and notch-root curves respectively (shown shaded as OGC and OD′E′ in Fig. 24(a); see also equation (12)) and is expressed as:

$$W_2 = W_1 K_t^2 \tag{43}$$

so that D' is the stress–strain state at the notch root and K_ε = OE′/OC. It is clear that eqn (43) will always predict lower values for K_ε than eqn (42).

Finally, the Stowell–Hardrath [173, 174] relation is written:

$$K_\varepsilon = \frac{K_\sigma(K_t - 1)}{K_\sigma - 1} \tag{44}$$

which may be rearranged as:

$$(K_t - 1)\varepsilon_t \sigma_{max} = (\sigma_{max} - \sigma)\varepsilon_{t\ max} \tag{45}$$

This last equation is expressed in terms of areas in Fig. 24(b), where the large vertical and large horizontal rectangles (shaded) are of equal areas. Thus the product of the pseudo-elastic strain difference and the notch-root stress is equal to the product of the actual stress difference and the notch-root strain.

Although Figs 24(a) and (b) have been drawn for the first loading, it is important to note that eqns (42), (43) and (44) are valid when peak-to-peak (i.e. stress and strain range) values are substituted, which will most often be the practical requirement. This was first shown by Crews and Hardrath [175] who applied eqn (44), taking the compressive tip as the origin (e.g. point σ_C in Fig. 2(b)).

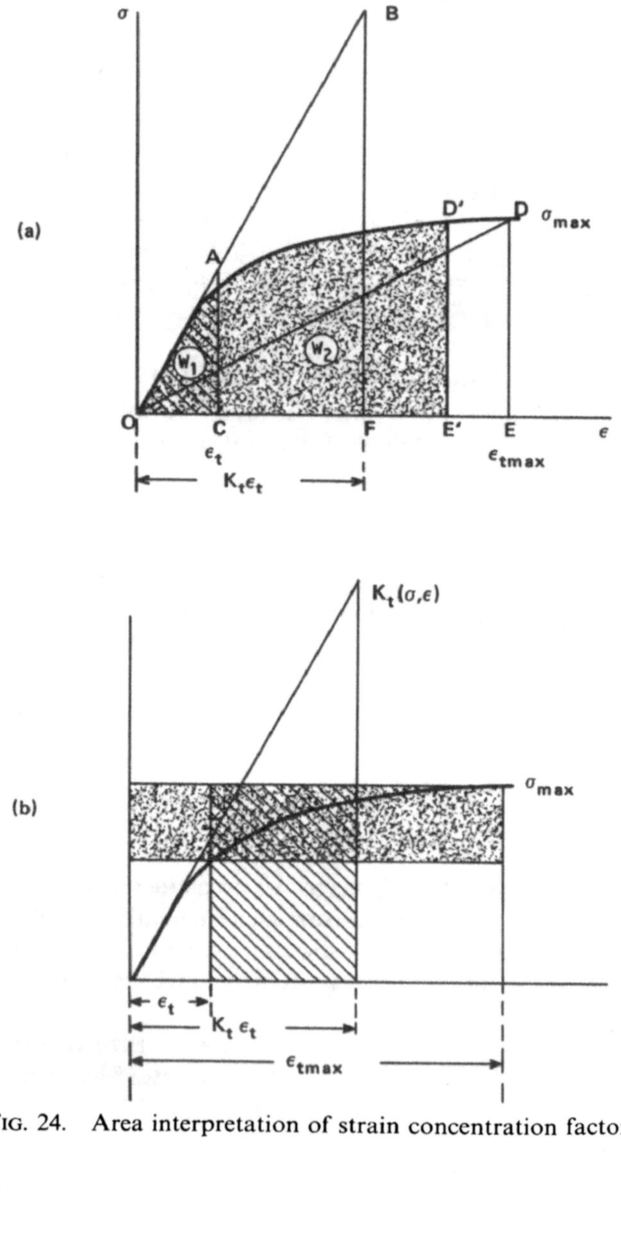

FIG. 24. Area interpretation of strain concentration factors.

2.11.1.2. Curve Construction

Figure 25 shows some stress–strain data (plotted in terms of amplitudes) for mild steel at ambient temperature [52, 175–178], so chosen because concentration factors for this material have been compared with more sophisticated analyses [175, 179–181]. It is assumed in the following, for a semicircular edge notch, that $K_t = 3$ and that the remote values of stress σ and strain ε are 160 MPa and 0·001 respectively. The method is to construct

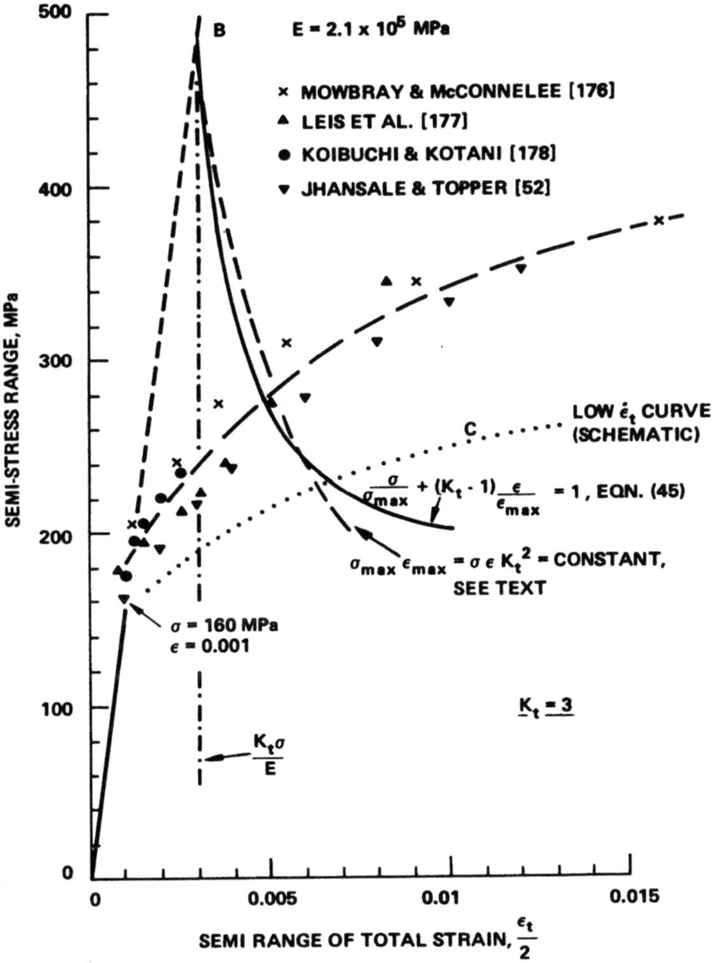

FIG. 25. Curve construction for strain concentration factors.

TABLE 11

EXPRESSIONS FOR STRAIN CONCENTRATION FACTOR K_ε

Expression	*Remote elasticity*	*Remote yield*
Neuber	$K_\varepsilon = K_t^{2\gamma/(\gamma+\beta)}\left(\frac{E^\gamma \varepsilon_t^{(\gamma-\beta)/\beta}}{A^\gamma}\right)^{1/(\gamma+\beta)}$	$K_\varepsilon = K_t^{2\gamma/(\gamma+\beta)}$
Stowell–Hardrath	$K_\varepsilon\left[1 - \frac{E\varepsilon_t^{(\gamma-\beta)/\gamma}}{A}\left(\frac{B}{K_\varepsilon}\right)^{\beta/\gamma}\right] = K_t - 1$	$K_\varepsilon\left[1 - \left(\frac{1}{K_\varepsilon}\right)^{\beta/\gamma}\right] = K_t - 1$
Equivalent energy	$\frac{1}{2E}(\Delta\sigma_{max}^2 - K_t^2\,\Delta\sigma^2) + \frac{1}{(1+\beta)A^{1/\beta}}[\Delta\sigma_{max}^{(1+\beta)/\beta} - K_t^2\,\Delta\sigma^{(1+\beta)/\beta}] = 0$	

OB in Fig. 25 such that the coordinates of B are (3 × 160 MPa, 3 × 0·001), i.e. $\sigma\varepsilon K_t^2 = 1{\cdot}44$ MPa.

(1) For the Neuber route, construct the hyperbola $\sigma_{max}\varepsilon_{t\,max} = \sigma\varepsilon K_t^2$ (see eqn (42)). The intersection of this curve with the stress–strain curve gives the notch-root condition (Fig. 25).
(2) Similarly, for the Stowell–Hardrath route, construct the curve whose coordinates are given by (σ_{max}, $\varepsilon_{t\,max}$) in eqn (45) and determine the point of intersection.
(3) The equivalent energy method uses the elastic energy $\sigma^2/2E$, and the plastic term given by eqn (16). Strain concentration factors are expressed in terms of total strain, and so direct substitution from eqn (7) is not possible. Further, use of eqn (8) in this instance is also not possible, since it does not describe the *shape* of the loading curve, the essence of the equivalent energy method. Consequently, it is recommended that the analytical procedure be adopted (see the next section).

In the example of Fig. 25, semi-range values of σ and ε have been taken, but the construction is equally valid for full-range values.

2.11.1.3. *Analytical Expressions*

The examples in this section use the full-range parameters $\Delta\sigma$ and ε_t.

In the Neuber and Stowell methods, use of eqn (8) is permissible over a given range of strain because, as Fig. 24 indicates, the exact curve shape is not required. Thus, using this equation together with eqn (3), and Hooke's law ($\Delta\sigma = E\varepsilon_t$ when there is no yield remote from the groove) and substituting into eqns (42) and (44) respectively, we find, after collecting terms, the expressions listed in Table 11. The equivalent energy expression is based on an analysis due to Polak who considered the tensile cycle only [172], and in our case is simply a stress-based relation comprising eqns (3), (16) (full-range) and (43). The following points may be noted concerning the expressions in Table 11:

(a) The rather complex Neuber relations arising when purely elastic conditions apply remote from the groove reduce to much simpler forms when yield occurs remotely, since the terms A and B disappear.
(b) The Neuber expressions may be solved explicitly but all others require numerical solution.
(c) There is no simplified form for the equivalent energy method. Further, the equation in Table 11 must be solved for $\Delta\sigma_{max}$, hence $\varepsilon_{t\,max}$ via eqn (2), thus giving K_ε indirectly.

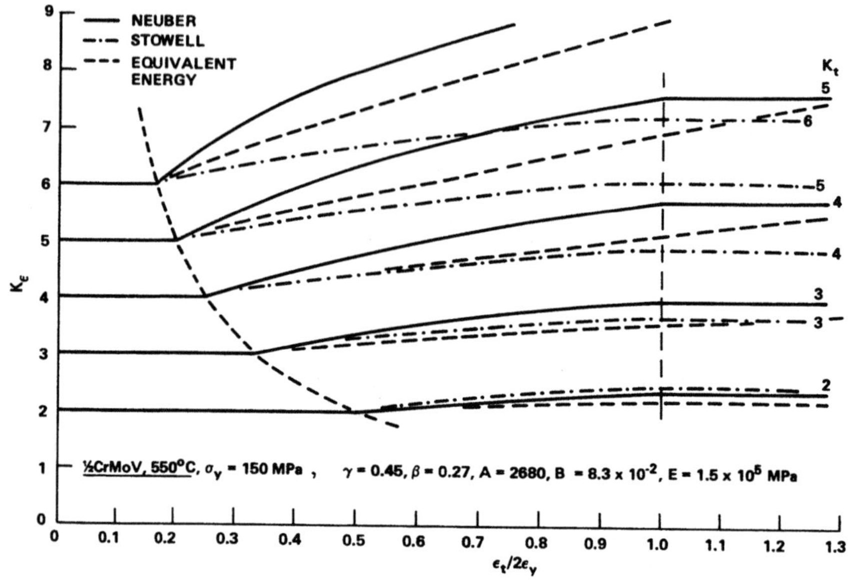

FIG. 26. (a) Analytical expression for strain concentration factor: $\frac{1}{2}$Cr–Mo–V steel at 550°C.

Taking typical values of parameters for 1Cr–Mo–V steel at 550°C [31] (see also Fig. 11) and $\frac{1}{2}$Cr–Mo–V steel at the same temperature [30, 47], Fig. 26 shows the variation of K_ε as the remote strain range is gradually increased to its yield value of $2\varepsilon_y$ and beyond, for several values of K_t. The Neuber and Stowell expressions predict a maximum in K_ε, independent of any further deformation once yield is exceeded remotely, unlike the energy method. It is observed that the higher strain hardening exponent β in the $\frac{1}{2}$Cr–Mo–V steel is responsible for a much more gentle increase in K_ε with increasing strain. It is also seen (Fig. 26) that crossovers in predicted K_ε occur between the Neuber and Stowell methods as the elastic concentration factor K_t is allowed to increase.

This section has examined the three most popular expressions for strain concentration factors: other more complicated variations have been proposed [182].

2.12. DISCUSSION

The foregoing calculations have been treated at some length because it appears that, at the elevated temperatures of interest in this book, there is

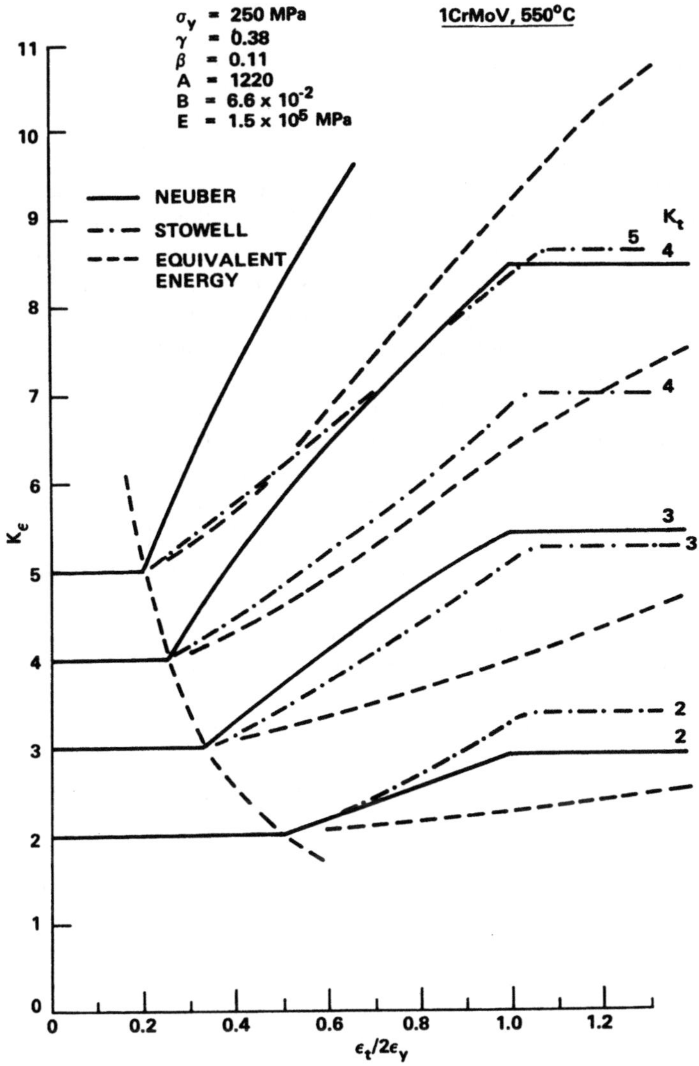

FIG. 26. (b) As for Fig. 26(a), 1Cr–Mo–V steel at 550°C.

no technique for measuring notch-root strains directly, as can be done at room temperature [176]. (However, a recent paper [183] claims that this has been achieved for 304 steel at 600°C.) Calculated strain ranges at the notch root are thus usually verified by a rather indirect method, demonstrating that the number of cycles to crack initiation locally are in agreement with those for plain specimens [92, 148, 184]. The calculations of local strain also require a biaxial correction [11, 176, 185] if transient thermal stressing is to be modelled. Further details lie beyond the scope of this chapter but it can be stated that, in the examples of Fig. 26, the relative order of the Stowell–Hardrath and Neuber methods is maintained, and that the effect of the correction is to increase K_ε at a given value of $\varepsilon_t/2\varepsilon_v$.

The Neuber method can be shown to overestimate local strains, both measured and obtained by finite element analysis [176, 180, 182]. It is claimed that the equivalent energy method gives a better approximation than that of Neuber [172].

All three methods depicted in Fig. 26 were derived under conditions of time-independent deformation and therefore apply only to continuous cycling at high temperature. It is not clear whether the equations apply during creep–fatigue. A limiting solution might be to use the stress–strain curve drawn through the relaxed data after a specified time interval. Since this curve lies below the peak curve, as shown in Fig. 10, the constructions of Fig. 25 all predict an increase in K_ε compared with continuous cycling. This is shown by the intersections at curve C in Fig. 25 which could, for example, represent a stress–strain curve obtained at a very low strain rate at high temperature.

Stress concentration factors in the presence of a temperature gradient, for example in the sudden heating of a rotor shaft, have been derived [186] and are claimed to be equivalent to those in bending [187]. However, the field appears rather controversial, many solutions being obtained only by finite element analysis.

The work of Peterson [188] is a standard handbook for elastic stress concentration for features at room temperature. A similar venture for the elasto-plastic stress and strain concentrations as encountered at elevated temperatures is long overdue.

2.13. CONCLUSION

The value of Young's modulus for metals, being a physical constant, is reproducible, well characterised with temperature, and is not always sensitive to alloying. On the other hand, properties such as fatigue endurance at a given strain level can show notable scatter, since both crack initiation and growth are involved. Straddling these two concepts are the cyclic stress–strain data of the present review, and it is remarkable that in a specimen which may typically contain 10^7 grains, elasto-plastic behaviour is sometimes not very reproducible. This may partly be due to the evolutionary stage: Nahm *et al.* [68] report a large initial specimen-to-specimen difference but nevertheless common saturation behaviour. Once saturation is attained, sequential effects may be involved, as discussed in Sections 2.6 and 2.8.6 respectively. There is no doubt, however, that there are large variations in the values of A and β for a given material, some of which are summarised in Tables 8(a)–(c).

These tables are not intended to be comprehensive data sheets: much more information is available in the references of this chapter, and in seeking the likely behaviour of a particular alloy as many sources as possible should be consulted to obtain the cast-to-cast variation in behaviour.

It is evident from the results obtained so far that there are not enough stress–strain data (a) at low strains or (b) at low strain rates (including dwell), both typical of service experience. It is also evident that the multiple-step test is here to stay, some results indeed giving good agreement with single-specimen tests [31, 167]. Many would argue that the transition from low to high strains and back again in this test produces results which simulate service behaviour. A recent variation [189] is to incorporate several 'minor' cycles within one major hysteresis loop. The worrying aspect is that at critical parts of the loop (notably zero strain) the minor cycles appear 'softer' for a given total strain, i.e. produce more plastic strain than reference data. This has led to large reductions in endurance [189].

Such variable strain tests are clearly valuable for interpreting sources of data scatter and assist in setting a lower bound for design purposes. Hysteresis loop 'probing', i.e. stopping at various parts of the loop in order to separate plastic and creep components, is now becoming quite common [118].

There are several simple empirical relations observed in the laboratory that could, with profit, be used in design. One (already encountered) assumes that a given alloy instantaneously adopts a characteristic stress value on changing the strain. This is probably a safe assumption in the ascending mode, but not so on decreasing the strain. Other examples are: (a) in $\frac{1}{2}$Cr–Mo–V steel, the range of relaxed stress is independent of the initial strain [60]; and (b) the range of softening stress is likewise independent for a 9Cr–1Mo steel [62]. Such rules clearly need justifying for the general case. Otherwise, it would be considered that the precise use of evolutionary and sequential effects in a design assessment would be reserved for the most sophisticated and expensive analyses, in which the margin of safety for the plant item was being seriously eroded. Various limit analyses using (a) monotonic data, (b) saturation data, or (c) saturation data on service-exposed material should satisfy most requirements.

The Ramberg–Osgood equations (2) and (3) [12] have formed the mainstay of this chapter, but there is a sting in the tail. Osgood himself reviewed many alternative expressions [13] and, while claiming that his own was the easiest to manipulate, conceded that others could give a better fit under some circumstances. For example, the equivalent to eqn (3) is provided by eqn (20) of that paper:

$$\varepsilon_p = A'\left(\cosh\frac{\beta'\,\Delta\sigma}{E} - 1\right) \tag{46}$$

Typical values of A' and β' are 10^{-4} and 10^3 respectively. Equation (46) gives a curve on a log $\Delta\sigma$ vs. log ε_p plot which could allow for the apparent increase in β (see equation (3) and Section 2.7), at low strains.

To avoid the confusion which still occurs in interpreting plots of saturation, and to a lesser extent evolutionary, stress–strain behaviour there is now a strong case for an agreed standard in presenting these data. Despite all these considerations, it is surprising that such data have only just found their way into Design Codes (for example the French code RCC-MR [190]); possibly because it is felt that monotonic curves represent lower-bound behaviour typifying the situation after a prolonged dwell. It is nevertheless hoped that the present survey will kindle interest in a topic which deserves wider attention from materials—and engineering—scientists alike.

REFERENCES

1. Coffin, L. F., 1954, A study of the effects of cyclic thermal stresses on a ductile metal, *Trans. ASME* (*Ser. A*), **76**, 931–50.
2. Murphy, M. C., 1981, The engineering properties of wrought copper, *Fatigue Engng Mat. Struct.*, **4**, 199–234.
3. Abdel-Raouf, H., Plumtree, A. and Topper, T. H., 1974, Temperature and strain rate dependence of cyclic deformation response and damage accumulation in OFHC copper and 304 stainless steel, *Metall. Trans.*, **5**, 267–77.
4. Jones, W. B. and van den Avyle, J., 1980, Substructure and strengthening

mechanisms in $2\frac{1}{4}$Cr–1Mo steel at elevated temperature, *Metall. Trans.*, **11A**, 1275–86.
5. Jaske, C. E., Leis, B. N. and Pugh, C. E., 1975, Monotonic and cyclic stress–strain response of annealed $2\frac{1}{4}$Cr–1Mo steel, *Structural materials for service at elevated temperatures in nuclear power generation*, ed. A. O. Schaefer, New York, ASME MPC-1, pp. 191–212.
6. Pineau, A., 1983, High temperature fatigue in relation to microstructure, in *Fatigue at high temperature*, ed. R. P. Skelton, London, Elsevier Applied Science, pp. 305–64.
7. Klesnil, M. and Lukas, P., 1980, *Fatigue of metallic materials*, Amsterdam, Elsevier.
8. Landgraf, R. W., 1977, Cyclic stress–strain responses in commercial alloys, in *Work hardening in tension and fatigue*, ed. A. W. Thompson, New York, AIME, pp. 240–59.
9. Conway, J. B., Stentz, R. H. and Berling, J. T., 1975, *Fatigue tensile and relaxation behaviour of stainless steels*, TID-26135, US Atomic Energy Commission.
10. Ellison, E. G. and Paterson, A. J. F., 1976, Creep fatigue interactions in a 1Cr–Mo–V steel, *Proc. Inst. Mech. Engrs*, **190**, 321–50.
11. Dawson, R. A. T., 1967, Factors to be considered in the design and operation of turbines to prevent failure by thermal fatigue, *Thermal and high strain fatigue*, London, The Metals and Metallurgy Trust, pp. 40–54.
12. Ramberg, W. and Osgood, W. R., 1943, Description of stress–strain curves by three parameters, NACA Tech. Note No. 902.
13. Osgood, W. R., 1946, Stress–strain formulas, *Proc. J. Aeronaut. Sci.*, **13**, 45–8.
14. Thomas, G. and Dawson, R. A. T., 1980, The effect of dwell period and cycle type on high strain fatigue properties of a 1%Cr–Mo–V rotor forging steel at 500–550°C, *2nd Int. Conf. Engng Aspects of Creep*, London, Inst. Mech. Engrs, Paper C335/80, pp. 167–73.
15. Hales, R., 1983, Fatigue testing methods at elevated temperatures, *Fatigue at high temperature*, ed. R. P. Skelton, London, Elsevier Applied Science, pp. 63–96.
16. Wareing, J., 1977, Creep–fatigue interaction in austenitic stainless steels, *Metall. Trans.* **8A**, 711–21.
17. Asada, Y. and Mitsuhashi, S., 1980, Creep fatigue interaction of 304 and 316 stainless steels in air and vacuum, *4th Int. Conf. Pressure Vessel Technol.*, Vol. 1, London, Inst. Mech. Engrs, Paper C48/80, pp. 321–7.
18. Berling, J. T. and Slot, T., 1969, Effect of temperature and strain rate on low cycle fatigue resistance of AISI 304, 316 and 348 stainless steels, *Fatigue at high temperature*, STP459, Philadelphia, ASTM, pp. 3–30.
19. Lundberg, L. and Larsson, L. E., 1980, Low cycle fatigue of 1Cr–0·5Mo at 530°C, Research Report SVF-86, Swedish Inst. for Metals, April 1980.
20. Skelton, R. P., 1985, Review: high strain fatigue testing at elevated temperature, *High Temperature Technol.*, **3**, 179–94.
21. Ellis, J. R., Jakub, M. T., Jaske, C. E. and Utah, D. A., 1975, Elevated temperature fatigue and creep–fatigue properties of annealed $2\frac{1}{4}$Cr–1Mo steel, *Structural materials for service at elevated temperature in nuclear power generation*, ed. A. O. Schaefer, New York, ASME MPC-1, pp. 213–46.

22. Reuchet, J. and Rémy, L., 1983, High temperature low cycle fatigue of MAR-M509 Superalloy I: the influence of temperature on the low cycle fatigue behaviour from 20 to 1100°C, *Mat. Sci. Engng*, **58**, 19–32.
23. Swindeman, R. W., Farrell, K. and Conway, J. B., 1983, Time dependent fatigue of type 304 stainless steel containing microvoids in the starting structure, *Thermal and environmental effects in fatigue: research design interface*, New York, ASME PVP Vol. 71, pp. 121–137.
24. Morrow, J. D., 1965, Cyclic plastic energy and fatigue of metals, *Internal friction and cyclic plasticity*, STP378, Philadelphia, ASTM, pp. 45–87.
25. Landgraf, R. W., Morrow, J. D. and Endo, T., 1969, Determination of the cyclic stress–strain curve, *J. Mat.*, **4**, 176–88.
26. Jaske, C. E., Mindlin, H. and Perrin, J. S., 1973, Cyclic stress–strain behaviour of two alloys at high temperature, *Cyclic stress–strain behaviour, analysis, experimentation and failure prediction*, STP519, Philadelphia, ASTM, pp. 13–27.
27. Wilson, C. J. and Robinson, A., 1977, Control of plastic extension in fatigue tests, *J. Phys. E: Sci. Instr.*, **10**, 129–32.
28. Fournier, D. and Pineau, A., 1977, Low cycle fatigue behaviour of Inconel 718 at 298 K and 823 K, *Metall. Trans.* **8A**, 1095–105.
29. Bressers, J., De Cat, R. and Fenske, E., 1984, Crack initiation and growth in high temperature low cycle fatigue, Report EVR 8808, European Concerted Action COST 50 Materials for Gas Turbines, Commission of European Communities, Luxembourg.
30. Skelton, R. P., 1985, Recent work at Central Electricity Research Laboratories, UK.
31. Marloff, R. H., Johnson, R. L. and Wilson, W. K., 1982, Biaxial low cycle fatigue of Cr–Mo–V steel at 1000°F (538°C), R&D Report 82-1E7-MULTI-P1, Westinghouse.
32. Jaske, C. E., 1977, Low cycle fatigue of AISI 1010 steel at temperatures up to 1200°F (649°C), *J. Press. Vessel Technol., ASME Ser. J*, **99**, 432–43.
33. Jaske, C. E., Frey, N. D. and Utah, D. A., 1979, Low cycle fatigue of heat-treated 1Cr–1Mo–$\frac{1}{4}$V and 2$\frac{1}{4}$Cr–1Mo steels and annealed type 304 stainless steel at high temperatures, *Applications of materials for pressure vessels and piping*, ed. G. V. Smith, MPC-10, ASME, pp. 175–210.
34. Brinkman, C. R. and Korth, G. E., 1974, Strain fatigue and tensile behaviour of Inconel 718 from room temperature to 650°C, *J. Test. Eval.*, **2**, 249–59.
35. Antolovich, S. D., Rosa, E. and Pineau, A., 1981, Low cycle fatigue of René 77 at elevated temperatures, *Mat. Sci. Engng*, **47**, 47–57.
36. Bressers, J. and Verhegge, B., 1980, The effect of creep–fatigue interaction on the life of Ni-base superalloys, Project CCR-2 Cost 50, Petten Establishment Report P/07/80/07.
37. Kanazawa, K. and Yoshida, S., 1975, Effect of temperature and strain rate on the high temperature, low cycle fatigue behaviour of austenitic stainless steels, *Creep and fatigue in elevated temperature applications*, London, Inst. Mech. Engrs, pp. 226.1–226.10.
38. Wareing, J., Vaughan, H. G. and Tomkins, B., 1980, Mechanisms of elevated temperature fatigue failure in type 316 stainless steel, *Creep–fatigue environment interactions, Proc. TMS–AIME Symp. Fall Meeting, Milwaukee*, ed. R. M. Pelloux and N. Stoloff, pp. 129–50.

39. Sanders, T. H., Frishmuth, R. E. and Embley, G. T., 1981, Temperature dependent deformation mechanisms of Alloy 718 in low cycle fatigue, *Metall. Trans.* **12A,** 1003–10.
40. Udoguchi, T. and Wada, T., 1971, Thermal effect on low cycle fatigue strength of steels, *Thermal stresses and thermal fatigue*, ed. D. J. Littler, London, Butterworth, pp. 109–23.
41. Lin, K. C., 1981, High temperature fatigue behaviour of unirradiated V–15Cr–5Ti tested in vacuum, *J. Nucl. Mat.*, **103/4**, 913–18.
42. Brinkman, C. R., Strizak, J. P. and King, J. F., 1978, Elevated temperature fatigue characterization of transition joint weld metal and heat affected zone in support of breeder steam generator development, *Fatigue testing of weldments*, STP648, Philadelphia, ASTM, pp. 218–34.
43. Jaske, C. E., 1976, Thermal mechanical, low cycle fatigue of AISI 1010 steel, *Thermal fatigue of materials and components*, STP 612, Philadelphia, ASTM, pp. 170–98.
44. Jaske, C. E. and Rice, R. C., 1976, Low cycle fatigue of two austenitic alloys in hydrogen gas and air at elevated temperatures, *ASME–MPC Symposium on Creep–Fatigue Interaction*, New York, ASME MPC, pp. 101–28.
45. Jaske, C. E. and Mindlin, H., 1971, Elevated temperature low cycle fatigue behaviour of $2\frac{1}{4}$Cr–1Mo and 1Cr–1Mo–$\frac{1}{4}$V steels, $2\frac{1}{4}$*Chrome* 1 *molybdenum steel in pressure vessels and piping*, New York, ASME, pp. 137–209.
46. Batte, A. D., Murphy, M. C. and Stringer, M. B., 1978, High-strain fatigue properties of a 0·5Cr–Mo–V steam turbine casing steel, *Metall. Technol.*, **5**, 405–13.
47. Skelton, R. P., 1979, The prediction of crack growth rates from total endurances in high strain fatigue, *Fatigue Engng Mat. Struct.*, **2**, 305–19.
48. Challenger, K. D., Miller, A. K., and Langdon, R. L., 1981, Elevated temperature fatigue with hold time in a low alloy steel: a predictive correlation, *J. Mat. Energy Syst.*, **3**, 51–61.
49. Eylon, D., Bartel, T. L. and Rosenblum, M. E., 1980, High temperature low cycle fatigue in beta processed Ti–5Al–5Sn–2Zr–2Mo–0·25Si, *Metall. Trans.*, **11A**, 1361–7.
50. White, D. J. and Horwood, G. P., 1971, Cyclic stress–strain relations from bending and push–pull tests, *J. Strain Anal.*, **6**, 99–107.
51. Masing, G., 1926, Eigenspannungen und Verfestigung beim Messing, *Proc. 2nd Int. Congr. Appl. Mech.*, Zurich, pp. 332–5.
52. Jhansale, H. R. and Topper, T. H., 1973, Engineering analysis of the inelastic stress response of a structural metal under variable cyclic strains, *Cyclic stress–strain behaviour: analysis, experimentation and failure prediction*, STP519, Philadelphia, ASTM, pp. 246–70.
53. Brinkman, C. R., Booker, M. K., Strizak, J. P. and Corwin, W. R., 1975, Elevated temperature fatigue behaviour of a $2\frac{1}{4}$Cr–1Mo steel, *Trans. ASME J. Press. Vessel Technol.*, **97**, 252–7.
54. Halford, G. R., 1963, The strain hardening exponent—a new interpretation and definition, *Trans. ASM*, **56**, 787–8.
55. Halford, G. R., 1966, The energy required for fatigue, *J. Mat.*, **1**, 3–18.
56. Laird, C., 1979, Mechanisms and theories of fatigue, *Fatigue and microstructure*, Ohio, ASM, pp. 149–203.
57. Kuhlmann-Wilsdorf, D. and Laird, C., 1979, Dislocation behaviour in fatigue

II. Friction stress and back stress as inferred from an analysis of hysteresis loops, *Mat. Sci. Engng*, **37**, 111–20.
58. Wilson, D. V., 1965, Reversible work hardening in alloys of cubic metals, *Acta Metall.*, **13**, 807–14.
59. Halford, G. R., 1966, Stored energy of cold work changes induced by cyclic deformation, Ph.D., Thesis, Univ. Illinois.
60. Murphy, M. C., Stringer, M. B. and Batte, A. D., 1979, The elastic–plastic behaviour of a $\frac{1}{2}$%Cr–Mo–V steam turbine casing steel during high strain thermal fatigue, *Proc. DVM Int. Symp. on Low Cycle Fatigue Strength and Elasto-Plastic Behaviour of Materials*, Stuttgart, pp. 53–62.
61. Jones, W. B., 1983, Effects of mechanical cycling on the substructure of modified 9Cr–1Mo ferritic steel, *Ferritic steels for high temperature applications*, Ohio, ASM, pp. 221–35.
62. Ebi, G. and McEvily, A. J., 1984, Effect of processing on the high temperature low cycle fatigue properties of modified 9Cr–1Mo steel, *Fatigue Engng Mat. Struct.*, **7**, 299–314.
63. Kanazawa, K., Yamaguchi, K. and Kobayashi, K., 1979, Temperature dependence of low cycle fatigue behaviour of martensitic stainless steels, *Mat. Sci. Engng*, **40**, 89–96.
64. Kanazawa, K., Yamaguchi, K. and Kobayashi, K., 1983, Temperature and strain rate dependence of low cycle fatigue properties of heat resisting steel SUH 660, *Trans. NRIM*, **25**, 28–36.
65. Coffin, L. F., 1971, The effect of frequency on the cyclic strain and low cycle fatigue behaviour of cast Udimet 500 at elevated temperature, *Metall. Trans.* **2**, 3105–13.
66. Wells, C. H. and Sullivan, C. P., 1965, Low cycle fatigue damage of Udimet 700 at 1400F, *Trans. ASM*, **58**, 391–402.
67. Brinkman, C. R., Korth, G. E. and Beeston, J. M., 1974, Comparison of the strain-controlled low cycle fatigue behaviour of stainless type 304/308 weld and base metal, *Int. Conf. on Creep and Fatigue in Elevated Temperature Applications*, London, Inst. Mech. Engrs, pp. 218.1–218.11.
68. Nahm, H., Moteff, J. and Diercks, D. R., 1977, Structural development during low cycle fatigue of AISI 304 stainless steel at 649°C, *Acta Metall.*, **25**, 107–16.
69. Sumner, G., 1967, The low endurance fatigue behaviour of a 20%Cr–25%Ni–0·7%Nb stainless steel at 25, 650 and 750°C, *Thermal and high strain fatigue*, London, Metals and Metallurgy Trust, pp. 295–311.
70. Kuwabara, K. and Nitta, A., 1977, Isothermal and thermal fatigue strength of Cr–Mo–V steel for turbine rotors, CRIEPI Report E277005.
71. Antolovich, S. D., Liu, S. and Baur, R., 1981, Low cycle fatigue behaviour of René 80 at elevated temperature, *Metall. Trans.*, **12A**, 473–81.
72. Leven, M. M., 1973, The interaction of creep and fatigue for a rotor steel, *Exptl Mech.*, **13**, 353–72.
73. Korth, G. E. and Harper, M. D., 1975, Fatigue and creep fatigue behaviour of irradiated and unirradiated type 308 stainless steel weld metal at elevated temperatures, *Properties of reactor structural alloys after neutron or particle irradiation*, STP 570, Philadelphia, ASTM, pp. 172–90.
74. Plumbridge, W. J., Dalski, M. E. and Castle, P. J., 1980, High strain fatigue of a type 316 stainless steel, *Fatigue Engng Mat. Struct.*, **3**, 177–88.

75. Burke, M. A. and Beck, C. G., 1984, The high temperature low cycle fatigue behaviour of the nickel base alloy IN-617, *Metall. Trans.*, **15A**, 661–70.
76. Nagata, N., Furuya, K. and Watanabe, R., 1979, Low cycle fatigue behaviour of blanket structural materials, *J. Nucl. Mat.*, **85/6**, 839–43.
77. Balina, V. S., Mitrofanov, E. A., Khein, E. A., Zhitkyavichene, V. P. and Medeksha, G. G., 1981, Investigation of low cycle fatigue of rotor and casing steels, *Teploenergetica*, **28**, 50–2.
78. Krempl, E. and Walker, C. D., 1969, Effect of creep rupture ductility and hold time on the 1000°F strain fatigue behaviour of a 1Cr–1Mo–0·25V steel, *Fatigue at high temperature*, STP 459, Philadelphia, ASTM, pp. 75–99.
79. Goodall, I. W., Hales, R. and Walters, D. J., 1980, On constitutive relations and failure criteria of an austenitic steel under cyclic loading at elevated temperature, CEGB Report No. RD/B/N 4916.
80. Berling, J. T. and Conway, J. B., 1970, Effect of hold time on the low-cycle fatigue resistance of 304 stainless steel at 1200°F, *1st Int. Conf. Pressure Vessel Technology, Delft, Holland*, Part II, New York, ASME, pp. 1233–46.
81. Rezgui, B., 1979, Interaction fatigue–fluage effet d'un temps de maintien sur la resistance à la fatigue d'un acier Z2CND 17-13 (Type 316L) à 600°C, CEA Report 4982.
82. Skelton, R. P., 1979, Crack growth with dwell in 316 and 9Cr1Mo steels at high temperature, *Time and load dependent degradation of pressure boundary materials*, IAEA Meeting, Innsbruck, Report No. IWG-RRPC-79/2, pp. 73–87. See also CEGB Note No. RD/L/N 216/78.
83. Huthmann, H., Menken, G. and Husslage, W., 1977, The fatigue behaviour of X6CrNi1811 under sodium corrosion conditions, *Specialists Meeting on Properties of Primary Circuit Structural Materials Including Environmental Effects, Bergisch Gladbach, Germany*, IAEA Report IWG FR/22, pp. 191–202.
84. Bressers, J., Roth, M., Tambuyser, P. and Fenske, E., 1982, The effect of time-dependent processes on the LCF life of gas turbine disc alloys, Final Report Project CCR2, Cost 50 Round II, Report EUR 8162.
85. Wareing, J., Tomkins, B. and Sumner, G., 1973, Extent to which material properties control fatigue failure at elevated temperatures, *Fatigue at elevated temperature*, STP520, Philadelphia, ASTM, pp. 123–38.
86. Nahm, H. and Moteff, J., 1976, Second phase formation and its influence on the fatigue properties of Incoloy 800 at elevated temperatures, *Metall. Trans.*, **7A**, 1473–7.
87. Kanazawa, K., Iwanaga, S., Kunio, T., Iwamoto, K. and Ueda, T., 1969, Investigation of low cycle fatigue at elevated temperatures (factors which govern number of strain cycles to failure), *Bull. JSME*, **12**, 189–99.
88. Udoguchi, T., Asada, Y. and Ichino, I., 1974, A frequency interpretation of hold time experiments on high temperature low cycle fatigue of steels for LMFBR, *Int. Conf. on Creep and Fatigue in Elevated Temperature Applications*, London, Inst. Mech. Engrs, pp. 211.1–211.9.
89. Yoshida, S., Kanazawa, K., Yamaguchi, K., Kobayashi, K., Sato, M., Suzuki, N. and Shiohara, M., 1978, *Elevated temperature fatigue properties of engineering materials. Part III*, Section 6, *Trans. NRIM*, **20**, 184–9.
90. *ibid.*, Section 7, pp. 190–7.
91. *ibid.*, Section 8, pp. 199–206.

92. Marloff, R. H. and Johnson, R. L., 1980, The influence of multiaxial stress on low cycle fatigue of Cr–Mo–V steel at 1000°F, *WRC Bull. 264*, pp. 1–21.
93. Ref. 89, Section 9, pp. 206–15.
94. Kuwabara, K., Nitta, A. and Kitamura, T., 1980, The evaluation of thermal fatigue strength of a $2\frac{1}{4}$Cr–1Mo steel under creep–fatigue interaction, CRIEPI Report E279007.
95. Wood, D. S., Baldwin, A. B. and Williamson, K., 1979, The creep/fatigue behaviour of 9Cr steel at 525°C, *Time and load dependent degradation of pressure boundary materials*, IAEA Meeting, Innsbruck, Report No. IWG-RRPC-79/2 pp. 88–101.
96. Yoshida, S., Kanazawa, K., Yamaguchi, K., Sasaki, M., Kobayashi, K. and Sato, M., 1977, Elevated temperature fatigue properties of engineering materials, Part 1, *Trans. NRIM*, **19**, 247–72.
97. Brinkman, C. R. and Korth, G. E., 1974, Low cycle fatigue and hold time comparisons of irradiated and unirradiated type 316 stainless steel, *Metall. Trans.*, **5**, 792–4.
98. Brun, G., Gauthier, J. P. and Petrequin, P., 1976, Etude de la fatigue oligocyclique d'un acier Z2CND 17-12 (type AISI 316L), *Mem. Sci. Rev. Metall.*, July/August, 461–83.
99. Yoshida, S., Kanazawa, K., Yamaguchi, K., Sato, M., Kobayashi, K. and Suzuki, N., 1978, Elevated temperature fatigue properties of engineering materials, Part II, Section 4, *Trans. NRIM*, **20**, 60–73.
100. *ibid.*, Section 5, pp. 74–83.
101. Raske, D. T., 1978, Low cycle fatigue and cyclic deformation behaviour of Type 16-8-2 weld metal at elevated temperature, *Fatigue testing of weldments*, STP 648, Philadelphia, ASTM, pp. 57–72.
102. Anon., 1983, *NRIM fatigue data sheet no.* 36, Nat. Res. Inst. for Metals, Tokyo, Japan.
103. Teranishi, H. and McEvily, A. J., 1981, On fatigue crack initiation and propagation at elevated temperature, *Advances in fracture research, 5th Int. Conf. on Fracture*, Vol. 5, Oxford, Pergamon Press, pp. 2439–47.
104. Huang, J. S. and Pelloux, R. M., 1980, Low cycle fatigue crack propagation in Hastelloy-X at 25 and 760°C, *Metall. Trans.*, **11A**, 899–904.
105. Nazmy, M. Y., 1983, High temperature low cycle fatigue of IN738 and application of strain range partitioning, *Metall. Trans.*, **14A**, 449–61.
106. Clavel, M., Levaillant, C. and Pineau, A., 1980, Influence of micromechanisms of cyclic deformation at elevated temperature on fatigue behaviour, *Creep fatigue–environment interactions, Proc. TMS–AIME Symp. Fall Meeting, Milwaukee*, pp. 24–45.
107. Conway, J. B., Berling, J. T. and Stentz, R. H., 1972, Low-cycle fatigue and cyclic stress–strain behaviour of Incoloy 800, *Metall. Trans.*, **3**, 1633–7.
108. Abdel Raouf, H., Plumtree, A. and Topper, T. H., 1973, Effects of temperature and deformation rate on cyclic strength and fracture of low carbon steel, *Cyclic stress–strain behaviour, analysis, experimentation and failiure prediction*, STP 519, Philadelphia, ASTM, pp. 28–57.
109. Solomon, H. D., 1976, High strain crack growth in Zircalloy-2 welds, *Nucl. Technol.*, **29**, 86–93.

110. Schmunk, R. E. and Korth, G. E., 1981, Tensile and low-cycle fatigue measurements on cross-rolled tungsten, *J. Nucl. Mat.*, **103/4**, 943–8.
111. Tsou, J. C. and Quesnel, D. J., 1983, Low cycle fatigue of aluminium at elevated temperatures, *Mat. Sci. Engng*, **59**, 99–113.
112. Hoffman, C., Eylon, D. and McEvily, A. J., 1982, Influence of microstructure on elevated temperature fatigue resistance of a titanium alloy, *Low cycle fatigue and life prediction*, STP770, Philadelphia, ASTM, pp. 5–22.
113. Sehitoglu, H. and Morrow, J. D., 1983, Characterization of thermomechanical fatigue, *Thermal and environmental effects in fatigue: research–design interface*, PVP Vol. 71, New York, ASME, pp. 93–110.
114. Anon., 1979, *ASME boiler and pressure vessel code*, Case N47-17, Class 1 components in elevated temperature service, Section III, Division 1, New York, ASME.
115. Coffin, L. F., 1965, The effect of quench ageing and cyclic strain ageing on low carbon steel, *J. Basic Engng Trans. ASME D*, **87**, 351–62.
116. Coffin, L. F., 1964, Cyclic strain and fatigue study of a 0·1%C–2·0%Mo steel at elevated temperatures, *Trans. Metall. Soc.*, *AIME*, **230**, 1690–99.
117. Pohl, K., Mayr, P. and Macherauch, E., 1981, Cyclic deformation behaviour of a low carbon steel in the temperature range between room temperature and 850 K, *Int. J. Fract.*, **17**, 221–33.
118. Buchmayr, B. and Hoffelner, W., 1982, Some interactions of creep and fatigue in IN738LC at 850°C, *Proc. Conf. High Temperature Alloys for Gas Turbines*, Dordrecht (Holland), D. Reidel, pp. 561–74.
119. Coffin, L. F., 1967, An investigation of the cyclic strain and fatigue behaviour of a low-carbon manganese steel at elevated temperature, *Thermal and high strain fatigue*, London, The Metals and Metallurgy Trust, pp. 171–97.
120. Jaske, C. E., Mindlin, H. and Perrin, J. S., 1975, Development of elevated temperature fatigue design information for Type 316 stainless steel, *Creep and fatigue in elevated temperature applications*, London, Inst. Mech. Engrs, pp. 163.1–163.7.
121. Tsuzaki, K., Hori, T., Maki, T. and Tamura, I., 1983, Dynamic strain ageing during fatigue deformation in Type 304 austenitic stainless steel, *Mat. Sci. Engng*, **61**, 247–60.
122. Villagrana, R. E., Kaae, J. L. and Ellis, J. R., 1981, The effect of ageing and cold working on the high temperature low cycle fatigue behaviour of Alloy 800H Part II: Continuous cyclic loading, *Metall. Trans.*, **12A**, 1849–57.
123. De Vries, M. I., Tjoa, G. L. and Elen, J. D., 1979, Effects of neutron irradiation on low cycle fatigue and tensile properties of AISI Type 304 stainless steel at 29 K, *Fatigue Engng Mat. Struct.*, **1**, 159–71.
124. Brinkman, C. R., Korth, G. E. and Beeston, J. M., 1973, Influence of irradiation on the creep/fatigue behaviour of several austenitic stainless steels and Incoloy 800 at 700°C, *Effects of radiation on substructure and mechanical properties of metals and alloys*, STP 529, Philadelphia, ASTM, pp. 473–92.
125. Beeston, J. M. and Brinkman, C. R., 1970, Axial fatigue of irradiated stainless steels tested at elevated temperatures, *Irradiation effects on structural alloys for nuclear reactor applications*, STP 484, Philadelphia, ASTM, pp. 419–50.
126. Grossbeck, M. L. and Liu, K. C., 1981, High temperature fatigue life of Type

316 stainless steel containing irradiation induced helium, *J. Nucl. Mat.*, **103/4**, 853–8.

127. Grossbeck, M. L. and Liu, K. C., 1982, Fatigue behaviour of Type 316 stainless steel irradiated in a mixed spectrum fusion reactor forming helium, *Nucl. Technol.*, **58**, 538–47.
128. Skelton, R. P., 1985, High strain fatigue testing in inert environments, *Techniques for high temperature fatigue testing*, London, Elsevier Applied Science, pp. 117–41.
129. Maiya, P. S. 1981, Effects of wave shape and ultra high vacuum on elevated temperature low cycle fatigue of Type 304 stainless steel, *Mat. Sci. Engng*, **47**, 13–21.
130. Maiya, P. S. and Majumdar, S., 1977, Elevated temperature low cycle fatigue behaviour of different heats of Type 304 stainless steel, *Metall. Trans.*, **8A**, 1651–60.
131. Massarelli, L., Ranucci, D. and Picco, E., 1980, Environmental effects on high temperature low cycle fatigue of cast nickel base alloys, Cost 50—Materials for Gas Turbines, CNR Report R80/3.
132. Furuya, K., Nagata, N. and Watanabe, R., 1980, Low cycle fatigue properties of Type 316 stainless steel in vacuum, *J. Nucl. Mat.*, **89**, 372–82.
133. Coffin, L. F., 1972, The effect of vacuum on the low cycle fatigue law, *Metall. Trans.*, **3**, 1777–88.
134. Smith, D. L., Natesan, K., Kassner, T. F. and Zeman, G. J., 1975, Effects of sodium on the low cycle fatigue behaviour of austenitic stainless steel, *Structural materials for service at elevated temperatures in nuclear power generation*, MPC-1, New York, ASME, pp. 290–301.
135. Natesan, K., Chopra, D. K., Zeman, G. J., Smith, D. L. and Kassner, T. F., 1977, Effect of sodium environment on the creep rupture and low cycle fatigue behaviour of austenitic stainless steel, *Specialists Meeting on Properties of Primary Circuit Structural Materials Including Environmental Effects, Bergisch Gladbach, Germany*, IAEA Report IWGFR/22, pp. 56–62.
136. Flagella, P. N., Denne, J., Kahrs, J., Whitlow, G. A., Schrock, S. L. and Ray, W. E., 1977, Effects of high temperature sodium exposure on the microstructural and time-dependent mechanical behaviour of type 316 stainless steel, *ibid.*, pp. 156–67.
137. Skelton, R. P., 1981, The effect of microstructure and tensile dwell on the growth of short fatigue cracks in 316 steel at 625°C, *Mechanical behaviour and nuclear applications of stainless steels at elevated temperatures*, Book No. 280, London, Metals Society, pp. 129–35.
138. Kuwabara, K. and Nitta, A., 1979, On high temperature fatigue and creep strength of a $2\frac{1}{4}$Cr–1Mo steel pipe material used for a long period, CRIEPI Report E279002.
139. Bernard-Connolly, M., Biron, A. and Bui-Quoc, T., 1978–79, Low-cycle cumulative damage with two strain repeated blocks on a stainless steel at high temperature, *Trans. Can. Soc. Mech. Engrs*, **5**, 173–8.
140. Bernard-Connolly, M., Biron, A. and Bui-Quoc, T., 1978–79, Low cycle fatigue behaviour and cumulative damage effect of SA-516-70 steel at room and high temperature, *4th Int. Conf. Press. Vessel Technol.*, Vol. 1, London, Inst. Mech. Engrs, pp. 297–301.

141. Kanazawa, K. and Yoshida, S., 1972, Effect of temperature and strain rate on the high temperature, low cycle fatigue behaviour of a 17Cr–10Ni–2Mo stainless steel, *Int. Conf. Mechanical Behaviour of Materials*, Vol. II, Society of Materials Science, Japan, pp. 530–9.
142. Wells, C. H. and Sullivan, C. P., 1967, The effect of temperature on the low-cycle fatigue behaviour of Udimet 700, *Trans. ASM*, **60**, 217–22.
143. Polak, J., Vasek, A. and Klesnil, M., 1985, Effect of elevated temperature on the low cycle fatigue of 2·25%Cr–1Mo% steel II. Variable amplitude loading, *Int. Conf. Low Cycle Fatigue—Directions for the Future*, ed. H. Solomon, Philadelphia, ASTM, to be published as STP 942.
144. Kuwabara, K. and Nitta, A., 1976, Effect of strain hold time of high temperature on thermal fatigue behaviour of type 304 stainless steel, *Symposium on Creep–Fatigue Interaction*, New York, ASME–MPC, pp. 161–77.
145. Wareing, J., 1983, Mechanisms of high temperature fatigue and creep–fatigue failure in engineering materials, *Fatigue at high temperature*, London, Elsevier Applied Science, pp. 135–85.
146. Teranishi, H. and McEvily, A. J., 1979, A comparison of the elevated temperature low cycle fatigue behaviour of $2\frac{1}{4}$Cr–Mo–V steel, *Low cycle fatigue strength and elasto-plastic behaviour of materials*, Stuttgart, DVM, pp. 25–38.
147. Hirota, Y., Kadoya, Y., Goto, T., Wake, M. and Fujii, H., 1982, Changes of material properties and life management of steam turbine components under long term service, Technical Report, Mitsubishi Heavy Industries Ltd, October.
148. Beech, S. M., 1982, Unpublished work.
149. Brinkman, C. R., Korth, G. E. and Hobbins, R. R., 1972, Estimates of creep–fatigue interaction in irradiated and unirradiated austenitic stainless steels, *Nucl. Technol.*, **16**, 297–307.
150. Abdel Raouf, H., Topper, T. H. and Plumtree, A., 1973, Effects of grain size and temperature on the cyclic strength and fracture of iron, *Fatigue at elevated temperatures*, STP520, Philadelphia, ASTM, pp. 300–10.
151. Challenger, K. D. and Moteff, J., Correlation of substructure with the elevated temperature low cycle fatigue of AISI 304 and 316 stainless steels, *ibid.*, pp. 68–79.
152. Challenger, K. D. and Moteff, J., 1972, Characterization of the deformation substructure of AISI 316 stainless steel after high strain fatigue at elevated temperatures, *Metall. Trans.*, **3**, 1675–8.
153. Garland, M. and Charrier, J., 1982, Configuration des dislocations engendres par fatigue oligocyclique à 800°C sur un acier austenitique, *Scripta Metall.*, **16**, 671–6.
154. Horton, C. A. P., Lai, J. K. L. and Skelton, R. P., 1983, The relationship between microstructure, fatigue and creep behaviour in a Type 316 stainless steel, *Mechanical properties of structural materials including environment*, Chester, Vol. 1, IAEA Report IWGFR-49, pp. 411–33.
155. Challenger, K. D. and Moteff, J., 1972, A correlation between strain hardening parameters and dislocation substructure in austenitic stainless steels, *Scripta Metall.*, **6**, 155–60.

156. Nahm, A. H. and Moteff, J., 1981, Characterization of fatigue substructure of Incoloy Alloy 800 at elevated temperature, *Metall. Trans.*, **12A**, 1011–25.
157. Rezgui, B., Petrequin, P. and Mottot, M., 1981, Hold time effects on low cycle fatigue properties of 316L stainless steel at 600°C and 650°C, *Advances in fracture research*, Vol. 5, Oxford, Pergamon, pp. 2393–402.
158. Leverant, G. R. and Sullivan, C. P., 1969, The low cycle fatigue of TD-nickel at 1800°F, *Trans. Metall. Soc. AIME*, **245**, 2035–9.
159. Plumbridge, W. J. and Bartlett, R. A., 1977, The metallography of 1Cr–Mo–V steels, Report No. 77/9, Dept Mech. Engng, University of Bristol.
160. Jablonski, D. A. and Sargent, S., 1981, Anisotropic fatigue hardening of a nickel base single crystal at elevated temperature, *Scripta Metall.*, **15**, 1003–6.
161. Merrick, H. F., 1974, The low cycle fatigue of three wrought nickel-base alloys, *Metall. Trans.*, **5**, 891–7.
162. Challenger, K. D., Miller, A. K. and Brinkman, C. R., 1981, An explanation for the effects of hold periods on the elevated temperature fatigue behaviour of $2\frac{1}{4}$Cr–1Mo steel, *J. Engng Mat. Technol., Trans. ASME*, **103**, 7–13.
163. Antolovich, S. D. and Jayaraman, N., 1982, Metallurgical instabilities during the high temperature low cycle fatigue of nickel-base superalloys, *Mat. Sci. Engng*, **57**, L9–L12.
164. Calderon, H. A., Weertman, J. R. and Fine, M. E., 1984, Effect of cyclic plastic deformation on elevated temperature Ostwald ripening of an alloy with coherent precipitates, *Scripta Metall.*, **18**, 587–92.
165. Challenger, K. D. and Vining, P. G., 1983, Substructure and back stress changes resulting from the cyclic loading of $2\frac{1}{4}$Cr–1Mo steel at 755 K, *Mat. Sci. Engng*, **58**, 257–67.
166. Pugh, C. E. and Purdy, C. M., 1977, Material properties requirements for LMFBR structural design: General considerations and data needs, *Specialists Meeting on Properties of Primary Circuit Structural Materials Including Environmental Effects, Bergisch Gladbach, Germany*, IAEA Report IWGFR/22, pp. 227–40.
167. Brinkman, C. R. and Korth, G. E., 1973, Heat-to-heat variations in the fatigue and creep–fatigue behaviour of AISI type 304 stainless steel at 593°C, *J. Nucl. Mat.*, **48**, 293–306.
168. Goodman, A. M., 1981, Design related aspects of creep and fatigue, *Creep and fatigue in high temperature alloys*, London, Elsevier Applied Science, pp. 145–86.
169. Corum, J. M., Greenstreet, W. L., Liu, K. C., Pugh, C. E. and Swindeman, R. W., 1974, Interim guidelines for detailed inelastic analysis of high temperature reactor system components, ORNL Report 5014 (December).
170. Neuber, H., 1961, Theory of stress concentration for shear-strained prismatical bodies with arbitrary non-linear stress–strain law, *Trans. ASME, Ser. E*, **28**, 544–50.
171. Molski, K. and Glinka, G., 1981, A method of elastic–plastic stress and strain calculation at a notch root, *Mat. Sci. Engng.*, **50**, 93–100.
172. Polak, J., 1983, Stress and strain concentration factor evaluation using the equivalent energy concept, *Mat. Sci. Engng*, **61**, 195–200.

173. Stowell, E. Z., 1950, Stress and strain concentration at a circular hole in an infinite plate, NACA Technical Note 2073 (April).
174. Hardrath, H. F. and Ohman, L., 1953, A study of elastic and plastic stress concentration factors due to notches and fillets in flat plates, NACA Report 1117.
175. Crews, J. H. and Hardrath, H. F., 1966, A study of cyclic plastic stresses at a notch root, *Exptl Mech.*, **6**, 313–20.
176. Mowbray, D. F. and McConnelee, J. E., 1973, Applications of finite element stress analysis and stress–strain properties in determining notch fatigue deformation and life, *Cyclic stress–strain behaviour—analysis, experimentation and failure prediction*, STP 519, Philadelphia, ASTM, pp. 151–69.
177. Leis, B. N., Gowda, C. V. B. and Topper, T. H., 1973, Cyclic inelastic deformation and the fatigue notch factor, *ibid.*, pp. 133–50.
178. Koibuchi, K. and Kotani, S., 1973, The role of cyclic stress–strain behaviour on fatigue damage under varying load, *ibid.*, pp. 229–45.
179. Mowbray, D. F. and McConnelee, J. E., 1971, Application of finite elastic–plastic stress analysis to notched fatigue specimens, *1st Int. Conf. Reactor Structural Mechanics*, Berlin, Paper M6/8, pp. 425–41.
180. Ohji, K., Ogura, K. and Takii, H., 1971, Elastic–plastic analysis of notched specimens using the finite element method, *14th Japan Cong. Materials Research*, Kyoto, Japan Soc. Mater. Sci., pp. 170–3.
181. Miki, N., Ohji, K. and Oguwa, K., 1983, Simple methods of estimating the stress and strain at the notch root, *Bull. JSME*, **26**, 331–9.
182. Seeger, T., Beste, A. and Amstutzh, 1977, Elastic–plastic stress–strain behaviour of monotonic and cyclic loaded notched plates, *Fracture 1977, ICF4*, Waterloo, Canada, Vol. 2, pp. 943–51.
183. Ohnami, M., Asada, Y., Sakane, M., Kitagawa, M. and Sakon, T., 1985, Notch effect on low cycle fatigue in creep–fatigue at high temperatures: experiment and FEM analysis, *Int. Conf. Low Cycle Fatigue—Directions for the Future*, Lake George, ASTM, Philadelphia, STP 942.
184. Sakane, M. and Ohnami, M., 1983, A study on the notch effect on the low cycle fatigue of metals in creep–fatigue interacting conditions at elevated temperature, *J. Engng Mat. Technol.*, **105**, 75–80.
185. Gonyea, D. C., 1973, Method for low cycle fatigue design including biaxial stress and notch effects, *Fatigue at elevated temperatures*, STP 520, Philadelphia, ASTM, pp. 678–87.
186. Gonyea, D. C., 1976, Thermal stress concentration factors in large shafts, *Thermal fatigue of materials and components*, STP 612, Philadelphia, ASTM, pp. 30–7.
187. Briner, M. and Beglinger, V., 1975, Thermal fatigue analysis for a large gas turbine rotor, *Creep and fatigue in elevated temperature applications*, London, Inst. Mech. Engrs, pp. 220.1–220.8.
188. Petersen, R. E., 1974, *Stress concentration design factors*, New York, Wiley.
189. Toyoda, H. and Mori, T., 1985, High temperature fatigue on a turbine rotor steel under variable strain ranges, *J. Soc. Mat. Sci., Jpn*, **34**, 208–13.
190. RCC-MR, 1985, *Technical appendix A3, Section II—Materials, design and construction rules for mechanical components of FBR nuclear test islands*, June 1985 edn, AFCEN.

APPENDIX 1: RELATION BETWEEN STRAIN RATE AND FREQUENCY

In the literature, continuous (triangular waveform) cycling is quoted indiscriminately either in terms of frequency ν or reversed total strain rate $\dot{\varepsilon}_t$, the relation between them being

$$\dot{\varepsilon}_t = 2\nu\varepsilon_t$$

Sometimes only the plastic strain range ε_p is given, in which case the corresponding total strain range ε_t may be calculated from eqn (7).

APPENDIX 2: COMPARISON BETWEEN LOCUS AND INDIVIDUAL LOOP SHAPE

The value of A in eqn (3) is the theoretical strength at unit plastic strain range. It follows from the crossovers shown in Fig. 5 that the constant A_i for an individual loop must be less than A_l, the value characterising the locus. The variation in A_i with respect to A_l thus defines the maximum deviation in stress and strain (Fig. 5). Henceforth denote all parameters with the suffix 'i' or 'l' as appropriate and for convenience put $1/\beta = n$.

A2.1. Calculation of β_i

Suppose in Fig. 5 at a given strain range that the locus and individual curves intersect at $\Delta\sigma'$, ε_t'. Then, from eqn (2), since $\varepsilon_{t_i} = \varepsilon_{t_l} = \varepsilon_t'$ and $\Delta\sigma_i = \Delta\sigma_l = \Delta\sigma'$, we find that

$$\beta_i = \beta_l \left[\frac{\ln(\Delta\sigma'/A_i)}{\ln(\Delta\sigma'/A_l)} \right] \tag{A.1}$$

A2.2. Maximum Strain Error

The largest deviation in total strain range occurs at some constant stress range $\Delta\sigma_i = \Delta\sigma_l = \Delta\sigma$. From eqn (2):

$$\delta_1 = \varepsilon_{t_l} - \varepsilon_{t_i} = \left(\frac{\Delta\sigma}{A_l}\right)^{n_l} - \left(\frac{\Delta\sigma}{A_i}\right)^{n_i}$$

and since the maximum deviation occurs when $\mathrm{d}\,\delta_1/\mathrm{d}\,\Delta\sigma = 0$,

$$\Delta\sigma^{n_i - n_l} = \frac{n_l A_i^{n_i}}{n_i A_l^{n_l}} \tag{A.2}$$

A2.3. Maximum Stress Error

The largest deviation in stress range occurs when the tangent moduli $(E_{\tan})_i$, $(E_{\tan})_l$ are equal, i.e. from the third column, bottom row of Table 4:

$$\frac{\Delta\sigma_i^{n_i-1}}{\Delta\sigma_l^{n_l-1}} = \frac{n_l A_i^{n_i}}{n_i A_l^{n_l}}$$

and since $\varepsilon_{t_l} = \varepsilon_{t_i}$ at this point, we have, from eqn (2):

$$\frac{\Delta\sigma_i}{E} + \left(\frac{\Delta\sigma_i}{A_i}\right)^{n_i} = \frac{\Delta\sigma_l}{E} + \left(\frac{\Delta\sigma_l}{A_l}\right)^{n_l}$$

Eliminating $\Delta\sigma_l$ from these two equations and putting $p = n_i - 1$, $q = n_l - 1$, and $Q = n_i A_l^{n_l}/n_l A_i^{n_i}$ for convenience, we find:

$$\frac{\Delta\sigma_i}{E} + \left(\frac{\Delta\sigma_i}{A_i}\right)^{n_i} - \frac{1}{E}(Q\,\Delta\sigma_i^p)^{1/q} - \frac{1}{A_l^{n_l}}(Q\,\Delta\sigma_i^p)^{n_l/q} = 0 \qquad \text{(A.3)}$$

which may be solved numerically for $\Delta\sigma_i$.

A2.4. Numerical Example

Figure 27 illustrates a typical locus (full line), plotted for eqn (2) taking $A_l = 1500$ MPa, $\beta_l = 0{\cdot}2$, $E = 1{\cdot}5 \times 10^5$ MPa. Now assume that $A_i = 1000$ MPa and that the individual loop intersects the locus at

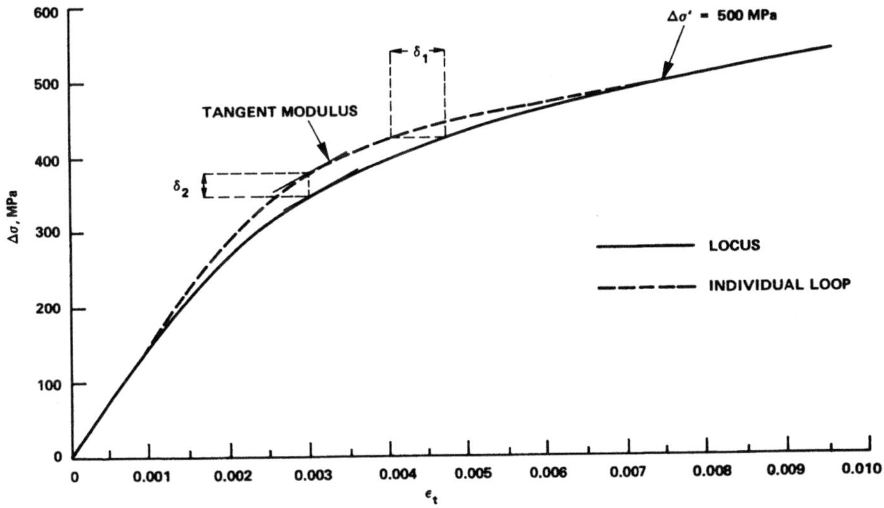

FIG. 27. Maximum deviation in stress, δ_2 and strain, δ_1.

$\Delta\sigma' = 500$ MPa. Then from eqn (A.1), $\beta_i = 0{\cdot}126$ and the corresponding curve, i.e. eqn (2), is shown plotted as the broken line.

The maximum strain error occurs at a stress given by eqn (A.2), that is, 428 MPa. Corresponding strain values from Fig. 27 are:

$$\varepsilon_{t_i} = 4{\cdot}04 \times 10^{-3}; \qquad \varepsilon_{t_1} = 4{\cdot}75 \times 10^{-3}$$

so that $\delta_1 = 7{\cdot}03 \times 10^{-4}$.

The maximum stress error occurs at a $\Delta\sigma_i$ value of 380 MPa, from eqn (A.3). The corresponding strain value from Fig. 27 is $3{\cdot}0 \times 10^{-3}$ and hence $\Delta\sigma_1 = 348$ MPa. The maximum stress error is thus $\delta_2 = (380 - 348) = 32$ MPa.

3

Materials Response to Thermal–Mechanical Strain Cycling

D. A. MILLER and R. H. PRIEST

Operational Engineering Division, Central Electricity Generating Board, Bristol, UK

SUMMARY

Service components which experience thermal fatigue are generally assessed using isothermal test data. A more accurate simulation for these situations is one involving both temperature and mechanical strain cycling. The techniques used to carry out these more complex tests are described in this chapter.

Thermal–mechanical strain cycling data from both continuously cycled and simultaneous cyclic/hold tests are presented for a range of engineering alloys. The response of these materials to thermal cycles is explained in terms of the damage mechanisms operative. The endurance of most materials under thermal–mechanical strain cycling are comparable with those obtained under isothermal conditions at a temperature equal to the maximum used in the thermal cycle. However, in situations where significant internal intergranular damage is induced (i.e. 'in-phase' tests with dwells), anisothermal testing can give lower endurances than those from nominally similar isothermal tests. A comparison of the amount of creep damage formed in these cycles indicates that in-phase cycling exacerbates creep cavity nucleation, thereby giving rise to lower endurances. It is important to establish if such differences in lifetime exist at long dwell times (i.e. >10 hours); if so, then existing design codes may be non-conservative.

The application of the strain range partitioning life prediction technique to anisothermal data is examined. It is shown that simplified versions of the strain range partitioning method can be successfully used to predict endurances from in-phase tests.

3.1. INTRODUCTION

Components in power generating equipment can be subject to a range of loading conditions. Many of these components (e.g. inlet valve chests, turbine blades and rotors) experience cyclic plasticity as a result of start-up and shutdown procedures. This cyclic plasticity is caused by the difference in thermal expansion between the surface of components exposed to temperature transients and the bulk material of those components. Under start-up conditions, compressive plastic yielding may occur in the surfaces of the components, subsequent tensile residual stresses being set up once the bulk material reaches the same temperature as the surface. This process is reversed on shutdown, such that a low-cycle fatigue situation occurs during each start-up, shutdown procedure. The important point to note is that the fatigue cycle does not, in general, arise under isothermal conditions.

In addition to cyclic plasticity, the interspersion of on-load periods at high temperature is likely to cause time-dependent creep-effects, the overall component strain history therefore being made up of simultaneous creep and fatigue processes. The problem of creep–fatigue damage is of particular importance because of the current increased tendency towards 'two-shifting' fossil-fired plant, i.e. adopting start-up and shutdown plant sequences on a daily basis.

Two thermal fatigue damage modes are commonly found in components (such as steam chests and rotors) subject to thermal cycling. These are due to either (1) upshock events or (2) downshock events:

(1) Upshock events occur when the inner surface of a thick-section component at nominally low temperature (room temperature to 200°C) is suddenly subjected to steam at 550°C (e.g. during a rapid start). As indicated above, this causes the surface material to be compressed as it tries to expand against the restraining (cooler) surrounding structure. This can lead to compressive yielding which, on equilibration of temperature through the wall, may result in tensile stresses at the surface during the on-load running period. The yielding process is reversed on shutdown. However, during the running period, on-load stresses give rise to creep damage. Creep damage thus accumulates during each cycle, resulting in creep-dominated crack initiation. A typical example of such a crack (found in a component) is shown in Fig. 1.

(2) Downshock is basically the reverse of the above description. In this

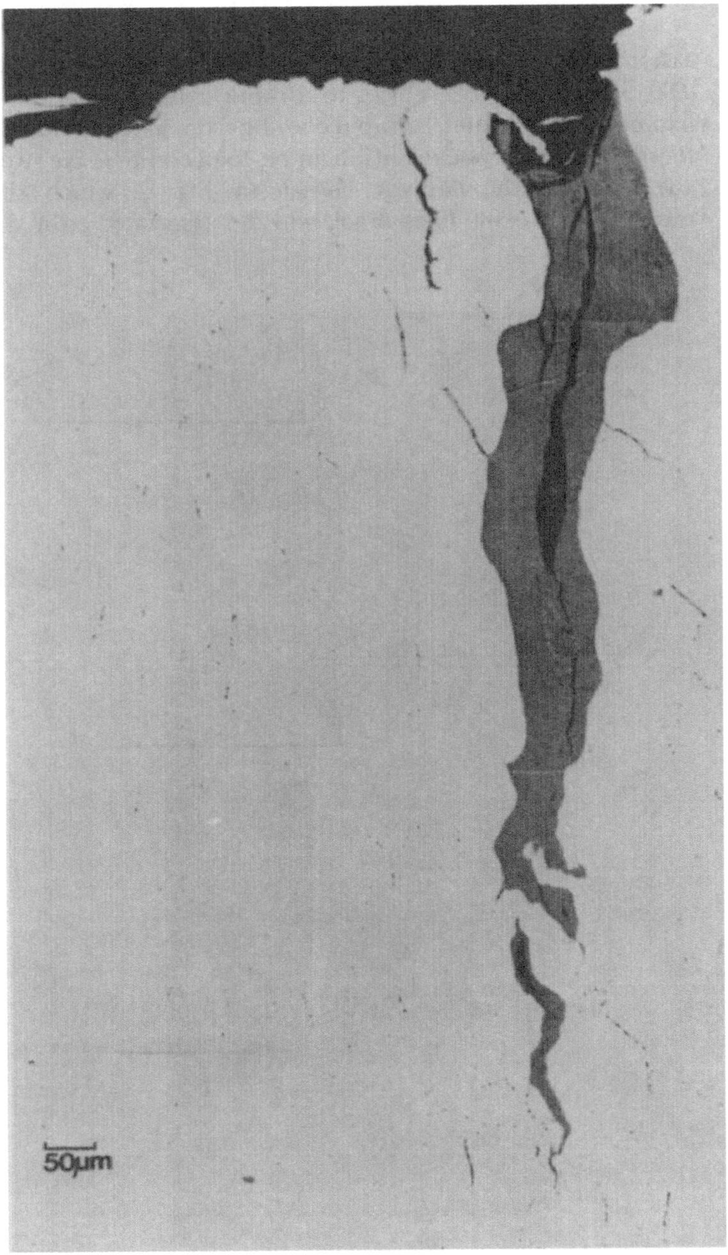

FIG. 1. Upshock damage showing a creep-dominated crack. (Courtesy Dr G. Wigmore, CEGB Operational Engineering Division.)

case, the thick-section component, operating at service temperature (500–570°C), may be subject to a rapid chilling event due to the impingement of water, or rapid cooling from a fast shutdown. This form of thermal cycle results in an on-load compressive stress, and gives rise to the damage shown in Fig. 2, which shows a transgranular oxide-filled crack with no associated creep damage.

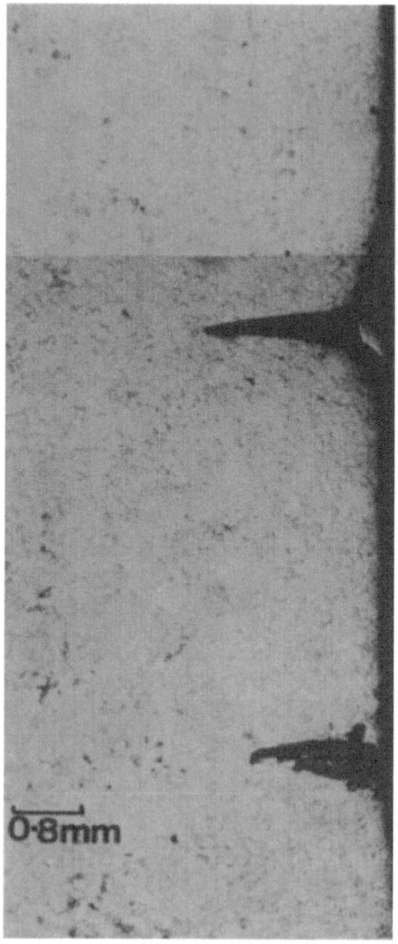

FIG. 2. Downshock damage showing a fatigue-dominated crack. (Courtesy Dr P. Neumann, CEGB Operational Engineering Division.)

In order that the designer can make reliable life predictions for components subject to thermal-cycling conditions, he needs to generate representative data to simulate the various damage mechanisms in operation during service. Such data not only give a basis for safety assessment calculations, but also aid the understanding of the damage processes. Historically, testing techniques which aim to simulate the deformation processes operative in service have usually been simplified. Uniaxial mechanical cycling under isothermal conditions is generally used to represent actual service conditions.

Nevertheless, the question remains as to how representative isothermal tests are of in-service thermal-cycling situations. The aim of this chapter is to examine data that have been generated under complex thermal-cycling test conditions, and to compare these with similar, though much simpler, isothermal data, in order to assess the necessity for anisothermal tests.

Thermal fatigue testing can be divided into two categories [1]. The first includes the case of plastic straining being caused by internal constraint and the second external constraint. Tapered disc specimens, which are alternatively immersed in high and low temperature fluidised beds, model the first category. Such tests provide a reasonably good simulation of the operational conditions encountered by components such as turbine blades, and can readily assess the relative thermal fatigue resistance of different materials. However, stress and strain gradients are set up in the test specimens during such tests and these cannot be measured directly, so that quantitative analyses are usually precluded, although use of finite element techniques has recently improved this situation.

Thermal straining by external constraint was first used by Coffin and Wesley [2], who applied temperature cycles to hourglass-type fatigue specimens held in a restraining jig. More recently [1, 3, 4], cyclic temperature tests with external constraint have been performed using more sophisticated servo-controlled fatigue machines, in which temperature and strain cycles can be applied independently using suitable programming ('thermal–mechanical strain testing'). These tests induce no stress/strain gradients in the test specimens, and experiments generally involve direct strain control using high temperature extensometers, with provision for automatic thermal strain compensation.

3.2. TESTING METHODOLOGY

The majority of high temperature cyclic/hold tests used to simulate service-induced thermal–mechanical strain cycles are performed isothermally.

Isothermal testing is now relatively commonplace. Such tests are usually performed uniaxially on solid cylindrical testpieces, axial extensometers being attached to a known specimen gauge length in order to monitor and control the cyclic excursions. For isothermal testing, steady elevated temperatures are usually obtained by the use of resistance wound furnaces. With this arrangement, heating rates can be slow, it taking tens of minutes, if not hours, to attain the required stable temperatures in the test specimens. However, in contrast to the isothermal tests, heating (and cooling) rates have to be relatively rapid for thermal–mechanical strain cycling tests. This requirement results in a number of restrictions being imposed on specimen design, heating/cooling arrangements and extensometers.

3.2.1. Specimen Design

3.2.1.1. Specimen Profile

In order to perform thermal cycles at a reasonable rate linearly between two set limits (avoiding temperature gradients in the specimens), many workers have opted for hollow specimen designs [3–7], the thickness of the specimen walls determining, to some extent, the speed of the temperature changes. It should be noted, however, that since thermal cycling data may need to be compared with isothermal data generated using solid specimens, the potential difference in response of hollow compared with solid specimens must be taken into account. Beauchamp [3] has observed reduced lives of ~20% and ~10% in his tests on 1Cr–Mo–V steel and 316 stainless steel, respectively, over similar tests performed on solid test specimens. These reductions are well within the expected scatter in the data, however, and are probably due to (a) the increased surface area available for cracking in the hollow specimens (i.e. specimen bore walls as well as external surfaces), and (b) faster failure processes in the tubular specimens, since there is less distance for the surface cracks to propagate before ultimate failure.

The use of hourglass specimens is often adopted in thermal cycling tests. With this specimen design, temperature control need only be restricted to the region around the minimum cross section, where maximum mechanical straining occurs, rather than along some parallel gauge length. Such a profile also ensures that specimens fail in the waisted region, making the use of diametral extensometers more reliable (see Section 3.2.3).

Some workers have indicated that fatigue life can be reduced due to premature crack initiation from surface machine marks [8]. Surface finishing effects are exacerbated by the use of hollow test specimens, and are particularly important in cyclically hardening materials such as stainless steels. To avoid such problems, specimen gauge lengths are usually polished before either isothermal or anisothermal testing, surface scratches of $<1\ \mu m$ deep being permissible [3, 7–9].

3.2.1.2. Gripping

The essential requirements of good specimen gripping are the same for both anisothermal and isothermal testing, namely:

(a) there is good alignment for axiality;
(b) there is positive gripping, so that no backlash occurs when going through zero load;
(c) the specimen is easily disassembled after test.

The two most popular methods of gripping are through the use of button-headed specimens with split collets [3, 5, 10], or thread-ended specimens with or without a taper [4, 6]. In the latter case, the thread takes the load while the taper ensures alignment. However, it is likely that after much use, good alignment may well be impeded with this method due to metal-to-metal contact on both thread and taper. The button-headed approach is therefore generally regarded as being the most reliable.

3.2.2. Temperature Control

3.2.2.1. Heating

There are a number of alternative heating techniques which have a rapid response and can be used for localised heating in thermal–mechanical tests.

Some workers [11] have chosen the technique of heating the insides of hollow test specimens, leaving the outsides free for the attachment of an extensometer. Others [12–14] have used direct resistance methods, in which a high alternating current at low voltage is applied directly to the specimen. Unfortunately, with both these methods there is increased testing complexity since cables must be attached to the specimens, the specimen grips needing insulation from the rest of the testing rig. A more serious disadvantage with the direct resistance method is that once the specimens crack, it is likely that localised heating will occur at the crack tip. This may cause error in the results due to premature failure or spoiled metallographic fracture surfaces.

Perhaps the most popular method of heating for thermal–mechanical strain testing is induction heating. With this method, water-cooled induction coils are placed around the specimen and an a.c. current is passed through them, causing eddy currents to be set up in the specimens [15]. These currents are parallel to the expected crack system, so no local hot

TABLE 1
SUMMARY OF HEATING TECHNIQUES USED FOR ANISOTHERMAL TESTING

Method	*Ref.*	*Advantages*	*Disadvantages*
Internal SiC rod	11	Outside of specimen; free to mount extensometer	Specimen gripping complicated by wiring; thermal gradients
Direct resistance	12–14	Fast response times; outside of specimen free to mount extensometer	Testing complicated by attachment of leads/ insulation of specimen; localised crack-tip heating
Induction	15, 16	Fast response times; control of temperature gradients by coil design	Space for extensometer limited; possible electrical pick-up in control circuitry

spots around the cracks are likely. Direct induction heating also has the capacity for very high heating rates. Indeed, by changing coil design, higher or lower heating rates may be obtained. However, one disadvantage of this method is that once induction coils have been placed around the test specimen, only limited space is left for an extensometer. Further, care also has to be exercised to ensure no induction pick-up in any thermocouple or servo-control circuitry [16]. This problem may, of course, be reduced if an optical pyrometer is used for temperature control (see Section 3.2.2.3).

The relative merits/disadvantages of the various heating techniques are summarised in Table 1.

3.2.2.2. Cooling

Cooling rates usually determine the speed of cycling. However, there are a number of ways of increasing cooling rates in a controlled manner. For example, a specimen grip assembly with water-cooled channels can be used [3]. Alternatively, compressed cooling air can be passed through the centre of hollow specimens [7], although this may lead to shock cooling and unacceptable thermal gradients. Cool air manifolds external to the specimen gauge lengths have also been used [5].

3.2.2.3. Temperature Measurement

Thermocouples and optical pyrometers are the two methods of

temperature measurement most often used for anisothermal testing [5, 17]. It should be noted that some workers have observed premature crack initiation to occur at spot welds, particularly in cyclically hardening materials [3]. To avoid such problems, Beauchamp and Ellison [18] used a thermocouple which was first flattened and then pulled against the surfaces of their 316 steel specimens using a lightly loaded wire. The large surface area of contact ensured good heat transfer, while the flattened couple avoided any problems with convection currents set up around the specimen. The only disadvantage of this method is that valuable working space outside the specimen gauge length is taken up with accommodating the thermocouple support wires.

The use of optical pyrometers [5] avoids problems of space limitation, premature failure at thermocouples and, when used, RF pick-up from induction heaters. However, such methods give less accurate temperature measurement.

3.2.3. Extensometers

The requirements of a good extensometer are that it records displacements accurately and repeatedly. In addition, the attachment of the extensometer to the specimen should not affect the material behaviour in any way (see Chapter 6).

For workers who use hourglass specimen designs, the use of diametral extensometers is mandatory [3, 6, 14]. There appear to be three basic configurations of diametral extensometer. These are shown in Fig. 3. The major difference between the first two configurations (Figs 3(b) and (c)) lies in the relative positions of hinge, specimen and displacement measuring device [3, 19]. In both cases, however, a mechanical magnification of the displacements can be obtained, which is an advantage when measuring small diametral strains. The third type of diametral extensometer has no mechanical magnification and uses probes which are diametrically opposed and which move radially [20–22] (Fig. 3(d)).

The design of diametral extensometers must be such that unwanted thermal displacements in the extensometers themselves are not picked up. This is usually facilitated by the use of materials with low thermal coefficients of expansion, such as Invar or fused silica. A potential disadvantage of diametral extensometers is that errors may occur in diametral strain readings due to oxide build-up between the extensometer probes and the specimen surface. However, except at very high temperatures this would be expected to be small, and in any event would result in a small ratchetting effect on mean strain as the test progresses [16].

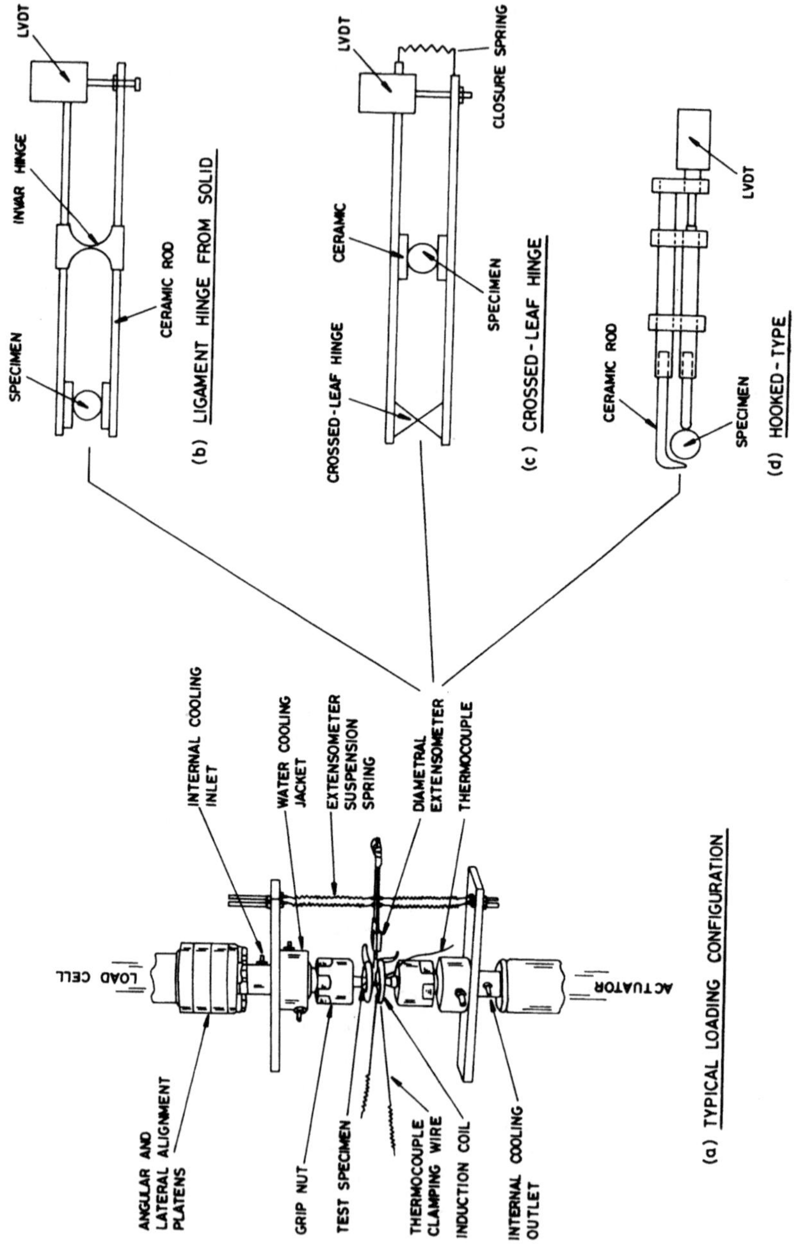

FIG. 3. Extensometer configuration. (After Refs 3 and 18.)

Despite the vast range of diametral extensometers, the majority of isothermal low-cycle fatigue data are obtained from tests performed under axial strain control. It is therefore advantageous if diametral strain measurement can be converted to equivalent axial strain values to enable direct comparisons of the data. This is usually achieved using analogue circuits [3, 6, 19].

The hollow test specimens employed by Hopkins [5] had parallel gauge lengths. These specimens incorporated internal ridges for extensometer attachment, displacements being measured between ridges using a quartz rod and tube arrangement attached to an LVDT. Although such an arrangement is complicated to mount, it has the advantage of keeping the extensometer out of the way of the induction heating coils. Kuwabara and Nitta [17] and Fujino and Taira [7] also used parallel gauge lengthed hollow specimens, a contact-type axial pi-gauge extensometer being attached to the outer surface of the specimens.

3.2.4. Thermal Cycles

Isothermal testing generally includes a component of creep which is introduced into the normal fatigue cycle by means of a constant strain tensile (and/or compressive) dwell or by reducing the tensile (or compressive) strain rate in the cycle; sequential fatigue–creep tests can also be performed (see Chapter 4).

For thermal–mechanical cycling rates, temperature and mechanical

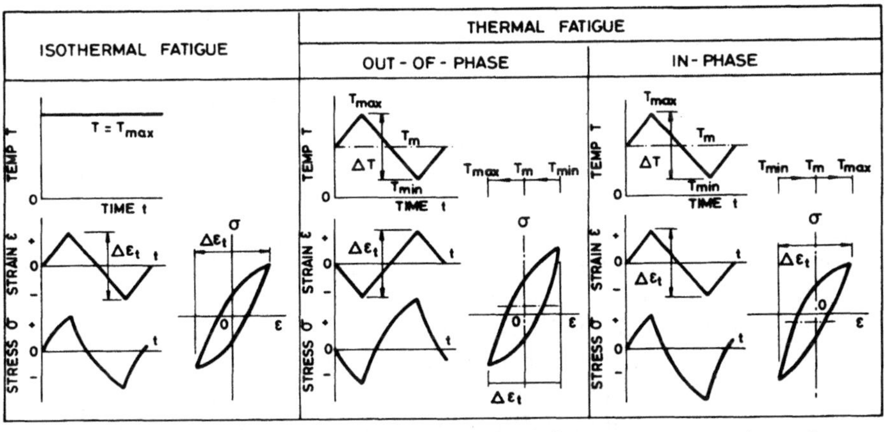

FIG. 4. Schematic diagram showing waveforms of temperature, strain and stress in thermal and isothermal fatigue tests [4]. (Reproduced with permission of ASME.)

strain cycles are applied simultaneously to the test specimens. Tests can either be performed 'in-phase' (IP), in which maximum tensile straining occurs at maximum temperature, or 'out-of-phase' (OP), in which maximum compressive straining occurs at maximum temperature. Figure 4 indicates these cycles [4]. As with isothermal low-cycle fatigue testing, hold periods can also be introduced at specified points around the thermal–mechanical fatigue loops. Thus, IP dwell tests include the hold periods at peak tensile strain and maximum temperature, whereas OP dwell tests include hold periods at peak compressive strain and maximum temperature. Data obtained from thermal–mechanical strain tests are reviewed in the next two sections and are presented as functions of either total or inelastic strain range. For simplicity, the test results have been subdivided as follows:

(1) thermal–mechanical fatigue (i.e. no dwells);
(2) thermal–mechanical creep–fatigue (i.e. with dwells).

3.3. THERMAL–MECHANICAL FATIGUE TEST DATA

3.3.1. Ferritic Steels

Studies in the thermal fatigue properties of ferritic steels [23–27] have revealed that over the inelastic strain range 0·1–1·0% IP tests exhibited the longest endurances compared with OP and isothermal tests. Indeed, the latter cycle types often produced similar endurances. Examples of this behaviour are shown in Figs 5 and 6.

Kuwabara and Nitta [23], in their study on 1Cr–1Mo–$\frac{1}{4}$V steel, observed that at strain ranges >0·50% the number of surface cracks in their OP test specimens was far greater than that found in the other two cycle types. In all tests surface defects were transgranular. Further, in the IP tests, internal intergranular damage was also observed. Although the total number of surface cracks in the isothermal and IP test specimens was almost the same, the number of large cracks (>0·2 mm) present was smaller in the IP cycle tests (Fig. 7). It was therefore postulated that fracture, and hence the increased endurance in the IP cycle tests, was due to the increased number of cycles required to cause linking between internal grain boundary defects and the small surface cracks observed in these tests. In contrast, the reduced endurances observed in the OP and isothermal tests was attributed to the rapid growth of the large transgranular cracks.

The nucleation of surface fatigue cracks appeared to be easier and

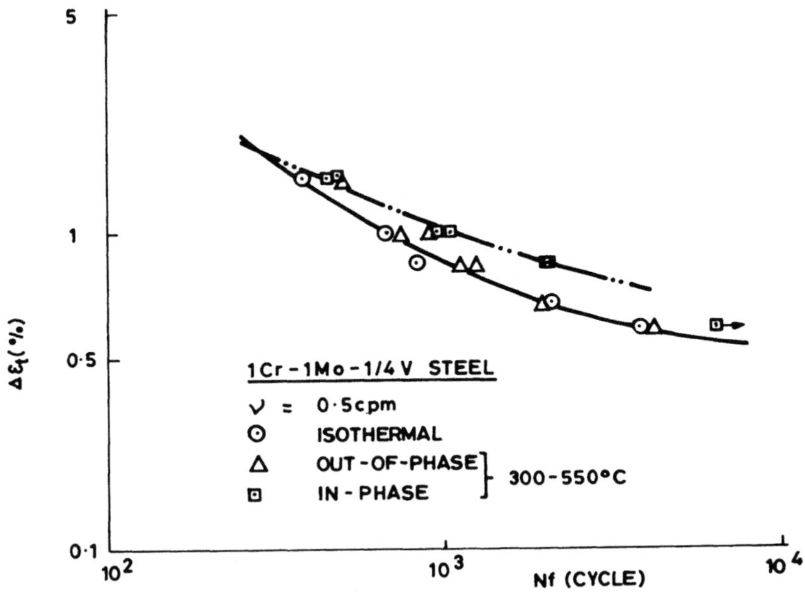

FIG. 5. Relationship between total strain range and failure life [23]. (Reproduced with permission of CRIEPI, Tokyo, Japan.)

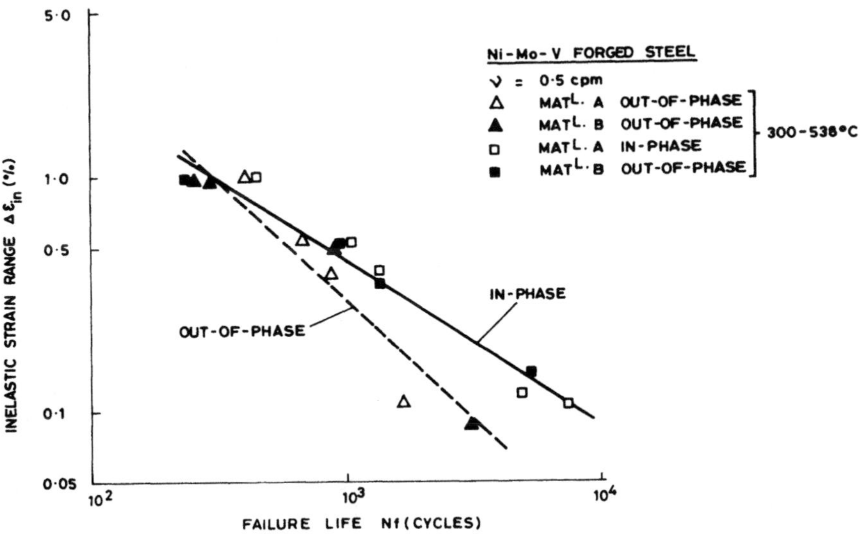

FIG. 6. Thermal fatigue life characteristics of Ni–Mo–V steel [24]. (Reproduced with permission of CRIEPI, Tokyo, Japan.)

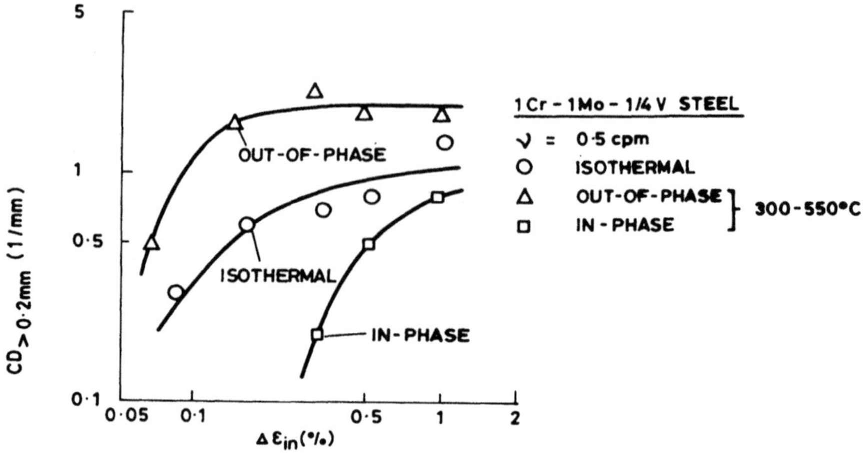

FIG. 7. Relationship between number of surface cracks of 0·2 mm and more depth per unit length and inelastic strain range at mid-life [23]. (Reproduced with permission of CRIEPI, Tokyo, Japan.)

occurred at a higher rate in the OP tests compared with the other two cycles types (Fig. 7). This was confirmed by work on a Ni–Mo–V rotor steel by Nitta *et al.* [24]. For strain ranges of 1·5% and 0·5%, the number of surface cracks per unit length was determined as a function of cycles for both OP and IP tests; these results are shown in Fig. 8. It can be seen that for a strain range of 1·5%, surface fatigue cracks were initiated at 20% of life for both IP and OP cycling, giving rise to comparable endurance values for both cycle types.

At 0·5% strain, the endurance of the OP test was approximately half that of the IP test. At this strain range, Fig. 8 shows that numerous surface cracks were present at 25% of life in the OP specimen, whereas initiation of surface cracks in the IP test was not observed until a number of cycles in excess of the OP test endurance.

Nitta *et al.* [24] have examined crack initiation life, N_i, for OP and IP tests, and concluded that N_i for the OP tests corresponded to 10% of life. However, N_i for the IP tests was equal to 70% of life at a strain range of 0·1% (Fig. 9). It was therefore concluded that the difference in endurance between the OP and IP thermal fatigue tests was a consequence of crack initiation behaviour. Indeed, no difference between crack propagation life, N_p $(= N_f - N_i)$, was found for the two different cycle types.

In order to examine the differences in surface fatigue crack initiation behaviour between IP and OP cycling, Nitta *et al.* [24] conducted

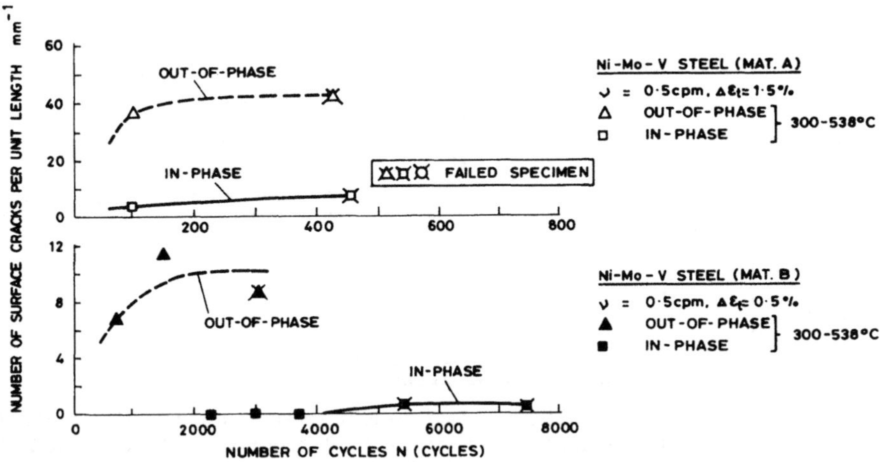

FIG. 8. Relationship between number of cycles and the surface cracks per unit length in the thermal fatigue of Ni–Mo–V steel [24]. (Reproduced with permission of CRIEPI, Tokyo, Japan.)

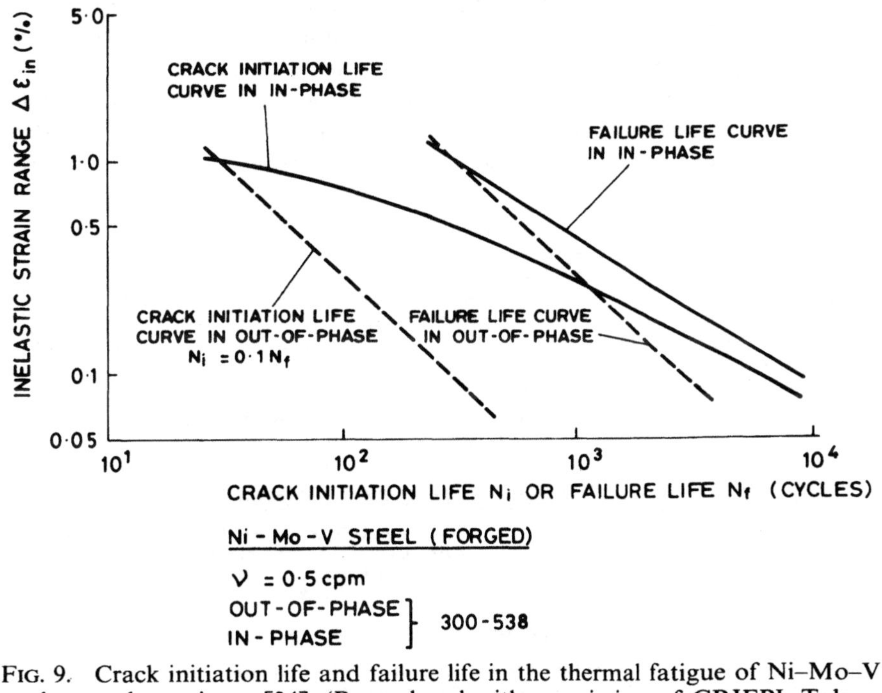

FIG. 9. Crack initiation life and failure life in the thermal fatigue of Ni–Mo–V steel, smooth specimen [24]. (Reproduced with permission of CRIEPI, Tokyo, Japan.)

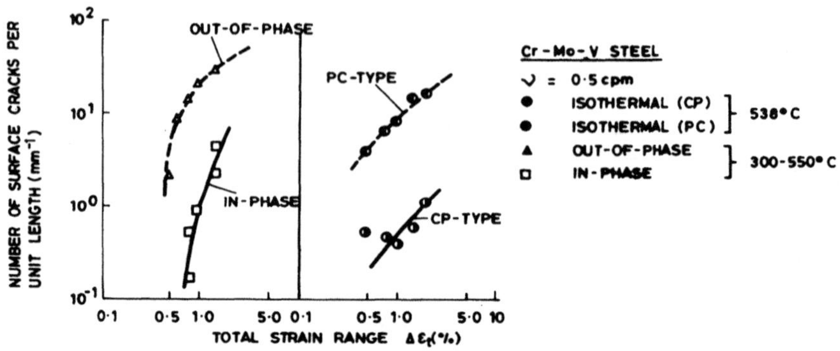

FIG. 10. Surface crack initiation characteristic in thermal fatigue and isothermal fatigue under saw-tooth strain waveform [24]. (Reproduced with permission of CRIEPI, Tokyo, Japan.)

isothermal tests on a Cr–Mo–V steel at 550°C in a scanning electron microscope. The tests involved unequal saw-tooth loading waveforms, i.e. slow–fast (CP) and fast–slow (PC). It was found that the isothermal saw-tooth tests gave similar surface fatigue crack initiation behaviour to that in non-isothermal cycling. For example, fast–slow (PC) cycles had a greater propensity to initiate surface fatigue cracks than slow–fast (CP) cycles (Fig. 10). These isothermal tests also enabled the role of environment (i.e.

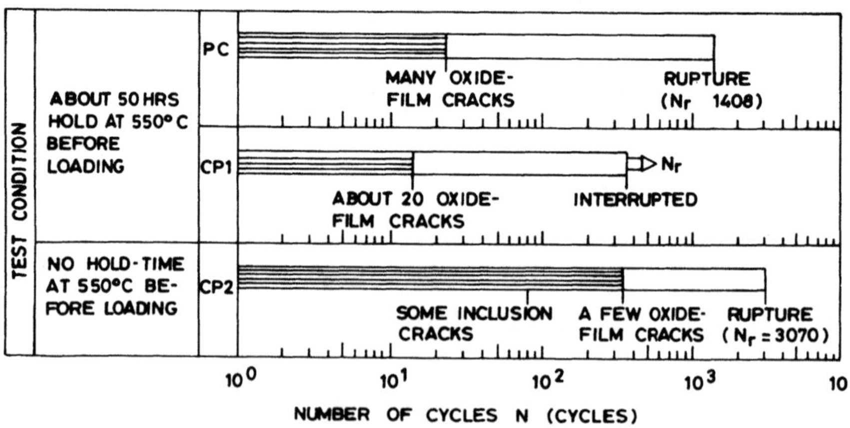

FIG. 11. Load control-type low-cycle fatigue test results of Cr–Mo–V steel, smooth specimen at 550°C within SEM [24]. (Reproduced with permission of CRIEPI, Tokyo, Japan.)

oxidation) on surface fatigue crack initiation to be examined. This is shown in Fig. 11, from which the following two points can be deduced:

(1) more oxide film cracks were present in the PC-type cycles;
(2) pre-oxidising induced more cracks at fewer cycles in the CP-type cycles.

In addition, it was also found that there was a good correlation between surface oxide cracking and parent metal cracking below the level of the oxide. This suggested that oxide cracking promoted cracks in the underlying parent metal. Thus, differences in endurance between IP and OP cycling were likely to be a consequence of a higher density of oxide film cracks in the OP tests. The cracking of the oxide film will be dependent on the applied stress, and since the OP and fast–slow (PC) type cycles possess a mean tensile stress, the oxide film cracking process in these cycles would have been exacerbated. An example of the mean tensile stress inherent in OP cycles is shown in Fig. 12, from which it is evident that the IP cycles have a mean compressive stress which would not promote oxide cracking.

A mechanism of environmentally assisted cracking related to oxide film cracking has been proposed to explain apparently anomalous isothermal low-cycle fatigue results in $2\frac{1}{4}$Cr–1Mo [28], in which compressive dwells were more damaging than tensile dwells. Such a mechanism could also be applied to explain the above observations.

It is worthwhile combining the available data in order to make a comparison of the effects of cycle type on endurance. An attempt to do this is shown in Fig. 13. Endurance data for the three cycle types are shown for 1Cr–1Mo–$\frac{1}{4}$V steel [23] and $2\frac{1}{4}$Cr–1Mo steel [25], together with the band of data obtained from isothermal tests on other ferritic steels [29]. It is clear that all the data fit well within the broad scatterband of $\times 2$ on mean endurance (see Sections 3.5 and 3.6).

3.3.2. Austenitic Steels

3.3.2.1. Crack Density Studies

Kuwabara and Nitta [9, 17] have presented data from anisothermal tests on 304 stainless steel ($\Delta T = 300$–$600°C$) and compared these with isothermal data on a similar steel generated at 600°C. These data from both isothermal and thermal cycle tests performed at a frequency of 0·5 cpm are shown in Fig. 14. It can be seen that for a total range of $\geq 0{\cdot}6\%$, isothermal tests are most damaging, followed by IP and, lastly, OP tests. Below this strain range there is a crossover in endurance.

By longitudinally sectioning failed test specimens, an attempt was made

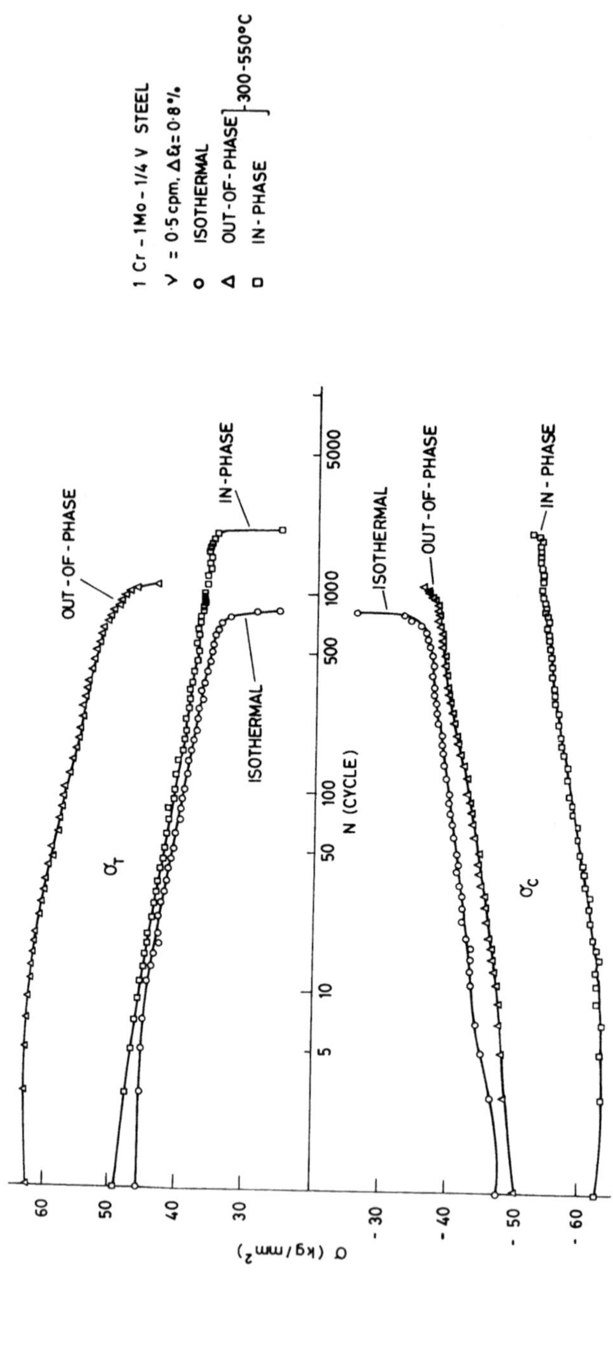

FIG. 12. Variation of peak tensile (σ_T) and compressive (σ_C) stress, as a function of cycles [23]. (Reproduced with permission of CRIEPI, Tokyo, Japan.)

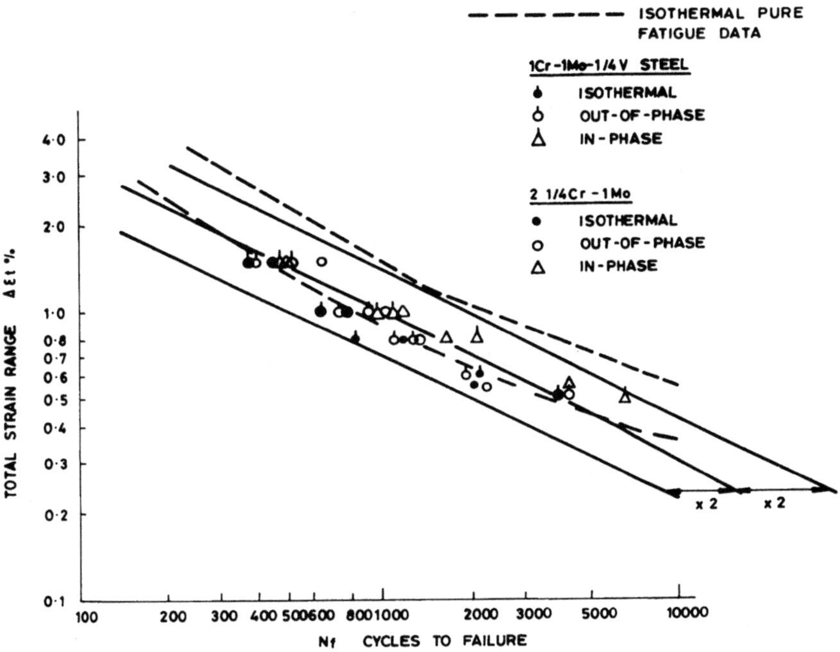

FIG. 13. Comparison of anisothermal and isothermal data from 1Cr–1Mo–$\frac{1}{4}$V steel and 2$\frac{1}{4}$Cr–1Mo steel (data from Refs 23, 25 and 29).

[9] to relate life reductions to failure modes. It was shown that the number of surface cracks was largest for the OP thermal cycles, followed by IP and then isothermal fatigue tests for strain ranges all in excess of 0·6%. Subsequently, to try and quantify the damage induced by these surface defects, the fraction of intergranular cracks to the total number of surface cracks, defined by

$$F_{IG} = (\eta_{IG} + \eta_{MIX})/N$$

where N is the total number of surface defects and η_{IG} and η_{MIX} are the numbers of intergranular and mixtures of intergranular and transgranular cracks, respectively, was plotted as a function of total strain range [9, 17] (Fig. 15). For F_{IG} values between 0·5 and 1·0, failures were mixed mode, though predominantly transgranular in nature. Values of $F_{IG} > 1{\cdot}0$ indicate failures that were mixed mode but predominantly intergranular; $F_{IG} \geq 1{\cdot}5$ indicates fully intergranular failures. Figure 15 shows that those tests likely to exhibit the strongest tendency to intergranular surface cracking were the

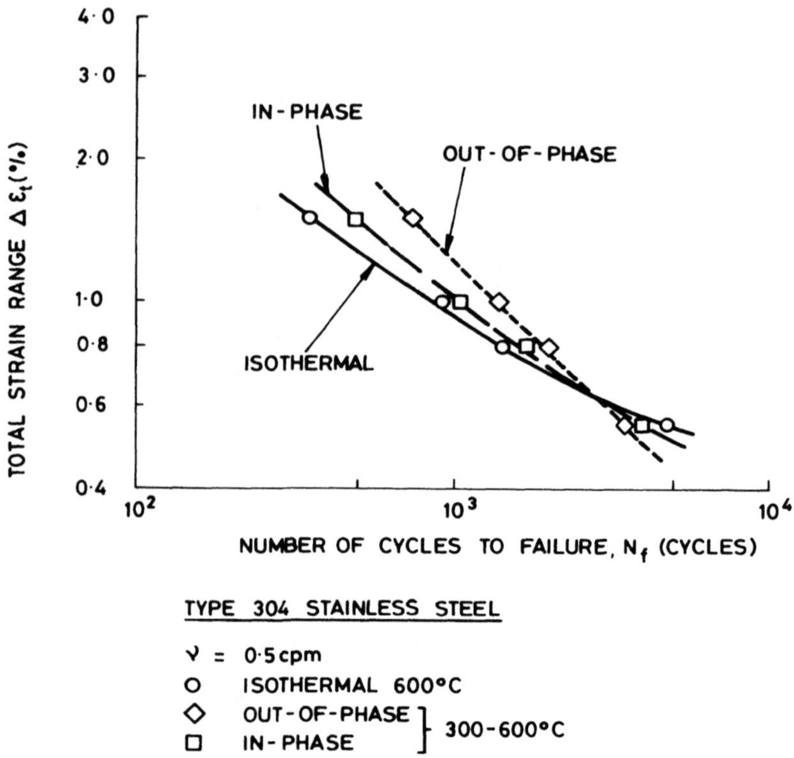

Fig. 14. Relationship between total strain range and number of cycles to failure [9, 17]. (Reproduced with permission of CRIEPI, Tokyo, Japan.)

IP thermal cycle tests, though for all tests predominantly transgranular behaviour would be expected. However, it is clear that as the total strain range is reduced, so F_{IG} increases, this increase being most rapid for the IP tests. Thus, the observation of the IP tests being more damaging than OP tests was explained on the basis of the materials tendency to produce intergranular surface cracks. The reasons why isothermal endurances were less than thermal cycle endurances for strain ranges >0·6%, or why OP endurances were smallest for strain ranges <0·6%, were not clearly explained, however.

Further thermal cycling tests have been performed by the same workers [9] on another batch of 304 stainless steel over the temperature range 200–550°C. Results from these tests indicated little difference in the lifetimes. However, analysis of the fraction of intergranular cracks, F_{IG}, as a function

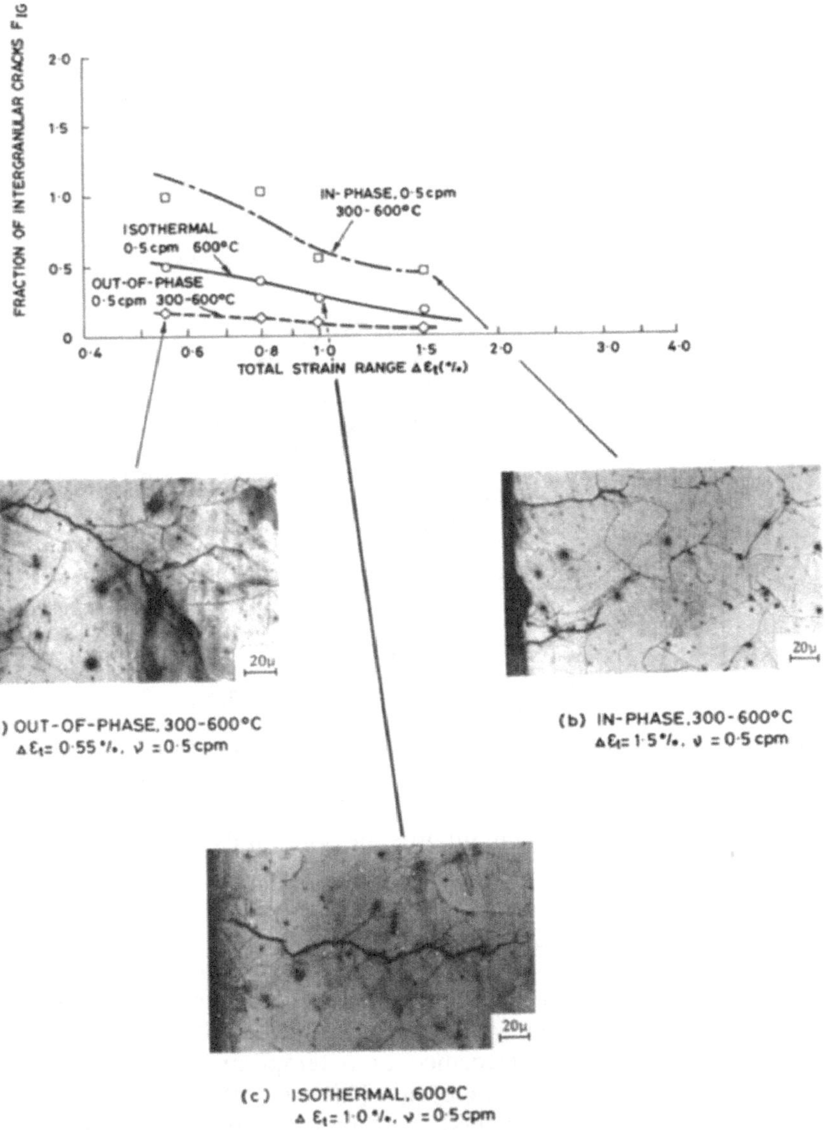

FIG. 15. Relationship between total strain range and fraction of intergranular cracks, 304 stainless steel. (After Refs 9 and 7.)

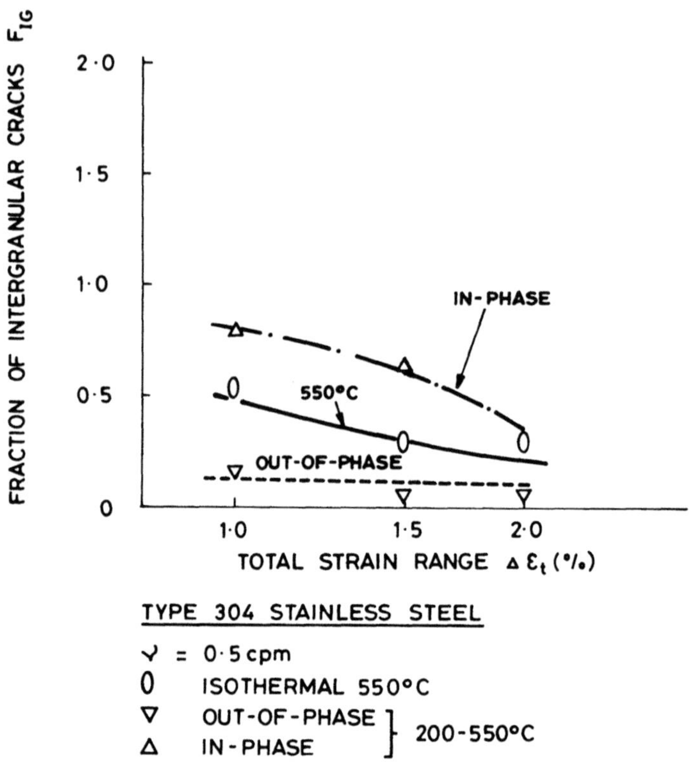

FIG. 16. Relationship between total strain range and fraction of intergranular cracks [9]. (Reproduced with permission of CRIEPI, Tokyo, Japan.)

of strain range once more highlights the potentially damaging nature of the IP thermal cycles (Fig. 16).

3.3.2.2. Grain Boundary Sliding Studies

Other tests on 304 stainless steel have been performed by Fujino and Taira [7]. They demonstrated that both IP and OP tests using a temperature range of 200–750°C gave lifetimes smaller than those obtained from isothermal tests performed at a temperature equal to the mean temperature of the thermal cycle. This was also the case even when isothermal test data obtained at a temperature corresponding to the maximum temperature used in the thermal cycle were compared with the IP thermal cycling data (Fig. 17(c)).

In order to explain their results, Fujino and Taira [7] also examined

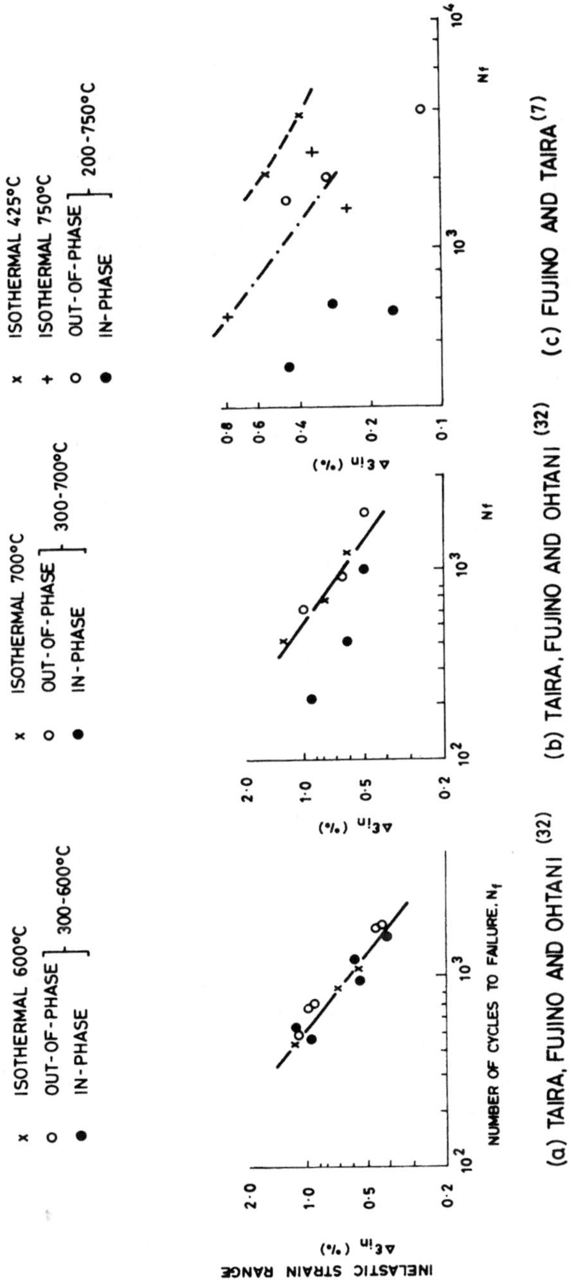

FIG. 17. Summary of thermal cycling data for 304 stainless steel.

potential increases in damage accumulation during thermal fatigue tests in terms of grain boundary sliding and cracking mechanisms. Measurements of grain boundary sliding, taken from the scribed surfaces of test specimens, showed that whereas for IP tests residual tensile grain boundary sliding accrued, little or no accumulation of residual sliding was evident for either isothermal or OP test cycles. Following the work of Davies *et al.* [30], it was proposed that tensile and compressive grain boundary sliding contributions depended on the strain rates and temperatures operative in each period. Thus, net tensile grain boundary sliding would be expected if tensile grain boundary sliding were reversed by compressive crystal slip, during an IP thermal cycle, for example. Further, the larger both the maximum temperature and the temperature range in a thermal cycle are, the more pronounced the residual grain boundary sliding and the more detrimental the effects, particularly if the amount of grain boundary sliding is proportional to grain boundary cracking [31]. Support for this is shown by the data of Taira and co-workers, in which large thermal cycles can indeed be seen to give most reduction in the endurance of IP tests (Fig. 17). Taira *et al.* [32] also showed that for 304 stainless steel and the austenitic alloy Hastalloy X, intergranular cracks were dominant in the IP thermally cycled specimens while transgranular cracks were dominant in OP and isothermal tests. These results are broadly in line with those of Kuwabara and Nitta [17].

3.3.2.3. Comparison of Data and Other Tests

Westwood [33] has also performed both isothermal (700°C) and thermal cycling tests (350–700°C) on 304 stainless steel. Although this programme of work was not as comprehensive as that of the Japanese investigators, several salient features were reported. In particular, for high strain range tests, good agreement was obtained between the purely cyclic endurance of isothermal and IP tests. It was noted, however, that one IP test at a total strain range of 1·35% did exhibit an endurance five times longer than that of an isothermal test at the same strain range. This is in contrast to the results of Taira *et al.* [32], who obtained marked reductions in the life of 304 stainless steel after IP thermal cycling over the range 300–700°C. Indeed, although Westwood's [33] tests were performed at strain ranges of $\geqslant 1\%$, in all cases anisothermal endurances were either the same or in excess of the lifetimes under isothermal conditions; further, the lower the strain ranges, the longer were the IP test endurances. These results do not seem to agree with the data of Kuwabara and Nitta [9, 17], who showed the opposite trend at low strain ranges.

Nevertheless, it is pertinent to compare the data obtained by Westwood [33] and Fujino and Taira [7] (see Fig. 18). A number of points can be seen from this figure. First, isothermal data from both sets of workers appear to be consistent with each other to within bounds of ×2 on life. Secondly, while at higher strain ranges IP cycling produced lifetimes comparable with the isothermal data, at lower strain ranges some effect of the IP cycling is indeed apparent. This effect is coupled with observations of little or no intergranular cracking in the higher strain range tests [33], whereas Fujino and Taira [7] link grain boundary sliding measurements made on their specimens tested at low strain ranges to the presence of such cracks, these cracks being responsible for the reduced life exhibited by their specimens. The higher thermal cycle temperature range used by Fujino and Taira [7] must also be noted, and may be responsible for the reduced IP lifetimes at lower strain ranges. However, it is also possible that the one IP result of Westwood [33] which showed an enhanced endurance compared with the isothermal results may simply reflect the scatter inherent in the materials data.

Data on another austenitic stainless steel (type 316) have also been presented by Beauchamp [3]. The tests performed included both isothermal tests at 625°C and IP and OP thermal cycling tests using the temperature range 400–625°C. Results obtained by Beauchamp [3] are shown in Fig. 19.

IP test specimens exhibited the same endurances as isothermal test specimens cycled at the same frequency. However, metallographic examination of tested specimens revealed that whereas isothermal specimens failed by intergranular nucleation and subsequent mixed-mode growth of surface cracks, there being little evidence of internal damage, the IP specimens revealed significant internal intergranular cracking in addition to surface crack formation. Both sets of specimens exhibited oxide formation on their external surfaces and oxide attack on the grain boundaries. OP thermal cycle tests exhibited lifetimes in excess of both IP and isothermal tests. Intergranular surface cracks were less common in these tests than in the IP tests, the amount of internal intergranular cracking also being reduced.

These surprising results have been rationalised by Beauchamp [3] in terms of competing mechanisms. He argues that since intergranular surface crack initiation is attributable to environmental effects, the reduced mean temperature of the thermal cycling tests means that surface crack initiation is also retarded. Thus, the feature determining endurance is internal cracking. In the case of IP tests, significant internal cracking was evident,

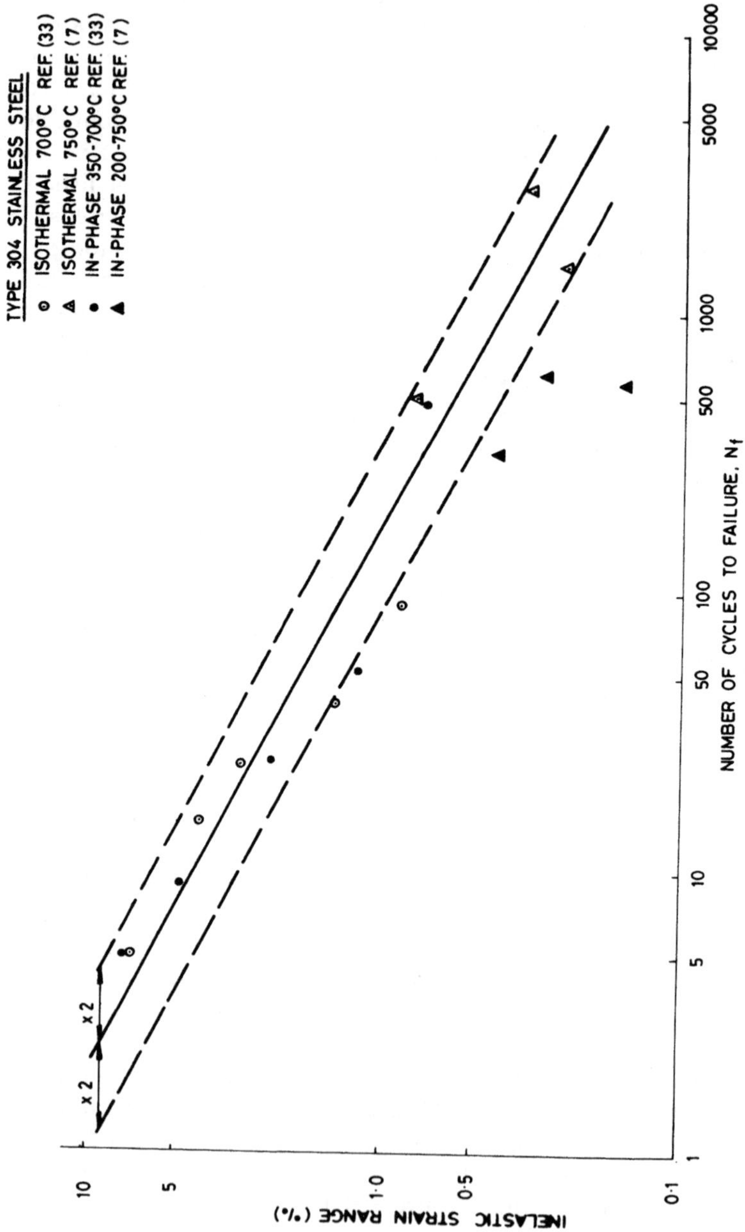

FIG. 18. Summary of cyclic data (isothermal vs. in-phase) for 304 stainless steel.

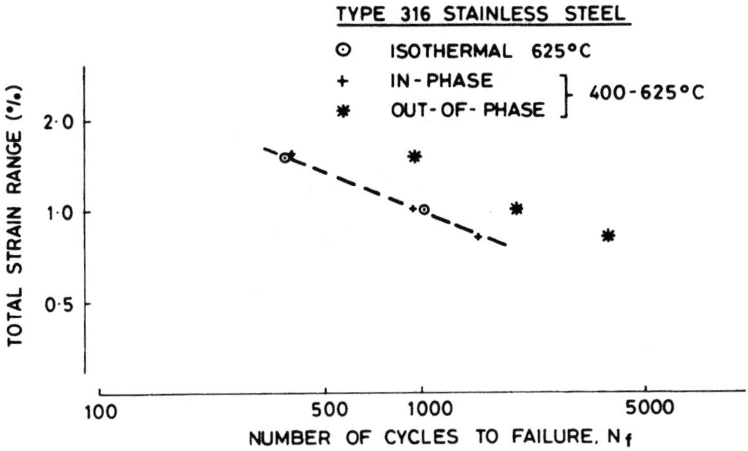

FIG. 19. Data on 316 stainless steel [3].

the effect of these cracks compensating for the reduced surface cracking to give lifetimes coincidentally equal to the isothermal endurances. In the OP tests, the absence of significant internal damage results in no such compensation, the enhanced lifetimes observed being due to suppressed surface crack initiation.

Other data on austenitic stainless steels have been obtained by Udoguchi and Wada [34]. Results from their isothermal and IP thermal cycle tests on AISI 347 stainless steel are shown in Fig. 20. It can be seen that, for all temperature ranges, IP cycling was more damaging than isothermal cycling at a temperature equal to the maximum value used in the thermal cycles. Further, the larger the temperature range used in the thermal cycles, the more damaging the cycle, in line with previous data (Fig. 17).

3.3.3. Nickel-base Superalloys

Nitta *et al.* [35] have examined the thermal fatigue characteristics of a number of Ni-base superalloys. The test conditions together with the materials examined are shown in Fig. 21. The results of the thermal and isothermal fatigue tests are also given in Fig. 21. Examination of these data reveals that there is no consistent trend in endurance between different cycle types and/or the material tested. However, these authors [35] have attempted to correlate the thermal fatigue data with the associated tensile strength and ductility (tensile and creep ductility for 100 h stress rupture test) of the range of alloys tested.

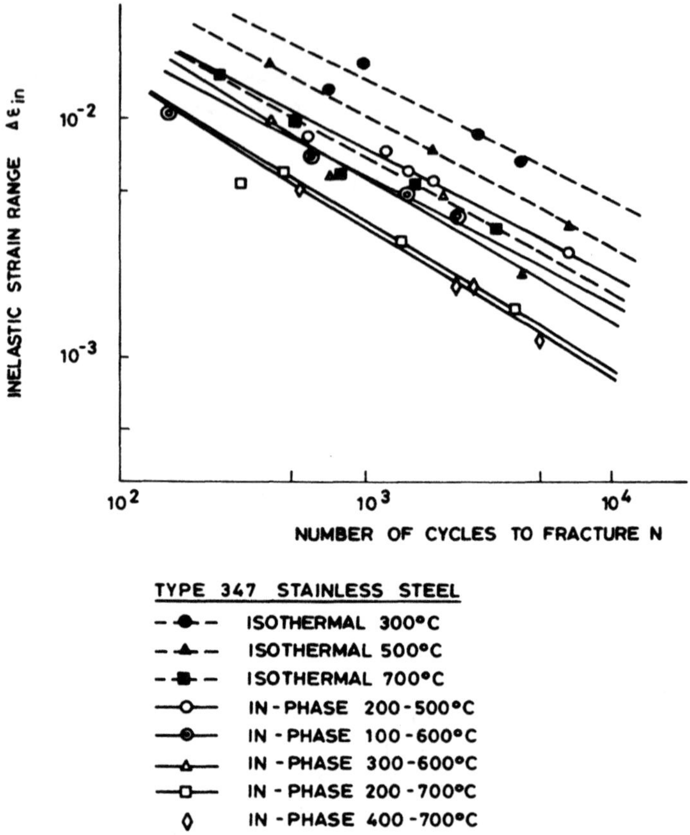

FIG. 20. Isothermal and thermal cycle fatigue tests on AISI 347 stainless steel [34]. (Reproduced with permission of Butterworths.)

It was found that the OP test endurances were dependent on the low temperature tensile ductility at high strain ranges (~1·0%) and on the tensile strength at lower strain ranges (~0·5%) (Fig. 22). However, it could be argued that for both strain ranges the data correlates well with tensile strength. It should also be noted that the OP tests exhibited a transgranular fracture mode.

For the IP tests, endurance was correlated with the tensile and creep ductility (obtained from 100 h stress rupture test), as shown in Fig. 23. It was suggested that, at higher strain ranges, the tensile ductility correlated well with endurance, whereas at the lower strain ranges creep ductility was the

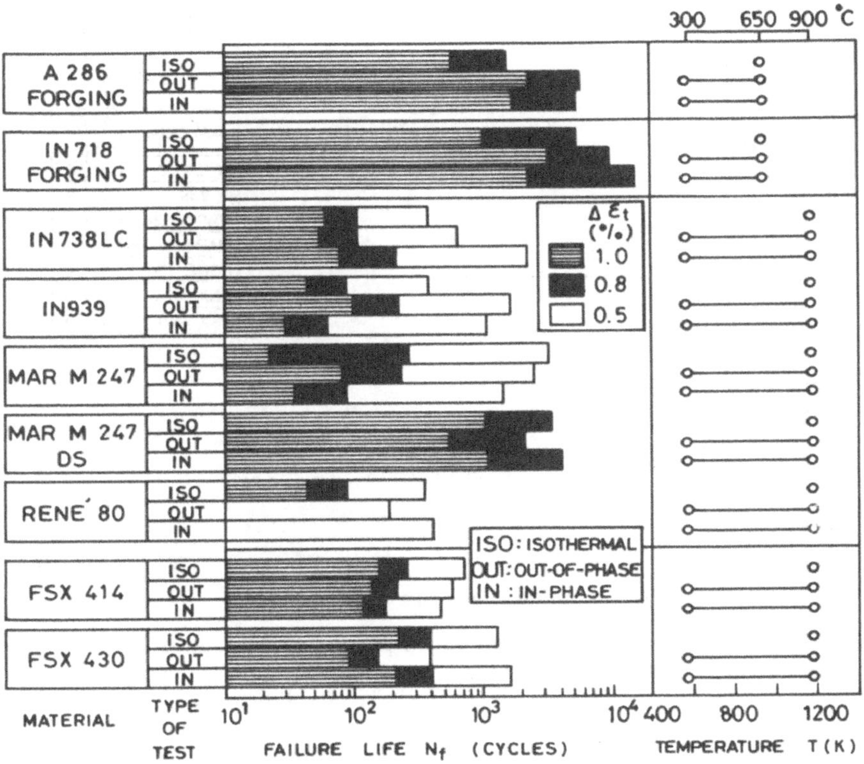

FIG. 21. The thermal and isothermal fatigue lives of the alloys tested [35]. (Reproduced with permission of CRIEPI, Tokyo, Japan.)

better correlating parameter. The in-phase specimens failed in an intergranular fracture mode. From Figs 22 and 23 it is evident that the thermal fatigue endurance of Ni-base superalloys is dependent on the inherent strength and ductility of each alloy.

3.3.4. Summary of General Trends in Thermal–Mechanical Fatigue

Experimental data presented in the previous sections on steels and Ni-base superalloys have highlighted the diverse responses that can be exhibited by materials subjected to thermal–mechanical strain cycling. This is not surprising given the complex nature of the various possible damage mechanisms which can interact to cause failure (such mechanisms include creep cavitation, fatigue crack nucleation/growth, environmental attack, microstructural ageing (coarsening and reprecipitation), cyclic strain

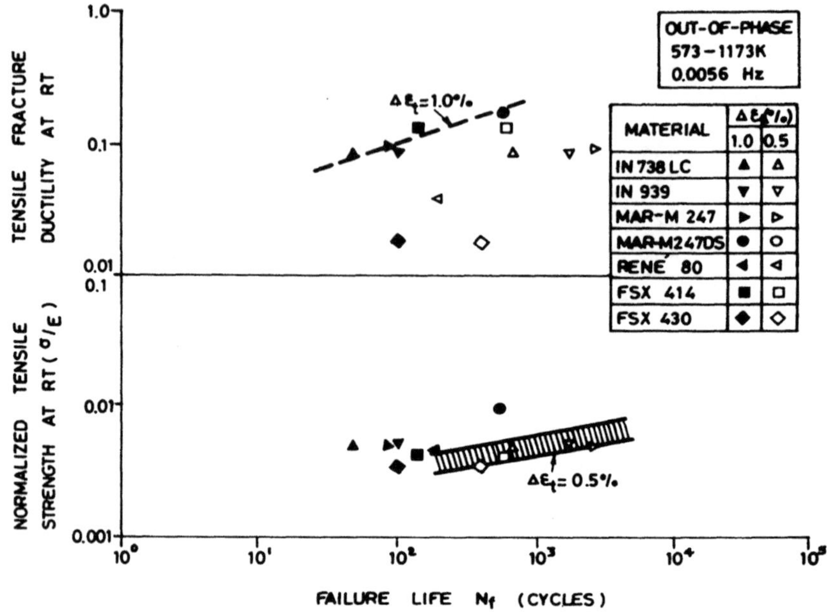

FIG. 22. The relationships between out-of-phase thermal fatigue life and tensile properties at *RT* (room temperature) for cast alloys [35]. (Reproduced with permission of CRIEPI, Tokyo, Japan.)

softening and hardening; see Chapters 2 and 4). Nevertheless, despite this complexity, it is possible to demonstrate that some trends exist in the above materials' behaviour.

In order to classify the different types of response to thermal cycling caused by the various damage mechanisms, Kuwabara *et al.* [4] have defined the four categories shown in Table 2. Representative examples of these four classes of behaviour are shown in Fig. 24 [4]. A rationale in terms of the various controlling mechanisms is now considered.

The thermal fatigue characteristics of Type O have been observed mainly in low-alloy steels, such as 1Cr–1Mo–$\frac{1}{4}$V or Ni–Mo–V steels. This is thought to be due to the difference in the number of cycles taken to crack initiation during IP and OP cycling for these materials (see Section 3.3.1). Early crack initiation was observed in the OP cycles, and this has been shown to be related to the behaviour of surface oxide scale. During OP cycling, high tensile stresses are produced at low temperatures, causing the surface oxide to crack more readily than under IP cycling conditions. These

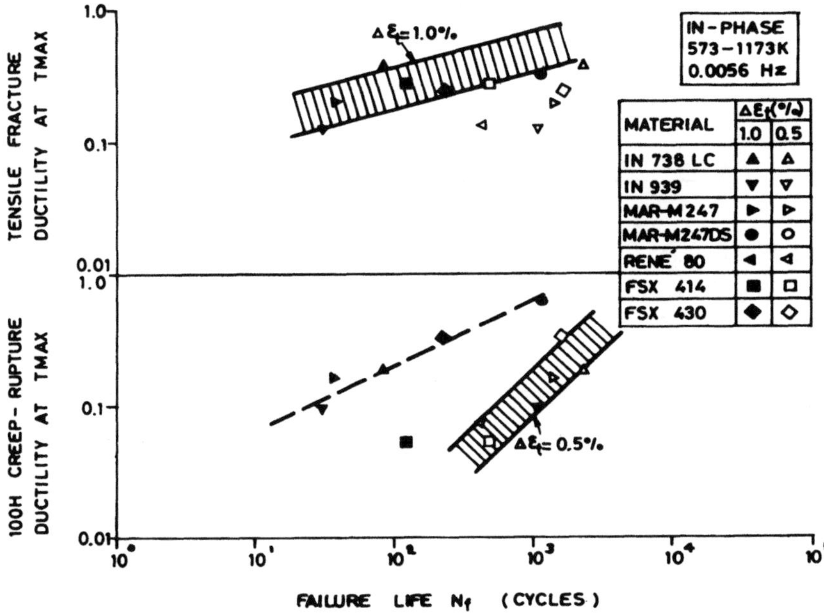

FIG. 23. The relationships between in-phase thermal fatigue life and ductilities at T_{max} for cast alloys [35]. (Reproduced with permission of CRIEPI, Tokyo, Japan.)

oxide cracks subsequently accelerate fatigue crack nucleation, thereby causing reduced endurances. Thus, the life characteristics of Type O failures can be explained for low-alloy steels by an environmental effect.

Type E failures clearly indicate little or no effect of creep and/or environment on endurance. This type of behaviour has been observed in 304 stainless steel when tested over the thermal cycle range 200–550°C. The maximum temperature in such a cycle is probably too low for any significant creep or environmental effects in 304 stainless steel to influence endurance in OP or IP cycling, therefore giving rise to the results shown in Fig. 24.

In the case of Type E′ behaviour, which has only been observed in the Ni-base superalloy IN 718, it appears that creep damage reduces the endurance in the IP cycles by giving rise to intergranular fracture. However, at low strain ranges the IP and OP endurances are the same, the creep component no longer dominating. This behaviour is in direct contrast to the Type I data. In the latter case, IP tests are more damaging than OP tests at lower strain ranges (Fig. 24).

TABLE 2

CLASSIFICATION OF THERMAL-CYCLING FAILURES [4]

(i)	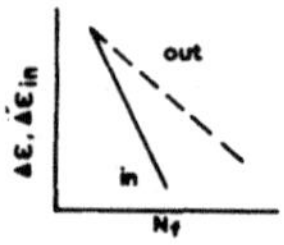	Type I: in-phase endurance is shortest at lower strain ranges
(ii)	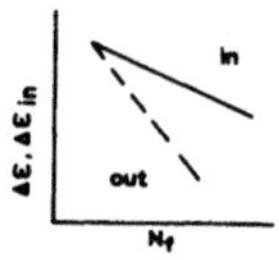	Type O: out-of-phase endurance is shortest at low strain ranges
(iii)	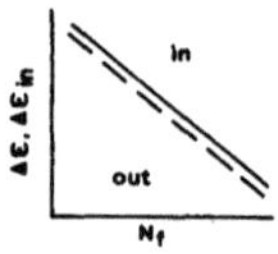	Type E: in-phase and out-of-phase endurances nearly equal
(iv)	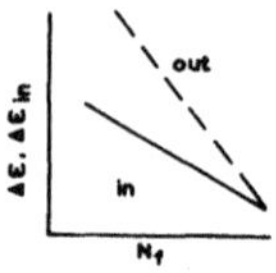	Type E′: in-phase endurances shortest at higher strain ranges, but nearly equal to out-of-phase endurances at low strain ranges

A general trend can be identified in the above material responses. For a given material, as the maximum temperature, T_{max}, in a thermal cycle is increased, thermal fatigue life characteristics shift from Types O or E to Type I, giving shorter lives under IP thermal fatigue conditions. In these cases, the fracture mode of the IP test specimens also tend to be intergranular in nature. A common feature of IP testing is the observation not only of intergranular surface fatigue cracks but also of internal creep damage [23]. Since the presence of such damage in isothermal fatigue tests is very rare, it is evident that IP cycles appear to accelerate the formation of creep damage. Under certain conditions (i.e. when environmental effects are limited and the material has low creep ductility) the presence of this damage will result in lower endurances in the IP cycles.

A summary of the various damage mechanisms and endurance behaviour associated with the three cycle types discussed above is shown in Fig. 25. However, the actual differences in endurance between the three cycle types may not be large. Indeed, it has been demonstrated for ferritic and austenitic stainless steels that on comparison of all the data (Figs 13 and 18) a ×2 scatterband on mean endurance encompasses all the test results. Therefore, even though there are mechanistic reasons for the

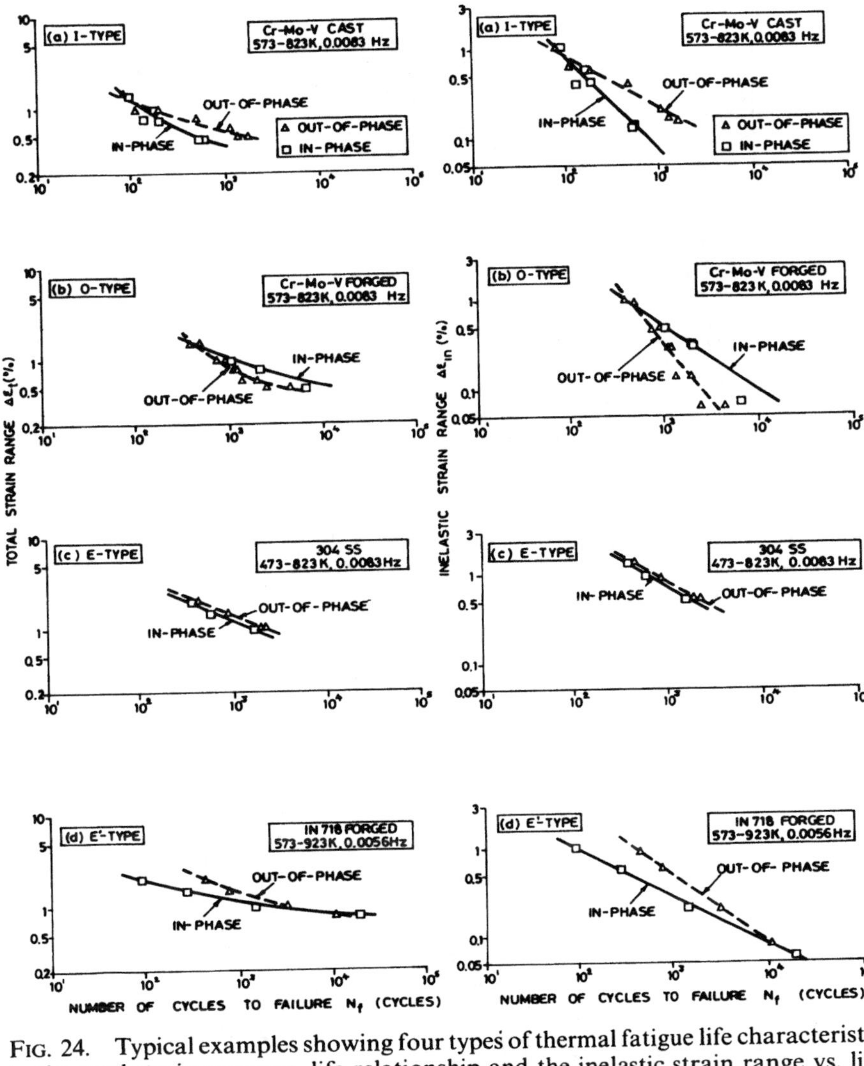

FIG. 24. Typical examples showing four types of thermal fatigue life characteristic in the total strain range vs. life relationship and the inelastic strain range vs. life relationship [4]. (Reproduced with permission of ASME.)

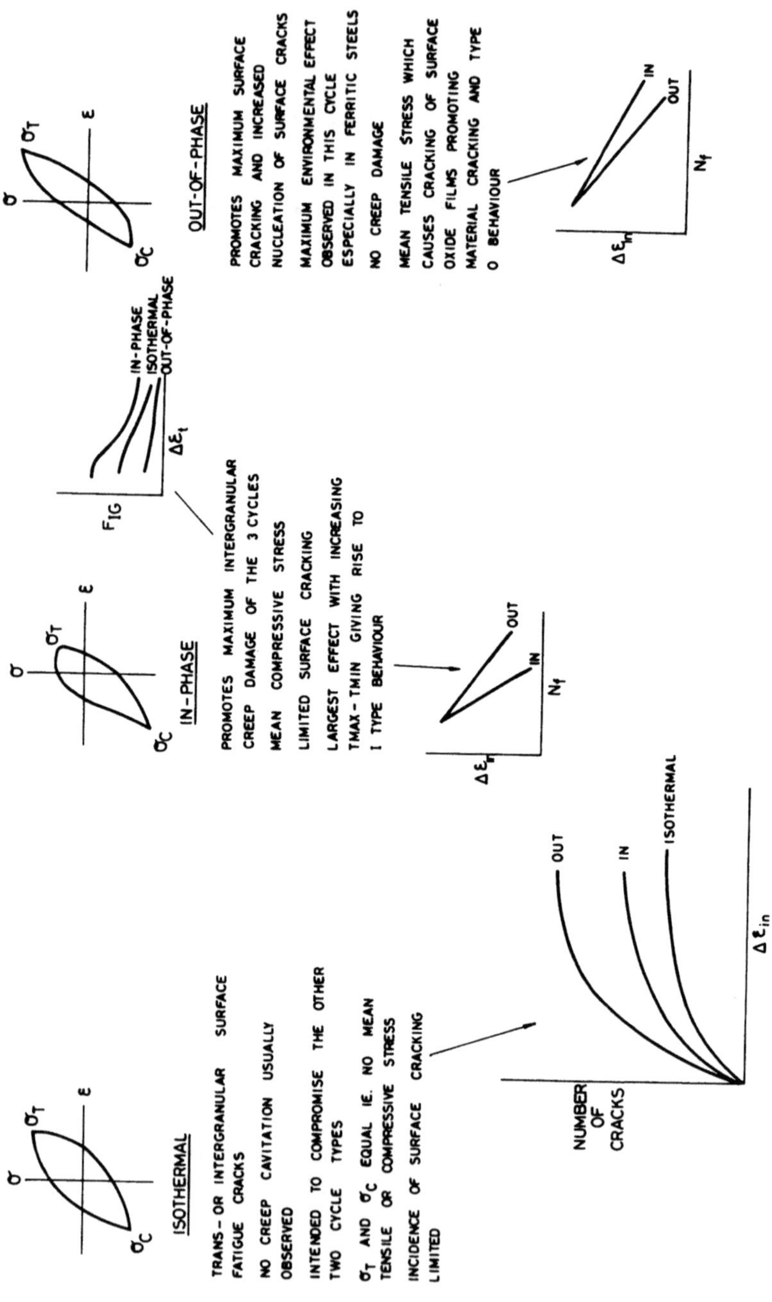

FIG. 25. Summary of thermal fatigue behaviour.

observed endurance behaviour, over the testing conditions examined these do not give rise to large variations in endurance.

3.4. THERMAL–MECHANICAL CREEP–FATIGUE TEST DATA

3.4.1. Ferritic Steels

Kuwabara and Nitta [23] have investigated the properties of a 1Cr–1Mo–$\frac{1}{4}$V steel and indicated that, on comparison with continuous cycling endurance data, the incorporation of a strain hold at maximum temperature in the OP and isothermal cycles (for dwells of 2–10 min) resulted in no significant reduction in endurance. In contrast, the IP thermal fatigue cycle tests exhibited a large reduction in endurance on the introduction of strain hold times (2–60 min). The effect of strain holds is clearly demonstrated by normalising the data with respect to continuous cycling data. To do this, Kuwabara *et al.* [23] defined a fatigue life ratio as the ratio of N_f (the endurance for cycles with strain holds) to N_{fo} (the endurance for continuous cycles at the equivalent inelastic strain range). Fatigue life ratio data are shown as a function of hold time for 1Cr–1Mo–$\frac{1}{4}$V steel in Fig. 26. The results clearly show that for IP thermal fatigue endurances decrease with increasing tensile strain hold time. Therefore, IP cycling with strain hold times are more damaging than isothermal low-cycle fatigue tests. This is an important observation, since the latter cycle is frequently used to simulate the more complex IP thermal–mechanical cycle. Indeed, data from such tests form the basis of design and life prediction techniques used at present.

Metallography of the 1Cr–1Mo–$\frac{1}{4}$V steel revealed that the only cracking

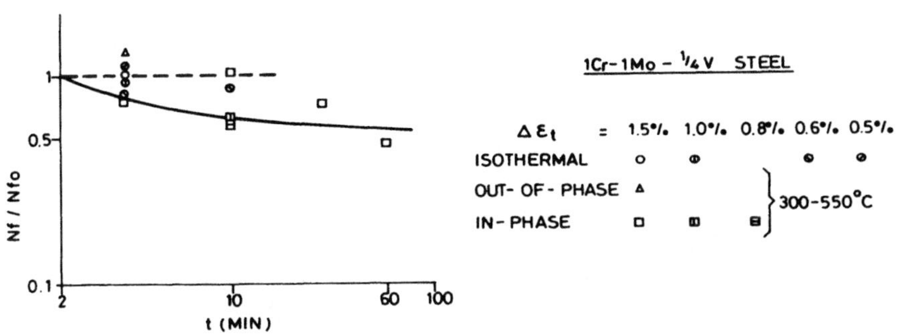

FIG. 26. Relationship between failure life ratio and strain hold time [23]. (Reproduced with permission of CRIEPI, Tokyo, Japan.)

mode found in OP testing was transgranular [23]. This was also the case for isothermal cycles, except for those with 10 min dwells at an inelastic strain range of 0·8%, which exhibited internal grain boundary cracking. By comparison, IP specimens without hold periods tested at an inelastic strain range of 0·8%, and 2 min hold period tests at an inelastic strain range of 1·5%, contained grain boundary damage. This evidence suggests that initiation of grain boundary damage is enhanced in the IP cycle tests compared with the isothermal tests. However, it should be noted that the data for isothermal strain hold tests in this study [23] only covers short dwell periods. It is well established [29] from isothermal tests on ferritic steels that by increasing dwell periods and/or decreasing strain ranges, failures due to creep damage accumulation can be promoted. It is therefore possible that for long dwell periods the difference in endurance between isothermal and IP cycles may become negligible, as has been observed in tests on 304 stainless steel [4]. This assumption clearly warrants further study for ferritic steels; the mechanism of creep damage enhancement in the in-phase cycle needs to be considered.

Kuwabara *et al.* [25] have also carried out a limited programme on the effect of dwells in $2\frac{1}{4}$Cr–1Mo steam pipe material after service exposure. However, the test conditions and number of tests conducted are so limited that it is difficult to draw any conclusions from this study.

Westwood and co-workers [1, 36, 37] have examined the influence of hold periods on endurance after carrying out both thermal–mechanical and isothermal cycling on $2\frac{1}{4}$Cr–1Mo and $\frac{1}{2}$Cr–Mo–V ferritic steels. However, unlike Kuwabara and co-workers, who employed a temperature cycle range of 300–550°C, the latter being the service temperature, Westwood *et al.* [36, 37] adopted a temperature acceleration approach. This resulted in a temperature cycle range of 300–667°C, with 10 min hold periods. The hold time and maximum temperature were selected using the Larson–Miller parameter for the material in question, and were intended to be equivalent to 16 h of full-load plant operation at the maximum design temperature of 594°C (i.e. simulated two-shift operation). For both steels examined, the IP cycle tests exhibited lower endurance than the equivalent isothermal cycles. This observation is in agreement with Kuwabara and Nitta's [23] study. However, the experiments of Westwood and co-workers [1, 36] induced large specimen shape changes during their tests at high strain ranges. Even at similar strain ranges to those used by the Japanese workers [23], the failure modes observed were different. Those given by Westwood and co-workers [1, 36] are listed in Table 3.

It is suggested that the above observations of specimen necking and

TABLE 3

SUMMARY OF FAILURE MECHANISMS OBSERVED BY WESTWOOD AND CO-WORKERS [1, 36]

Material	*Strain range*	*In-phase (IP)*	*Out-of-phase (OP)*	*Isothermal*
$2\frac{1}{4}$Cr–1Mo	High	Barrelling and tensile rupture	Off-centre necking; shear fracture	Multiple intergranular microcracking
	Low		Surface fatigue cracking	Growth of surface cracks and final tensile rupture
$\frac{1}{2}$Cr–Mo–V	High	Necking, off-centre barrelling		Pronounced necking, intergranular cracking; final ductile rupture
	Low	Transgranular cleavage, secondary intergranular cracking and grain growth		Intergranular crack growth through 40% of the wall thickness; fast shear

barrelling which strongly influenced the failure mode of the specimens are a consequence of the high strain range (up to 9%) and the accelerated temperature conditions used. At temperatures of 667°C the ductility of the material will be far greater than that exhibited at 530°C (service temperature), and will therefore limit the creep damage formed, enabling other mechanisms to dominate.

3.4.2. Austenitic Steels

3.4.2.1. Type 304: Frequency and Crack Density Effects

From data obtained on 304 stainless steel over a range of cyclic frequencies, it has been shown that the lower the frequency, the more damaging the cycle [9, 17]. This observation was interpreted by Kuwabara and Nitta [9, 17] in terms of the fraction of intergranular cracks parameter, F_{IG} (see Section 3.3.2.1). This fraction was shown to increase as cyclic frequency (or strain rate) was reduced. Indeed, the more pronounced the reduction in endurance, the larger was F_{IG}. Although a frequency effect similar to that observed in the isothermal tests was also apparent for IP cycling, OP data did not appear to conform to this trend.

Kuwabara and Nitta [9] have also presented data on 304 stainless steel in which test cycles also included dwell periods at 600°C. These hold period tests included isothermal, IP and OP (hold periods at maximum compressive strain) cycles with dwell durations of ⩽18 min.

To clarify the effect of frequency and hold time, the relationship between the time for one cycle and endurance was used [9] (Fig. 27). A number of points can be seen from Fig. 27. For example, for tensile hold periods included in isothermal tests performed at a strain range of 1·5%, the reduction in life obtained was equal to that observed from reduced frequency testing. The effect of compressive dwells in the isothermal tests

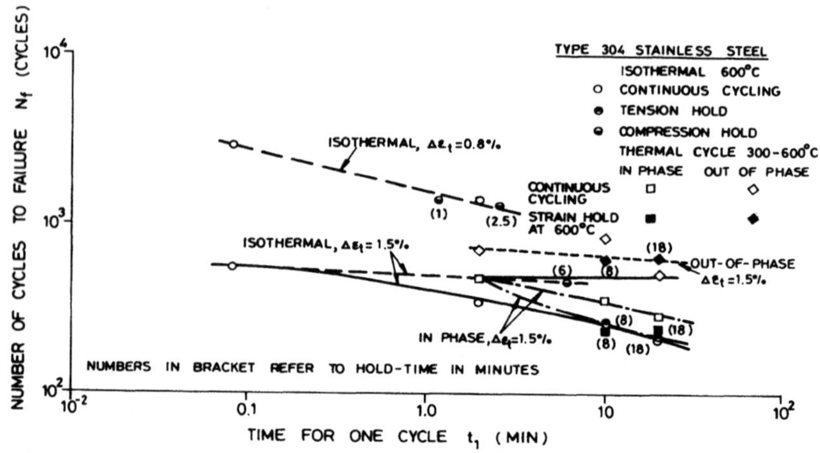

FIG. 27. Relationship between time for one cycle and number of cycles to failure [9, 17]. (Reproduced with permission of CRIEPI, Tokyo, Japan.)

appeared to enhance life at high strain ranges and reduce life slightly at lower strain ranges (~0·8%). Indeed, at a strain range of 0·8%, the effect of compressive dwells on endurance was comparable with the frequency effect.

For the thermal cycling data, it can be seen that the interspersion of a hold period at 600°C in the IP tests considerably reduced endurance, producing an effect comparable with the frequency effect noted previously. However, in the OP tests, hold times at 600°C and any frequency effects were small. Nevertheless, for both IP and OP tests, it is pertinent to note that endurances for cycles including 8 min hold periods were less than those which included 18 min hold periods.

Once more, these results were rationalised in terms of the F_{IG} index [17]. The relationship obtained between F_{IG} and the time for one cycle is shown

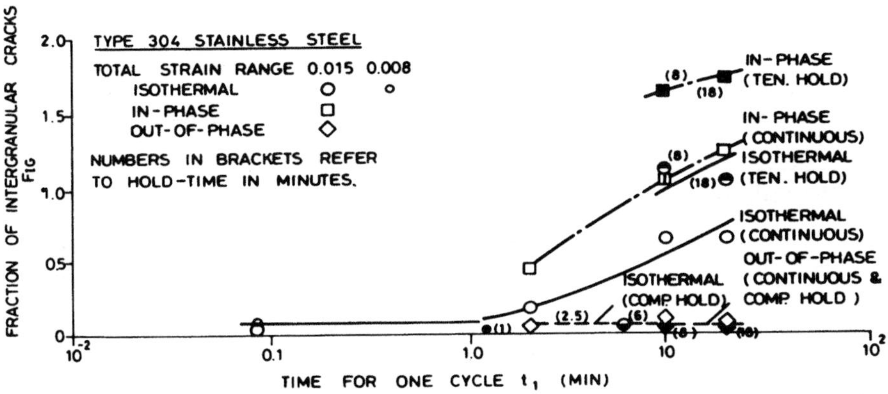

FIG. 28. Relationship between time for one cycle and fraction of intergranular cracks [9, 17]. (Reproduced with permission of Pergamon Press.)

in Fig. 28. This figure indicates a tendency towards intergranular surface cracking as hold periods are increased or cyclic frequencies reduced. However, intergranular cracks appear to occur more readily in hold period tests than in low-frequency tests [17]. This tendency to intergranular cracking correlates with the lifetime reductions indicated in Fig. 27.

Internal intergranular cracks were observed in the IP thermal cycle tests, which included 18 min hold times. Similar cracks were also observed in the OP hold period test specimens. However, increasing the duration of the hold periods had little effect on the proportion of intergranular surface cracks in the OP specimen (Fig. 28). Internal cracks were also evident in the isothermally tested specimen which experienced 6 min compressive hold periods, although the number of these cracks was much reduced compared with the IP dwell tests [38].

3.4.2.2. The Work of Westwood

Westwood [33] has also produced hold-time data from isothermal (700°C) and IP (700–350°C) tests on 304 stainless steel: these data are listed in Table 4. From the isothermal data, it can be seen that there was a marked reduction in endurance caused by the inclusion of a 60 min tensile dwell, though a single test including an 840 min (12 h) tensile dwell gave an endurance approaching that of the purely cyclic tests.

The inclusion of a 60 min dwell period in the IP tests gave endurances similar to those of the 60 min dwell isothermal tests at high strain ranges. However, at the lower strain range of 1%, the endurance of the 60 min dwell IP test was significantly larger than the comparable isothermal test,

TABLE 4
HOLD-TIME DATA FOR 304 STAINLESS STEEL [33]

Temperature (°C)	*Total strain range (%)*	*Inelastic strain range (%)*	*Hold time (min)*	*Cycles to failure*
Isothermal, 700	4·80	4·15	—	15
	4·00	3·00	—	24
	2·20	1·48	—	40
	2·20	1·53	—	40
	7·75	7·05	—	5
	1·45	0·88	—	89
	3·00	2·80	60	11
	6·70	6·70	60	3
	1·10	1·00	60	25
	2·70	2·60	840	26
In-phase (IP), 700/350	6·00	4·84	—	9
	2·30	1·22	—	52
	1·35	0·75	—	469
	3·10	2·37	—	25
	8·40	7·61	—	5
	1·80	1·23	60	21
	5·20	5·20	60	5
	1·00	0·49	60	153
	3·00	2·80	840	15

although still less than the purely cyclic IP endurance. Further, as with the isothermal tests, the single IP test performed with a hold time of 840 min (12 h) once more gave an endurance which approached the purely cyclic data line. Thus, Westwood [33] concluded that a minimum fatigue life occurred at hold times of 60 min, even though only two test results were obtained with longer hold times. Notwithstanding restricted data, Westwood plotted endurance versus hold time for both isothermal and IP cycling at a total strain range of ~1·6% (Fig. 29), and concluded that while IP and isothermal lifetimes were approximately equal for purely cyclic deformations, in dwell tests IP endurances were always lower.

Westwood performed a metallographic analysis of his test specimens to interpret his dwell test data [33]. He observed that the depth and width of the surface cracks in his specimens increased with increasing hold times under both isothermal and IP testing. Further, significantly more internal intergranular cracking was observed in the 60 min dwell specimens than in those tested under purely cyclic conditions or with 840 min dwells.

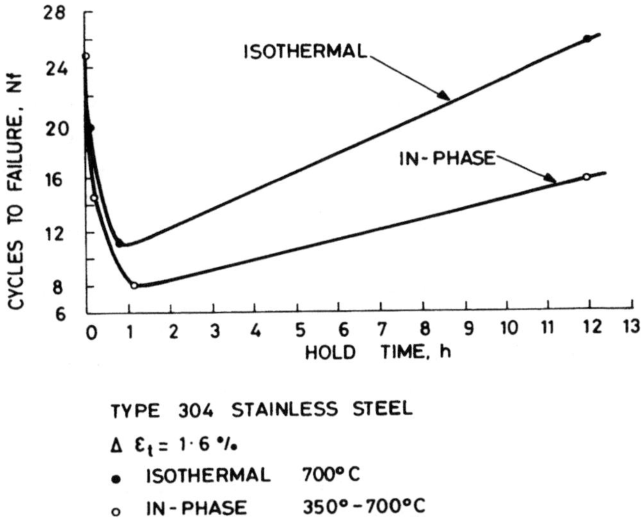

FIG. 29. Fatigue life vs. hold time for 304 stainless steel [33]. (Reproduced with permission of Pergamon Press.)

From an assessment of stress relaxation behaviour and grain boundary sliding-induced cavitation, it was concluded that the minimum fatigue resistance observed for 60 min dwell tests was associated with grain boundary cavitation damage, leading to increased cyclic growth rates of intergranular fatigue cracks. Increased fatigue resistance with long hold times was subsequently attributed to microstructural changes which effectively inhibited the cavitational damage processes. In this context, it is interesting to speculate whether the endurance minimum observed by Kuwabara and Nitta [9] in their 304 stainless steel thermal cycle tests with hold periods of 8 min is attributable to a similar mechanism. A similar trend to that found by Westwood [33] (Fig. 29) has also been observed in 316 stainless steel by Wareing [39], the improvement in endurance at long dwell times again being related to microstructural ageing effects.

3.4.2.3. Type 316 Steel

Beauchamp [3] performed isothermal tests (625°C) on 316 stainless steel at two strain rates in order to identify any cyclic frequency effect on specimen endurance. A reduction in life of some 50% was observed due to the slower cycling. In addition, metallographic examination of the tested specimens revealed that whereas, in the quickly cycled specimens, failure was caused by transgranular surface cracks with no evidence of surface oxide formation, the lower strain rate failures exhibited mixed-mode surface cracking with noticeable oxide formation on the external surfaces of the specimens. Indeed, in the latter tests, surface-crack initiation was intergranular, the lower endurance of these tests being attributed by Beauchamp [3] to oxide attack on the grain boundaries. Levaillant *et al.* [40] have observed a similar environmental effect during isothermal tests on 316 stainless steel.

Limited dwell data have also been obtained on 316 stainless steel by Beauchamp [3]. For 30 min dwells included during IP tests (400–625°C) there was only a slight reduction in endurance compared with similar tests with no hold periods. While there appeared to be much more oxidation on the surfaces of the dwell test specimens, much smaller amounts of internal intergranular cracking were observed compared with the purely cyclic IP specimens. This was thought to be linked with stress relaxation behaviour during the dwell.

The slightly reduced life of the dwell tests over the purely cyclic tests was thus attributed to the enhanced environmental effects operative in the (low-frequency) dwell tests.

3.4.3. Nickel-base Superalloys

Thomas and Varma [41] have reported results obtained from IP tests on IN 597, which included 5 min dwells at the maximum tensile strain. The tests were carried out under diametral strain control with a maximum temperature in the cycle of 850°C and minimum temperatures in the range 400–850°C. The resultant endurance data are given as a function of inelastic strain range in Fig. 30. Also included in Fig. 30 are isothermal data generated under axial strain control at 850°C. These tests exhibited reduced endurances compared with the IP tests [41]. It is evident that the behaviour of IN 597 is similar to that observed in ferritic steels [23] in that, in most cases, IP dwell tests exhibit a lower endurance than isothermal dwell tests. Further, the difference between isothermal and IP dwell test endurances also appears to increase as the difference between the maximum and minimum temperature in the cycle (T_{max} and T_{min}) increases.

To explain these results, Thomas and Varma [41] divided their data into two categories:

(a) Tests in which the minimum temperature (T_{min}) in the cycle was $\geqslant$700°C. Endurances were similar to those in isothermal tests.
(b) Tests in which 400°C $\leqslant T_{min} \leqslant$ 600°C. Specimens exhibited a reduced life in comparison with isothermal tests.

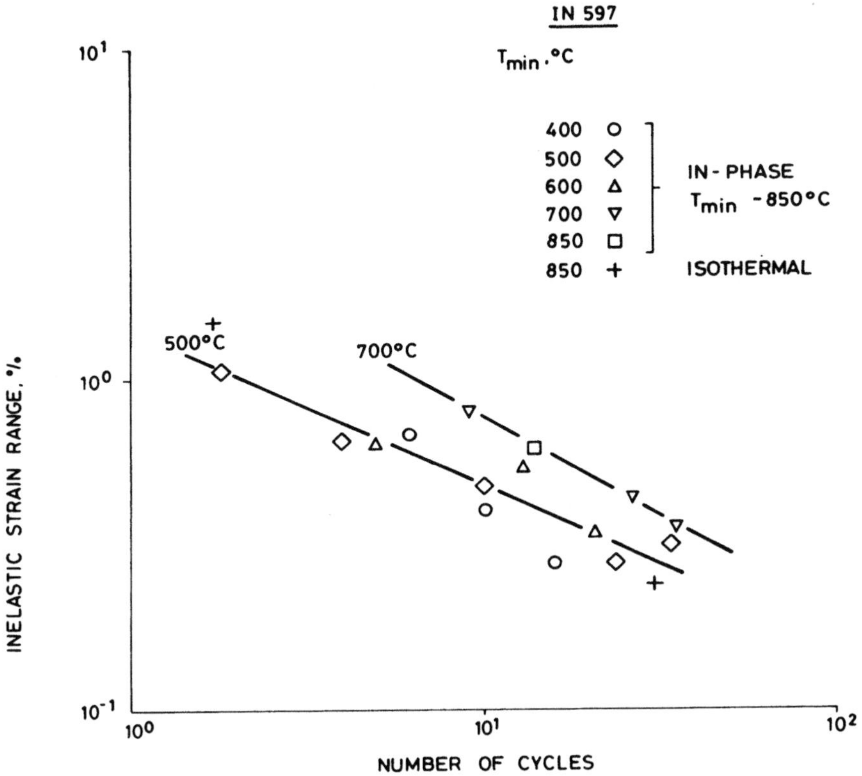

FIG. 30. Plot of inelastic strain range vs. endurance for IN 597 [41]. (Reproduced with permission of the National Physical Laboratory.)

After testing, specimens were examined metallographically and the associated creep damage was measured. It was found that cavity density generally increased with decreasing T_{min}, i.e. a larger temperature range ($T_{max} - T_{min}$) resulted in a more rapid production of cavities. This effect is shown in Fig. 31, in which cavity density is plotted against the product of inelastic strain range and cycles to crack initiation (i.e. the effective accumulated inelastic strain). With the exception of one result from a specimen tested with $T_{min} = 500°C$, the general trend is for the rate of cavitation as a function of accumulated strain to be greater the lower the values of T_{min}. These data are therefore in agreement with the behaviour of ferritic and austenitic stainless steels, in that it appears that thermal–mechanical creep–fatigue tests (i.e. IP with tensile dwells) enhance the

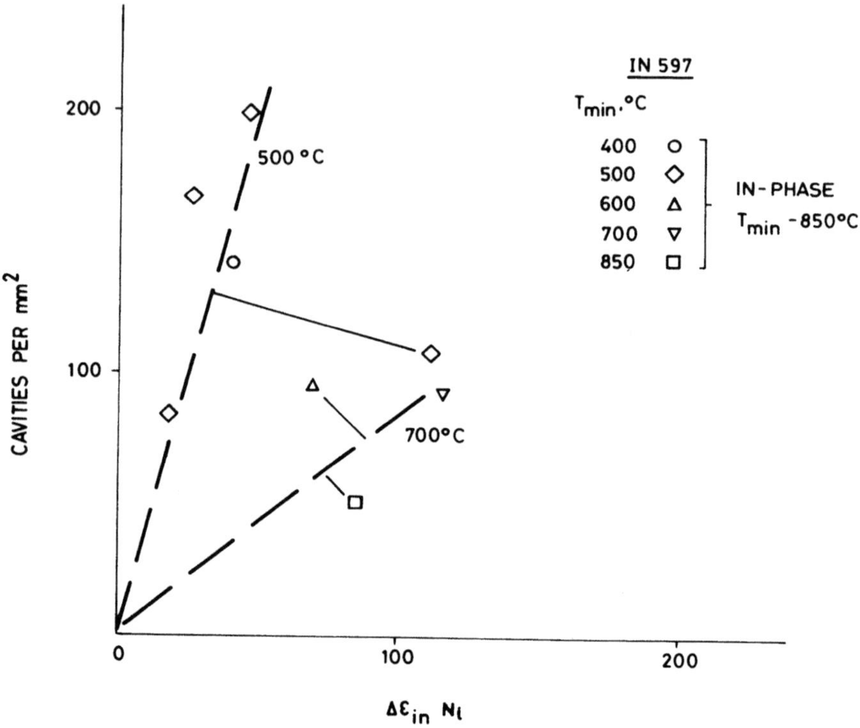

FIG. 31. Cavity density vs. $\Delta\varepsilon_{in}N_i$ [41]. (Reproduced with permission of the National Physical Laboratory.)

development of intergranular creep damage; this observation will be discussed further in Section 3.5.

3.4.4. Summary of General Trends in Thermal–Mechanical Creep–Fatigue

In this section a comparison is made between the damage mechanisms and associated endurance behaviour of isothermal and thermal–mechanical tests involving dwell periods. The influence of tensile and compressive dwells on *isothermal* endurance and failure mode is summarised schematically in Fig. 32, for ferritic and austenitic stainless steels (see also Chapter 5). In general, it is found that compressive dwells are not very damaging since the creep damage component in these tests is negligible; however, it should be noted that in some cases environmental effects have been reported as causing a reduction in endurance in compressive dwell tests [28]. This is not the case for tensile dwell tests.

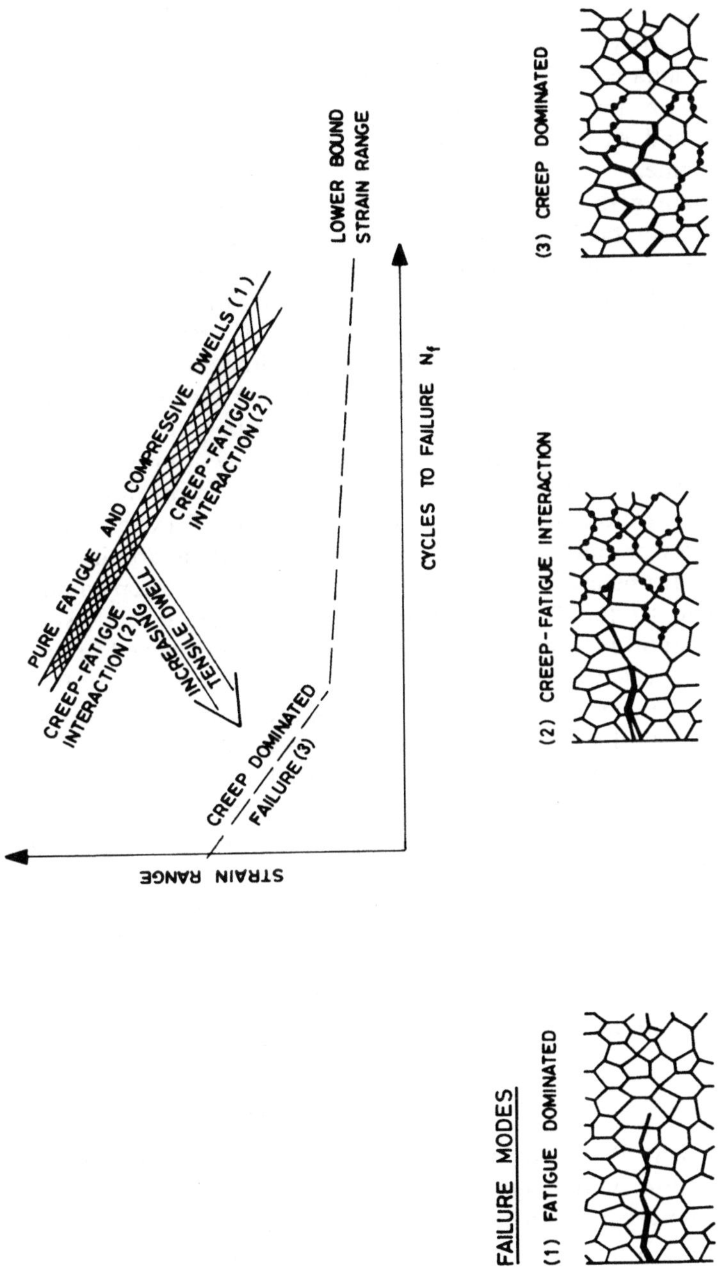

FIG. 32. Generalised isothermal dwell behaviour.

From Fig. 32 it is shown that by increasing tensile dwell periods, the endurance of cyclic tests decreases, failure modes changing from fatigue- to creep-dominated; a small creep–fatigue interaction region occurs at intermediate dwell times. High strain ranges favour fatigue-dominated failures, and as the strain range is decreased creep-dominated failures predominate.

The damage mechanisms associated with tensile dwell cycles are also shown schematically in Fig. 32. During the dwell, creep cavity nucleation and growth occur in each cycle; cyclically reversed plasticity gives rise to fatigue crack nucleation and growth. The growth of creep cavities during dwell periods has been modelled previously [42], resulting in life prediction methods for creep-dominated failure. Depending on the test conditions (i.e. dwell time, strain range, temperature, frequency and creep ductility), either fatigue or creep damage will be dominant or they may interact.

It is now possible to compare isothermal behaviour with that found in thermal–mechanical tests. It has been indicated in previous sections that the incorporation of compressive dwells into OP cycles did not affect endurances or failure modes (transgranular fatigue-dominated). This is similar to the behaviour found in isothermal tests, i.e. creep damage did not contribute to the failures. However, in the OP dwell tests a larger environmental (oxidation) effect may have been expected in comparison with those tests without dwells, given the increased time at T_{max} and the presence of mean tensile stresses. The apparent absence of any such effect was probably due to the short dwell times employed.

The data reviewed on the influence of dwells on IP endurances for ferritic steels, austenitic stainless steels and Ni-base superalloys all show that the incorporation of a dwell into the IP cycles causes a reduction in endurance and promotes an intergranular fracture mode. However, it is more important to note that, generally, IP dwell endurances are less than the equivalent isothermal tensile dwell endurances. A further important observation is that creep cavity formation is accelerated in the IP dwell tests compared to the isothermal dwell tests. The mechanistic reasons for these differences are now considered.

3.5. INTERPRETATION

It is evident that the largest reduction in endurance and the most accumulation of damage is observed in the IP creep–fatigue tests. At present, however, isothermal tensile dwell data are employed in design codes. Since it has been shown in the previous section that these data are not always conservative, it is important to establish, on a mechanistic basis, the reasons for such differences in endurance.

The marked reduction in endurance observed in IP tests is at first somewhat surprising, since the only difference between these and isothermal tests is the temperature variation during reversed plasticity. In fact, comparison of the relaxation behaviour (where cavity growth occurs during the dwell) for IP and isothermal cycles shows very similar behaviour [23]. Therefore, the above differences in endurance and enhanced cavity formation cannot be explained by differences in creep-relaxation behaviour. It is evident that the thermal strain cycle must, in some way, enhance creep cavity formation compared with the isothermal strain cycle. Indeed, this is confirmed from the tests with no dwells, in which maximum internal creep damage was found in the IP cycle. For example, Fig. 28 shows that, for 304 stainless steel, the F_{IG} factor is higher for an IP cycle dwell test than for a comparable tensile dwell isothermal test. Furthermore, it has been shown that cavity formation increases as the differences between T_{max} and T_{min} increases [41]. Consequently, the difference in endurance between IP and isothermal dwell tests also increases as $T_{max} - T_{min}$ increases.

3.5.1. Model I: Cavity Nucleation by Grain-boundary Sliding

Possible explanations for these effects are given in Fig. 33. It has already been noted that IP cycling (without dwells) in austenitic stainless steel results in a net tensile grain boundary sliding component, whereas in

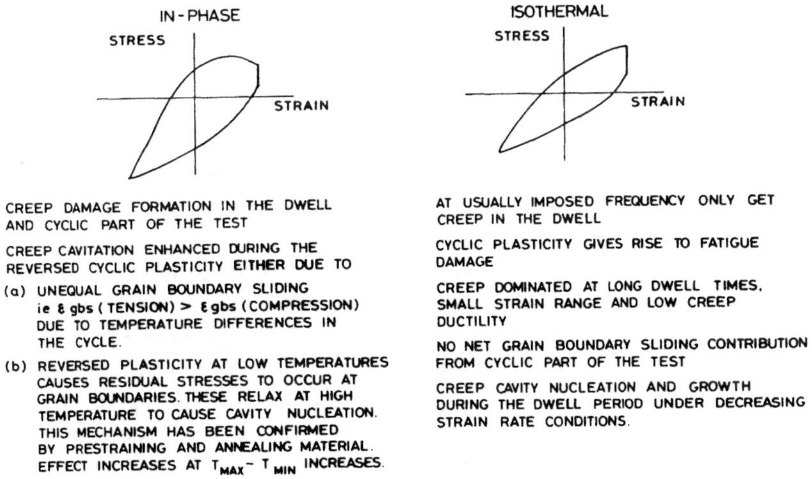

FIG. 33. Creep damage formation in isothermal and in-phase tensile dwell tests.

isothermal tests little or no net accumulation of grain boundary sliding occurs (Section 3.3.2). This arises due to the imbalance of deformation mechanisms (not present in isothermal tests) produced by the temperature difference around the loop in IP tests. In the IP tests, creep deformation ($\varepsilon_{gb} + \varepsilon_{matrix}$) occurs during the tensile part of the cycle. Components of this deformation (ε_{gb}) will not be reversed by the lower temperature deformation that occurs during the compressive part of the cycle. This therefore leads to a net tensile component of grain boundary sliding (ε_{gb}) which accumulates per cycle. It is pertinent to note that a number of different cavity nucleation mechanisms have been proposed [43, 44] which model grain boundary sliding as the cause of stresses which result in cavity nucleation. If such mechanisms are operative in thermal cycling situations, then the net tensile grain boundary sliding component in IP testing will promote cavity nucleation, and give rise to the intergranular damage found in IP cycles. Further, the inclusion of a tensile dwell gives time for growth of the nucleated creep cavities, leading to a higher incidence of creep damage in the IP cycles compared with the isothermal cycles.

3.5.2. Model II: Cavity Nucleation by Matrix Deformation

Recent experimental evidence [45, 46] suggests that cavity nucleation can also be caused by matrix deformation. Dyson and co-workers [45, 46] have demonstrated that, by prestraining and subsequently annealing a number of Ni-base alloys, submicron cavities can be formed at all grain boundaries. On applying a uniaxial creep stress to these alloys, only those cavities on boundaries perpendicular to the stress axis were seen to grow. It has been shown [45] that the mechanism responsible for cavity nucleation is related to matrix deformation, in that slip bands impinging on grain boundaries set up residual stresses which, during high temperature annealing, relax and cause submicron cavity nucleation. Similar observations have been reported by Weertman and co-workers [47, 48].

It is possible to extend this mechanism of prestrain and annealing (which enhances cavity nucleation) to the IP thermal cycle testing. During strain reversal, the material is subject to plastic deformation at low temperatures which will cause slip band impingement at grain boundaries. This will lead to the formation of grain boundary cavities during the subsequent tensile reversal at high temperature. Indeed, supportive evidence for this type of mechanism during IP cycling is given in recent work by Dyson and Wei [49]. In this study, torsional creep–fatigue tests were conducted on Nimonic 101 in two material conditions: (a) as received, and (b) following 9% reversed prestrain.

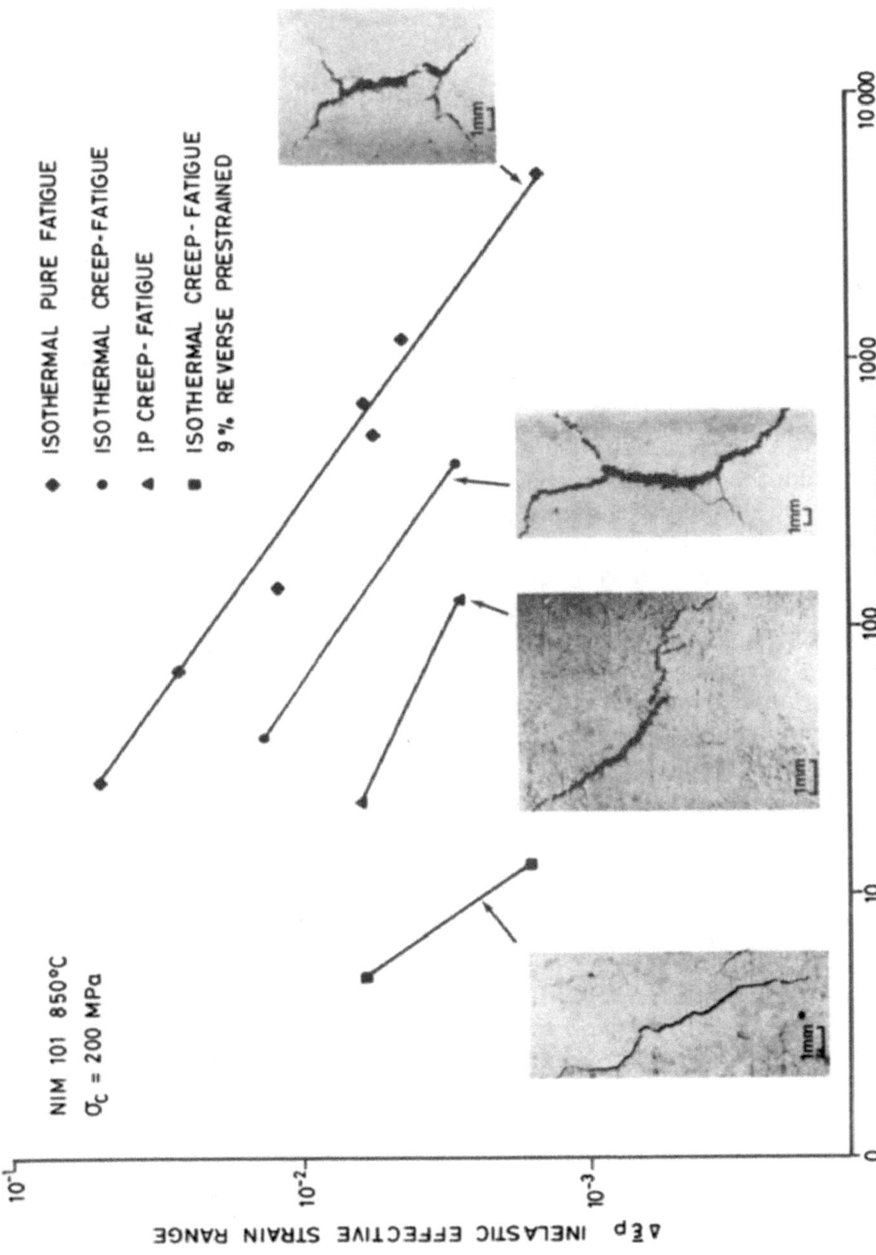

FIG. 34. Creep–fatigue data on (a) as-received and (b) 9% prestrained Nimonic 101 [49]. (Reproduced by courtesy of Ke Wei and Dr B. F. Dyson.)

The following tests were conducted.

For material (a):

(1) isothermal pure fatigue;
(2) isothermal creep–fatigue (peak tensile dwell);
(3) in-phase creep–fatigue (peak dwell).

For material (b):

(4) isothermal creep–fatigue (peak tensile dwell).

The results are shown in Fig. 34, together with the associated metallographic evidence of the damage present in the specimens after testing. It is evident that the endurances of (1) > (2) > (3) > (4). It was found that the isothermal pure fatigue specimens (1) failed in a pure fatigue manner, whereas the isothermal creep–fatigue specimens (2) failed by creep–fatigue interaction. However, the IP specimens (3) and the isothermal creep–fatigue specimens on material B (4) both failed in a creep-dominated manner. Furthermore, the isothermal creep–fatigue tests on the prestrained material (4) exhibited the lowest endurance. This is a consequence of two related factors. The prestraining has caused the nucleation of creep cavities, which results in the material having a lower creep ductility than the as-received material. The endurance of the prestrained material is therefore related to the growth of grain boundary cavities only. The difference in endurances between the as-received material (1) and prestrained material (4) is therefore a measure of the number of cycles required to nucleate creep cavities. Thus, on comparing the endurance of the isothermal creep–fatigue tests with IP tests, it is evident that the reduction in endurance is a consequence of the increased number of cavities nucleated per unit plastic strain reversal in the IP tests. Furthermore, it is likely that the number of cavities nucleated is temperature-dependent, more being produced at the lower temperatures in the IP cycle.

There is clearly a need to confirm the operation of this mechanism of cavity nucleation for IP tests. In particular, the influence of such a mechanism on endurance for long dwell periods requires further investigation, since it has important implications regarding the long-term isothermal data presently used in existing design codes.

3.6. LIFE PREDICTION

It has been shown that, for some thermal cycles such as Type O (Table 2), IP and OP data give almost identical lifetimes (Fig. 13). Indeed, for some ferritic steels and Ni-base superalloys [4], those isothermal data generated at a temperature equal to the maximum used in comparable thermal cycle tests, T_{max}, bound all the thermal cycle data within a scatterband of $\times 2$. Further, for $2\frac{1}{4}$Cr–1Mo steel, similarly generated isothermal data gave lower-bound predictions to the thermal cycling data [4]. Examples of Type E cycles have also shown how isothermal data obtained at T_{max} can be used to estimate thermal cycling data on a total strain range versus endurance basis [32] (Fig. 17(a)). Thus, for Types E and O material behaviour, life predictions for thermal cyling tests can reasonably be obtained from isothermal data generated at T_{max}.

This, however, is not the case for Type I material response. Results from thermal cycling tests on 304 stainless steel have indicated (Fig. 18) that data can fall below the $\times 2$ scatterband on isothermal results, leading to non-conservative predictions on the basis of isothermal data: hold-period thermal cycle tests would lead to further non-conservatisms. Thus, for situations where intergranular internal damage causes a noticeable decrease in endurance, some other life-prediction technique is required. Several techniques for life prediction are available [50–55]; all have been used primarily for the isothermal case [29, 56], but some workers have extended their applicability to thermal cycling situations [3, 4, 9, 27, 35, 40, 57]. The method that has been most often used for life prediction during thermal fatigue is strain range partitioning, SRP, considered in detail below.

3.6.1. Strain Range Partitioning (SRP)

3.6.1.1. Basic Method

It is assumed that the inelastic strain accumulated during one isothermal high temperature cycle can be partitioned into four possible kinds of strain component [52]. These depend on the direction of straining (tension or compression) and on the type of inelastic strain accumulated (creep or plasticity). These types of strain component are identified as ε_{cp}, ε_{pc}, ε_{pp} and ε_{cc}, where, for example, ε_{cp} is that part of the inelastic strain range which results from a tension-going creep strain balanced by a compression-going plastic strain (see also Chapter 8 in the earlier volume). The well-known Manson–Coffin relationship is then used to describe four separate relationships for high temperature applications:

$$\begin{aligned} N_{pp} &= A_1(\Delta\varepsilon_{pp})^{\alpha_1}, \qquad N_{cp} = A_2(\Delta\varepsilon_{cp})^{\alpha_2} \\ N_{pc} &= A_3(\Delta\varepsilon_{pc})^{\alpha_3}, \qquad N_{cc} = A_4(\Delta\varepsilon_{cc})^{\alpha_4} \end{aligned} \tag{1}$$

where the coefficients A and exponents α are experimentally determined material constants.

Thus, predicted lifetimes under cyclic/hold test conditions at elevated temperatures can be obtained by using the generic life lines described by eqn (1) along with the interactive damage rule:

$$1/N_{\mathrm{Pred}} = F_{\mathrm{pp}}/N_{\mathrm{pp}} + F_{\mathrm{cp}}/N_{\mathrm{cp}} + F_{\mathrm{cc}}/N_{\mathrm{cc}} + (F_{\mathrm{pc}}/N_{\mathrm{pc}}) \tag{2}$$

where F_{**} are the inelastic strain component fractions (e.g. $F_{\mathrm{pp}} = \Delta\varepsilon_{\mathrm{pp}}/\Delta\varepsilon_{\mathrm{in}}$) and the brackets emphasise that, for any given loop, it is only possible to have a 'cp' or a 'pc' damage component, not both.

The generic lifetime relationships correlating the various strain components to their life contributions are nominally temperature-insensitive, so the application of SRP to thermal cycling conditions is attractive. Further, by equating the varying temperatures that occur in thermal cycling tests under constant strain rate conditions to varying strain rates in an isothermal test [25, 58, 59], slow–fast tests, for example, can be used to obtain generic life relationships for the 'cp' strain component, etc. [9].

Although the generic endurance equations needed in the application of SRP to thermal cycling situations can be obtained from isothermal tests, the problem of partitioning actual thermal cycle tests into plasticity and creep components remains. The method generally used is the step-stress method [60], during which stress is held constant at different points around the (thermal cycle) hysteresis loops and steady-state creep rates determined. By integrating these rates throughout the period of one cycle, creep strain values can be evaluated, as shown in Fig. 35 [25].

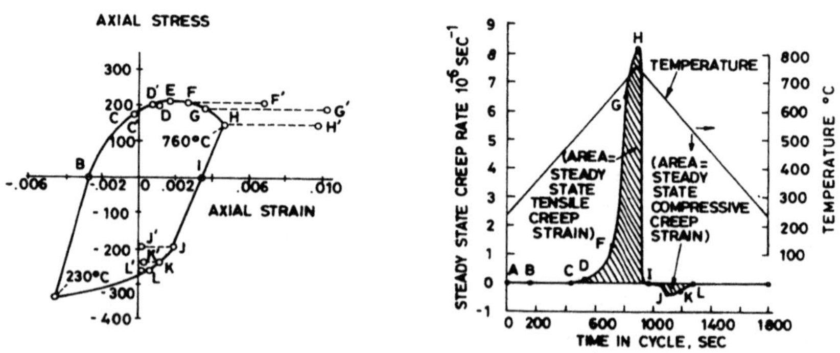

(a) STRESS HOLDING POSITION ON HYSTERESIS LOOP

(b) RESULTS OF OBTAINING CREEP STRAIN COMPONENT BY STEP-STRESS METHOD

FIG. 35. Procedure of obtaining the creep strain component by the step-stress method (example of in-phase thermal fatigue) [25]. (Reproduced with permission of CRIEPI, Tokyo, Japan.)

3.6.1.2. *Simplified Methods*

Some workers have found this technique difficult to apply, and have therefore adopted simplified step-stress approaches. First attempts [9, 17] made the assumption that creep strain could be evaluated by assuming a linear change in creep strain rate as a function of temperature. Thus, using the data of Kuwabara and Nitta [9] on 304 stainless steel, and assuming that for this material creep rate data are available from static creep tests at only two temperatures, say 550°C and 600°C, the variation in creep strain rate throughout the whole of the thermal cycle range (taken to be 300–600°C) would be estimated as shown in Fig. 36. Note that the creep strain component is assumed to be generated only in the half-cycle including temperature increments.

For the example shown in Fig. 36, it was found that the creep rate at 550°C was two to three orders of magnitude less than that at 600°C, and a further simplification to the step-stress method was introduced. It was assumed [27] that creep strain was generated only for temperature increments above some critical value, T_c, taken to be 500°C for low-alloy steels and 550°C for austenitic steels. Further, creep strain components

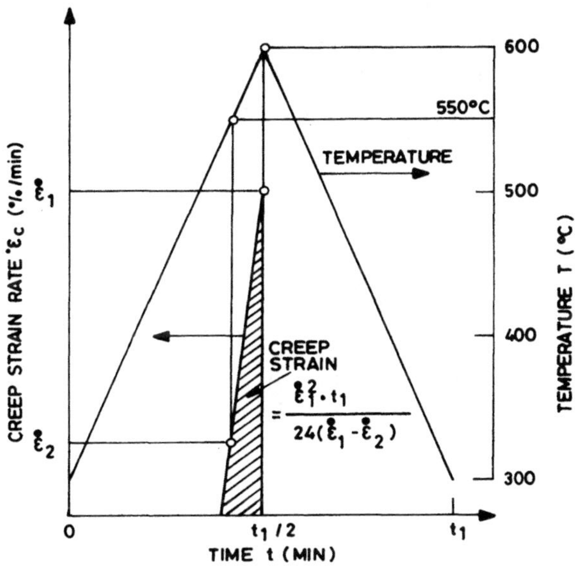

FIG. 36. Schematic diagram showing the method of separating the creep strain component in thermal fatigue [9, 17]. (Reproduced with permission of CRIEPI, Tokyo, Japan.)

were evaluated from the creep rates measured at any given stress at the maximum temperature, using the relationship

$$\dot{\varepsilon}_c = B\sigma^n \tag{3}$$

where $\dot{\varepsilon}_c$ is the steady-state creep strain rate, σ the stress, and B and n are constants. Thus, if σ_t is the maximum tensile stress at T_{max}, the creep strain component, for example, for IP cycles would be given by [27]

$$\Delta\varepsilon_{cp} = \frac{B\sigma_t^n t_1(T_{max} - T_c)}{4(T_{max} - T_{min})} \tag{4}$$

where t_1 is the time for one cycle, excluding hold times.

For dwell tests, the strain contribution evaluated from the relaxed stress during the dwell divided by Young's modulus is simply added to eqn (4).

An alternative to the step-stress method is the loop inversion technique [27], which is similar in concept to the half-cycle rapid load–unload method of partitioning strain ranges under isothermal test conditions [60]. Creep strain is once more assumed to occur during the temperature-ascending part of the thermal cycle, and not during the temperature-descending parts. Thus (Fig. 37), the segment of the hysteresis loop in the temperature-descending part of the cycle is simply inverted to the segment in the temperature-ascending part of the cycle, thereby enabling the purely cyclic, ε_{pp}, strain component to be evaluated. The creep strain component is then simply the inelastic strain range minus ε_{pp}.

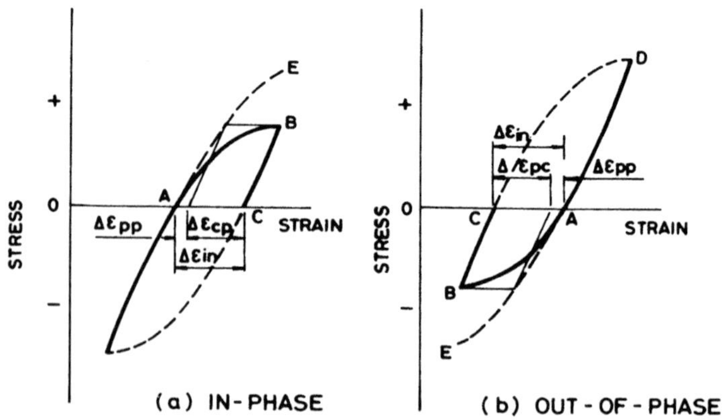

FIG. 37. Procedure of the loop inversion method for evaluating the creep strain component in thermal fatigue [27]. (Reproduced with permission of CRIEPI, Tokyo, Japan.)

3.6.2. Application of SRP

Figures 38 and 39 show the results obtained from both the simplified step-stress and loop inversion methods of strain range partitioning for a range of thermal cycle hold and no-hold tests performed on forged Cr–Mo–V steel and 304 stainless steel [27]. Using the simplified step-stress approach, SRP predictions were within a factor of two of the observed lifetimes for all except the OP tests performed with no-hold periods on the ferritic steel. For these tests, non-conservative predictions were obtained. However, by using the loop inversion method, OP tests with no dwells were predicted to lie within the ×2 scatterband on observed endurances for both steels although, for the remaining test cycles, conservative predictions ensued.

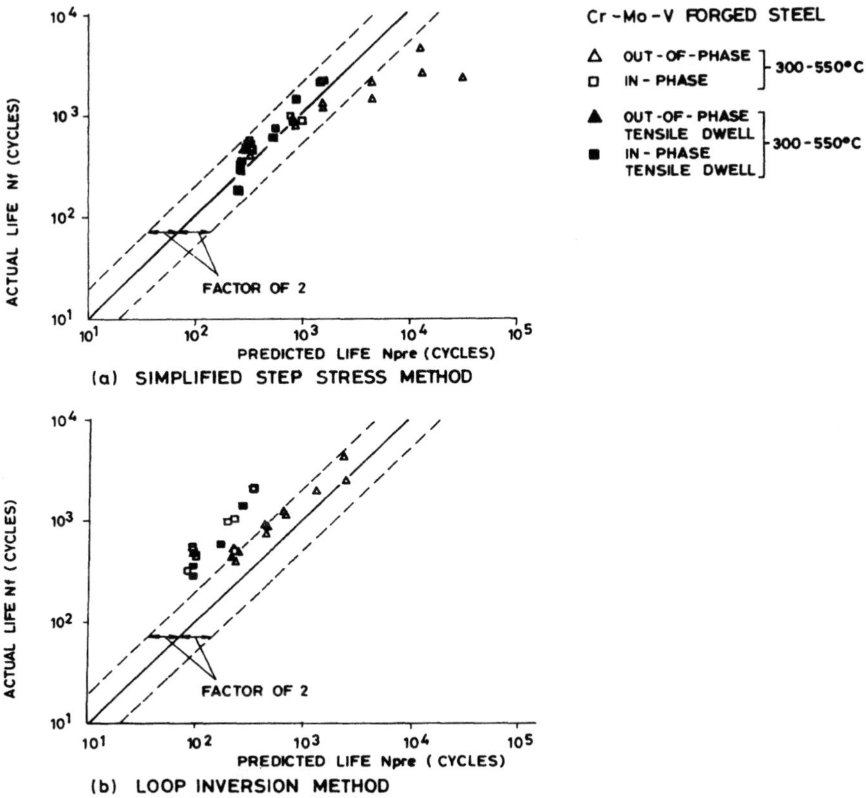

FIG. 38. Thermal fatigue life prediction for Cr–Mo–V forged steel [27]. (Reproduced with permission of CRIEPI, Tokyo, Japan.)

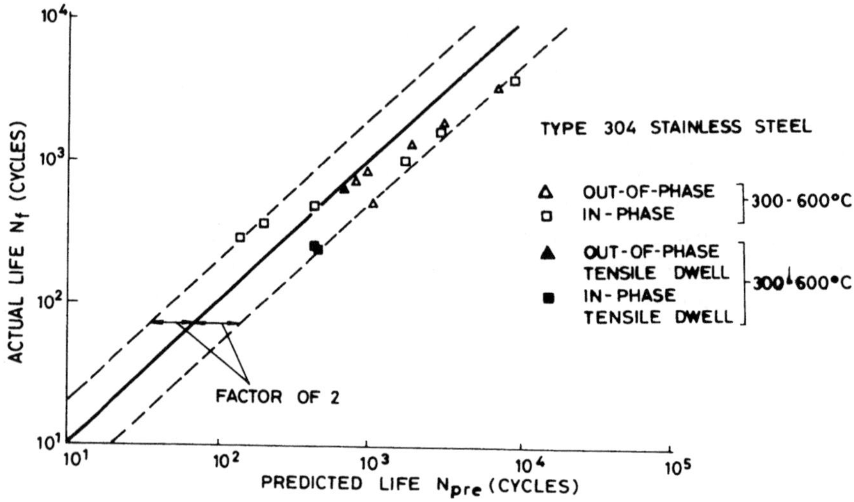

(a) SIMPLIFIED STEP STRESS METHOD

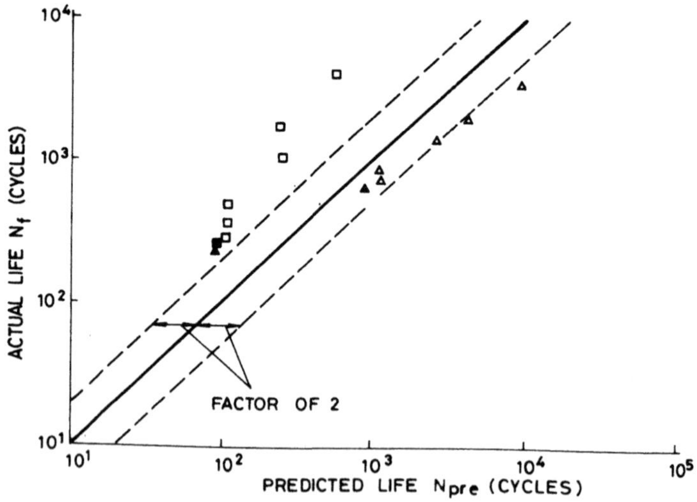

(b) LOOP INVERSION METHOD

FIG. 39. Thermal fatigue life prediction for type 304 stainless steel [27]. (Reproduced with permission of CRIEPI, Tokyo, Japan.)

Similar results were obtained from $2\frac{1}{4}$Cr–1Mo steel and Ni–Mo–V steel data [27]. Further, for type 321 austenitic steel [27], the simplified step-stress method yielded non-conservative SRP predictions for all test types. The loop inversion method enabled more accurate predictions, the data being scattered within a ×2 scatterband on life.

Beauchamp [3] was able to perform the full step-stress technique [60] when generating his thermal cycling data on 1Cr–Mo–V steel. However, this author attended to some of the inadequacies in the SRP approach, notably the inherent assumption that all the strain accumulated within, for example, a cyclic hold period was attributable to creep, irrespective of the rate at which this strain was accumulated [61]. He therefore modified the creep strain integration to sum only those contributions which occurred below a certain transition strain rate, and beyond which true creep damage was deemed to occur [54, 61]. Using this modified step-stress approach, he obtained results for no-hold thermal cycle test conditions as shown in Table 5.

TABLE 5
LIFE PREDICTIONS FOR 1Cr–Mo–V STEEL USING SRP [3]

Total strain range	*Cycle type*	*Predicted life (cycles)*	*Observed life (cycles)*
1×10^{-2}	In-phase, 565–315°C	1 234	1 170
$1{\cdot}48 \times 10^{-2}$	In-phase, 565–315°C	420	467
1×10^{-2}	Out-of-phase, 315–565°C	1 404	965
$1{\cdot}5 \times 10^{-2}$	Out-of-phase, 315–565°C	965	564

Predictions were thus good, though it was noted that by subdividing the apparent creep strain component (according to strain rate) into matrix and grain boundary damage, it was the matrix damage that contributed most to the failure process. Nevertheless, had the whole of the creep strain component been used to estimate ε_{cp} or ε_{pc}, predictions would have been more conservative.

In summary, the loop inversion method for obtaining SRP components, and subsequent endurance calculations, generally gives conservative predictions for thermal fatigue lifetimes for both hold and no-hold tests. However, although SRP has been shown to work for tests on these low strength/high ductility materials, the method may be inapplicable to those of high strength and low ductility. This is indicated in Fig. 40 which shows that, for one alloy, thermal fatigue lives can be shorter than those obtained

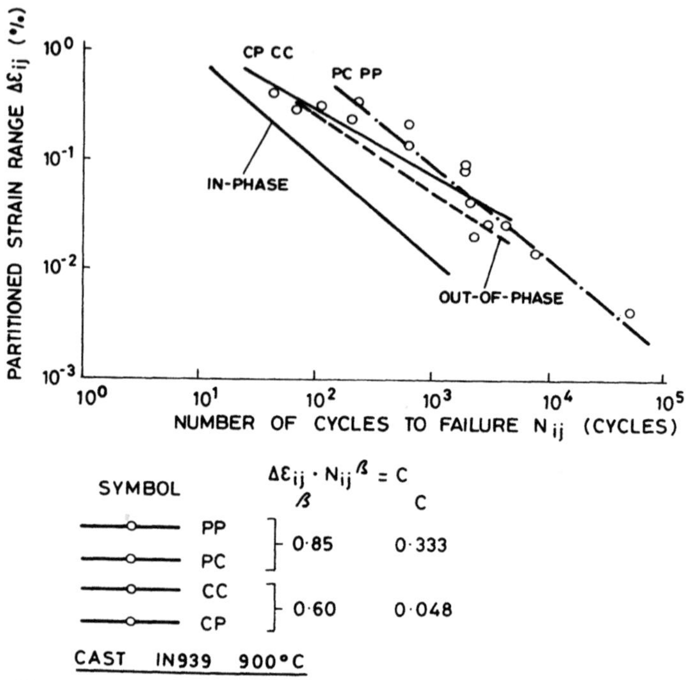

FIG. 40. Comparison of the partitioned inelastic strain range vs. life curve with thermal fatigue life curve for IN 939 Ni-base superalloy [4]. (Reproduced with permission of ASME.)

from the generic SRP life relationships [4]. Also, to date, only tests with relatively short hold periods have been performed under thermal cycling conditions. Isothermal tests on a range of materials have shown that SRP is inadequate for accounting for the damaging effect of long hold times [56]. Therefore, to obtain accurate life predictions for thermal cycling data at low strain ranges and/or long hold times, detailed understanding of the micromechanisms of failure must be obtained.

3.7. CONCLUDING REMARKS

In this chapter, an attempt has been made to review non-isothermal creep–fatigue data obtained from engineering alloys, in order to assess their relative importance. It has been shown that although thermal–mechanical fatigue tests have different endurances to those associated with isothermal fatigue tests, the variations between the test types are generally small. In fact, both in-phase and out-of-phase (IP and OP) test conditions have resulted in the majority of the data lying within a ×2 scatterband on the isothermal data. Thus, for the purpose of design codes, it would appear that it is generally adequate to use isothermal fatigue data.

However, under certain conditions, IP tests have been observed to exhibit much lower endurances than the equivalent isothermal and, further, these endurances decreased as the maximum temperature and/or temperature range used in the thermal cycles were increased. It has been possible to explain these differences in endurance on a mechanistic basis. In the IP tests, creep cavity formation is exacerbated by enhanced cavity nucleation caused by the reversed plasticity which occurs at low temperature. In addition, the data available suggest that, for plant which is subject to IP thermal cycles together with tensile dwell periods, creep damage also initiates earlier than would have been predicted from isothermal tensile dwell tests. Clearly, this may have important implications, since the isothermal data used in existing design codes may lead to non-conservative life predictions.

In order to fully confirm the magnitude of these effects, more extensive IP testing at low strain ranges and long dwell periods ($\gtrsim$16 h) is required to establish if these differences persist under conditions more representative of service.

Few of the currently popular life-prediction techniques used to assess creep–fatigue endurances can be easily applied to the non-isothermal case. To date, most effort has been concentrated on the strain range partitioning technique, since the generic life component equations are nominally temperature-insensitive. Even so, complex procedures have to be used to partition thermal cycles into their component strain ranges.

Notwithstanding these difficulties, SRP has been successful in describing thermal cycling data from a range of sources. However, as noted above, these tests have been restricted to relatively high strain ranges and short hold times. Previous analyses of SRP under isothermal test conditions have highlighted the potential uncertainties in this approach for long hold times/low strain ranges. Thus, it is concluded that, once more, the problem of generating longer-term thermal cycling data needs to be addressed if representative and reliable long-term predictions are to be made.

REFERENCES

1. Westwood, H. J. and Moles, M. D. C., 1979, Creep–fatigue problems in electricity generating plant, *Can. Metall. Q.*, **18**, 215–30.

2. Coffin, L. F. and Wesley, R. P., 1954, An apparatus for the study of the effects of cyclic thermal stress on ductile metals, *Trans. Am. Soc. Mech. Engrs*, **76**, 923–36.
3. Beauchamp, D. J., 1982, Thermal–mechanical strain cycling at elevated temperature, Ph.D. Thesis, Univ. Bristol.
4. Kuwabara, K., Nitta, A. and Kitamura, T., 1983, Thermal–mechanical fatigue life prediction in high-temperature component materials for power plant, ASME Int. Conf. on Advances in Life Prediction Methods, pp. 131–41.
5. Hopkins, S. W., 1976, Low-cycle thermal mechanical fatigue testing, *Thermal fatigue of materials and components*, STP 612, Philadelphia, ASTM, pp. 157–69.
6. Jaske, C. E., 1976, Thermal–mechanical, low cycle fatigue of AISI 1010 steels, *ibid.*, STP 612, Philadelphia, ASTM, pp. 170–98.
7. Fujino, M. and Taira, S., 1979, Effect of thermal cycling on low cycle fatigue life of steels and grain broundary sliding characteristics, *Proc. ICM3*, **2**, pp. 49–58, Pergamon Press.
8. Wareing, J. and Vaughan, H. G., 1979, Influence of surface finish on low cycle fatigue characteristics of type 316 stainless steel at 400°C, *Metall. Sci.*, **13**, 1.
9. Kuwabara, K. and Nitta, A., 1978, Thermal fatigue behaviour and strength estimation of power plant component materials under creep–fatigue interaction—experiments and analysis on type 304 stainless steel, CRIEPI Report E278003.
10. Gemma, A. E., Langer, B. S. and Leverant, G. R., 1976, Thermomechanical fatigue crack propagation in an anisotropic (directionally solidified) nickel base superalloy, *Thermal fatigue of materials and components*, STP 612, Philadelphia, ASTM, pp. 199–213.
11. Hirschberg, M. H., 1969, A low cycle fatigue testing facility, *Manual on low cycle fatigue testing*, STP 465, Philadelphia, ASTM, pp. 67–86.
12. Carden, A. E., 1970, Thermal fatigue evaluation, *ibid.*, STP 465, Philadelphia, ASTM, pp. 163–88.
13. Grussenkov, A. P. and Kasantsev, A. G., 1979, A kinetic deformation criterion for creep low-cycle fatigue under non-isothermal loading, *Proc. ICM3*, **2**, pp. 43–8, Pergamon Press.
14. Sheffler, K. D., 1976, Vacuum thermal–mechanical fatigue behaviour of two iron base alloys, *Thermal fatigue of materials and components*, STP 612, Philadelphia, ASTM, pp. 214–26.
15. Jaske, C. E. and Leis, B. N., 1979, Important considerations in thermal–mechanical fatigue evaluation of metallic materials, Paper 131, Int. Corrosion Forum, Atlanta, Georgia.
16. Ellison, E. G., 1981, Thermal–mechanical strain cycling and testing at higher temperatures, *Measurements of High Temp. Mech. Props. of Materials*, London, HMSO, Paper 13, pp. 204–18.
17. Kuwabara, K. and Nitta, A., 1979, Thermal–mechanical low cycle fatigue under creep–fatigue interaction on type 304 stainless steel, *Proc. ICM3*, **2**, pp. 69–78, Pergamon Press.
18. Beauchamp, D. J. and Ellison, E. G., 1982, Development of a temperature and strain cycling rig, *J. Strain Anal.*, **17**(1), 45–52.
19. Slot, T., Stentz, R. H. and Berling, J. T., 1969, Controlled strain testing procedures, *Manual on low cycle fatigue testing*, STP 465, Philadelphia, ASTM, pp. 100–28.

20. Cook, T. and Cavill, G., 1968, An experimental arrangement for direct high strain fatigue tests at elevated temperature, *J. Sci. Instr.*, **1**, 337–40.
21. Lord, D. C. and Coffin, L. F., 1969, High temperature material behaviour, *Manual on low cycle fatigue testing*, STP 465, Philadelphia, ASTM, pp. 129–48.
22. Perruchet, C., 1978, Contribution à l'evaluation de la methode SRP à l'alliage base nickel waspaloy, AGARD Conf. Proc. 243, Paper g.
23. Kuwabara, K. and Nitta, A., 1977, Isothermal and thermal fatigue strength of CrMoV steel for turbine rotors, CRIEPI Report E277005.
24. Nitta, A., Kuwabara, K. and Kitamura, T., 1982, Creep–fatigue damage in power plant materials—Part 2. The behaviour of elevated temperature low cycle fatigue crack initiation and propagation in steam turbine rotor steels, CRIEPI Report E282002.
25. Kuwabara, K., Nitta, A. and Kitamura, T., 1980, The evaluation of thermal fatigue strength of a $2\frac{1}{4}$Cr–1Mo steel under creep–fatigue interaction, CRIEPI Report E279007.
26. Kuwabara, K. and Nitta, A., 1979, On high-temperature fatigue and creep strength of a $2\frac{1}{4}$Cr–1Mo steel pipe material used for a long period, CRIEPI Report E279002.
27. Nitta, A., Kuwabara, K. and Kitamura, T., 1983, Prediction of thermal fatigue life in high temperature component materials for power plant, CRIEPI Report E282015.
28. Challenger, K. D., Miller, A. K. and Brinkman, C. R., 1981, An explanation for the effects of hold periods on the elevated temperature fatigue behaviour of $2\frac{1}{4}$Cr–1Mo steel, *J. Engng Mat. Tech., Trans. ASME*, **103**, 7–14.
29. Miller, D. A., Priest, R. H. and Ellison, E. G., 1984, A review of material response and life prediction techniques under fatigue–creep loading conditions, *High Temp. Mat. Proc.*, **6**(3 & 4), 155–94.
30. Davies, P. W., Stevens, R. N. and Wilshire, B., 1966, Grain boundary sliding in polycrystalline aluminium, *J. Inst. Metals*, **94**, 49–55.
31. Kitagawa, M., 1972, A microscopic approach to intergranular fatigue fracture of chemical lead at room temperature, Ph.D. Thesis, Univ. Illinois.
32. Taira, S., Fujino, M. and Ohtani, R., 1979, Collaborative study on thermal fatigue properties of high temperature alloys in Japan, *Fatigue Engng Mat. Struct.*, **1**, 495–508.
33. Westwood, H. J., 1979, Tensile hold time effects on isothermal and thermal low cycle fatigue of 304 stainless steel, *Proc. ICM3*, **2**, pp. 59–67, Pergamon Press.
34. Udoguchi, T. and Wada, T., 1969, Thermal effect on low cycle fatigue strength of steels, *Proc. Int. Conf. on Thermal Stresses and Thermal Fatigue, Berkeley, England*, London, Butterworth, pp. 109–23.
35. Nitta, A., Kuwabara, K. and Kitamura, T., 1983, The characteristics of thermal–mechanical fatigue strength in superalloys for gas turbine, CRIEPI Report E282011.
36. Westwood, H. J. and Lee, W. K., 1981, Creep–fatigue crack initiation in $\frac{1}{2}$CrMoV steel, *Proc. Int. Conf. on Creep and Fracture of Engng Mat. and Struct., Swansea*, Pineridge Press, pp. 517–30.
37. Lee, W. K., Westwood, H. J., Moles, M. D. C. and Lawson, W. H. S., 1981, Fracture characteristics of $2\frac{1}{4}$Cr–1Mo welds under thermal-fatigue conditions, *Proc. Fifth Can. Fract. Conf., Winnipeg, Canada*, pp. 85–95.

38. Kuwabara, K. and Nitta, A., 1979, Thermal–mechanical low-cycle fatigue under creep–fatigue interaction on type 304 stainless steels, *Fatigue Engng Mat. Struct.*, **2**, 293–304.
39. Wareing, J., 1983, Mechanisms of high temperature fatigue and creep–fatigue failure in engineering materials, *Fatigue at high temperature*, ed. R. P. Skelton, London, Elsevier Applied Science, pp. 135–86.
40. Levaillant, C., Rezgui, B. and Pineau, A., 1979, Effects of environment and hold times on high temperature low cycle fatigue behaviour of 316L stainless steel, *Proc. ICM3*, **2**, pp. 163–72, Pergamon Press.
41. Thomas, G. B. and Varma, R. H., 1984, Assessment of LCF and creep–fatigue behaviour of Ni–Cr base superalloys, COST 50—Materials for Gas Turbine, UK 11—Final Report Round 3 DMA(A)90.
42. Miller, D. A., Hamm, C. D. and Phillips, J. L., 1982, A mechanistic approach to the predication of creep-dominated failure during simultaneous creep–fatigue, *Mat. Sci. Engng*, **53**, 233–44.
43. Goods, S. H. and Brown, L. M., 1979, The nucleation of cavities by plastic deformation, *Acta Metall.* Overview No. 1, **27**, 1–15.
44. Svensson, L. E. and Dunlop, G. L., 1979, The role of interfacial dislocations in the nucleation of intergranular creep cavities, *Can. Metall. Q.*, **18**, 39–47.
45. Dyson, B. F., Loveday, M. S. and Rodgers, M. F., 1976, Grain boundary cavitation under various states of applied stress, *Proc. R. Soc., Lond.*, **A349**(1657), pp. 245–59.
46. Dyson, B. F. and Henn, D. E., 1973, The effect of room temperature prestrain on grain boundary cavitation in Nimonic 80A, *J. Microsc.*, **97**, 165–70.
47. Shiozawa, K. and Weertman, J. R., 1982, The Nucleation of grain boundary voids in a nickel-base superalloy during high temperature creep, *Scipta Metall.*, **16**(6), 735–9.
48. Shiozawa, K. and Weertman, J. R., 1983, Studies of nucleation mechanisms and the role of residual stresses in the grain boundary cavitation of a superalloy, *Acta Metall.*, **31**, 993–1004.
49. Dyson, B. F. and Wei, K., 1986, To be published.
50. Taira, S., 1962, Lifetime of structures subjected to varying load and temperature, *Creep in structures*, Academic Press, pp. 96–124.
51. Coffin, L. F., 1976, The concept of frequency separation in life prediction for time dependent fatigue, ASME–MPC Symp. on Creep Fatigue Interactions, pp. 346–63.
52. Manson, S. S., 1973, The challenge to unify treatment of high temperature fatigue—a partisan proposal based on strain range partitioning, *Fatigue at elevated temperatures*, STP 520, Philadelphia, ASTM, pp. 744–82.
53. Majumdar, S. and Maiya, P. S., 1979, Creep–fatigue interactions in an austenitic stainless steel, *Can. Metall. Q.*, **18**, 57–64.
54. Priest, R. H. and Ellison, E. G., 1981, A combined deformation map – ductility exhaustion approach to creep–fatigue analysis, *Mat. Sci. Engng*, **49**, 7–17.
55. Wareing, J., 1977, Creep–fatigue interaction in austenitic stainless steels, *Metall. Trans.*, **8A**, 711–21.
56. Priest, R. H. and Ellison, E. G., 1982, An assessment of life analysis techniques for fatigue/creep situations, *Res Mech.*, **4**, 127–50.

57. Priest, R. H., Beauchamp, D. J. and Ellison, E. G., 1983, Damage during creep–fatigue, *ASME Int. Conf. on Advances in Life Prediction Methods*, pp. 115–22.
58. Coffin, L. F., 1976, Instability effects in thermal fatigue, *Thermal fatigue of materials and components*, STP 612, Philadelphia, ASTM, pp. 227–38.
59. Taira, S. and Fujino, M., 1978, A damage analysis in high temperature thermal fatigue, *J. Soc. Mat. Sci., Jpn*, **27**, 440–46.
60. Manson, S. S., Halford, G. R. and Nachtigall, A. J., 1975, Separation of the strain components for use in strain range partitioning, NASA Report TMX-71737.
61. Priest, R. H. and Ellison, E. G., 1980, Estimation of cyclic creep damage by strain and strain rate considerations, *Int. Conf. on Engng Aspects of Creep, Sheffield, England*, **1**, pp. 185–92.

4

Metallography of High Temperature Fatigue

W. J. PLUMBRIDGE*

Department of Metallurgy and Materials Engineering,
University of Wollongong, Australia

SUMMARY

The vast majority of metallographic studies of fatigue have been concerned with room-temperature cycling and, generally, with elementary materials such as pure metals and simple alloys. In this chapter, using ambient temperature fatigue as a baseline, recent developments in the observation and understanding of high temperature fatigue are explored, up to and including crack initiation. Both the effects of microstructure on cyclic response and the influence of cyclic strain on microstructure are examined, to highlight the requirements for improved fatigue resistance. Emphasis has been placed on realistic, but complex, cycles where interactions between fatigue and creep are possible, and also on commercial alloys, in particular those based on aluminium and titanium. Data in these areas are limited, and it is concluded that Metallography has yet to reach its full potential in assisting life prediction of components and structures in service.

4.1. INTRODUCTION

Over the past three decades, the subject of fatigue has received considerable attention and, as information has been generated at a seemingly exponential rate, the number of reviews has also reached unmanageable levels. To attempt to undertake yet another and claim to be comprehensive would be a monumental task, beyond this author and, indeed, the purview

* This chapter was written whilst the author was Reader in Metallurgy in the Department of Mechanical Engineering, University of Bristol, UK.

of this book. To avoid disappointment, it is therefore appropriate at this stage to define 'objectives' and 'limits' of the present contribution.

In this chapter we are limited to two principal aspects of fatigue; (a) metallography and (b) high temperature, defined in this context as any temperature in excess of ambient. Even with this restriction there still remains a great volume of data gathered over the past decade which has necessitated further selectivity, as outlined later. The major objective of this review is to assess the developments in our knowledge of high temperature fatigue gained over the past few years. 'Metallography' has been interpreted in its broadest sense to include both microstructural observations and the elucidation of mechanisms arising from these observations. It is strongly held that Metallography can only make its most effective contribution to the prediction of materials behaviour and to the development of improved materials, if a thorough *understanding* of the processes involved is obtained. For its real value to be universally appreciated, especially by the Engineering fraternity, Metallography must be seen to be achieving these goals, and it is hoped that the present chapter will reveal some of its recent successes.

4.2. THE FATIGUE PROCESS—OVERVIEW AND APPROACH

4.2.1. Traditional Description

While we know a great deal more about the detail of the events leading to fatigue failure, the basic 'initiation/stage I/stage II' description of Forsyth [1] still holds. Stage II growth has been further subdivided to include pre- and post-continuum growth regions; yet the precise definition and significance of crack initiation remains as hotly disputed as ever [2]. Temperature effects have perhaps become predominant in the past ten years since they accentuate time-dependent phenomena associated with strain rate, environment and creep. These are particularly important at low strain rates, and where continuous cycling is replaced by cycles containing dwells so that interactions between fatigue and creep result.

To establish a perspective for this article it is perhaps helpful to outline the traditional fatigue sequence, which is as follows:

(1) *Cyclic deformation processes.* These are defined as recoverable phenomena which may occur on a local or global scale, producing fatigue hardening, softening or stability (see Chapter 2). They involve interactions between dislocations and precipitate particles, inclusions, grain boundaries and, most significantly, other dislocations.

(2) *Crack initiation.* On the exhaustion of the local reversed-strain accommodating capacity, initiation takes place at a discontinuity (or stress-concentrating feature), either within the matrix or along a grain boundary. No precise definition of the event is offered except to note that such damage may not be repaired by annealing, unlike the reversible processes occurring in the pre-initiation phase. The classical mechanism for smooth surfaces involves intrusions and extrusions.

(3) *Stage I crack growth.* Extension occurs along slip planes of the 'initiated' defect under mode II (in-plane shear) sliding conditions. Slight changes in growth direction are governed by crystallography on traversing grain boundaries. Under tensile–compressive loading, shear processes become more difficult as the crack penetrates into the interior of the material, and this leads to macroscopic crack rotation.

(4) *Stage II crack growth.* Extension occurs along a path perpendicular to the applied stress axis and is generally non-crystallographic. Early and final periods of growth may be sensitive to environment and microstructure but the intermediate, or continuum, region (described by the Paris equation during LEFM conditions) is much less so. When the net section stress attains a critical level, total fracture ensues.

4.2.2. Scope

The salient point for the designer is the number of cycles to failure, N_f, which is the sum of the cycles devoted to initiation, N_i, plus those expended in propagation, N_p. Unless the argument that initiation is complete after the first cycle is accepted, a quantitative definition of that event remains as elusive as ever. There has been increasing emphasis in recent years on initiation and early-stage growth, and this review willl reflect that attitude by omitting a consideration of long crack growth (see Chapter 5 of the earlier volume). Further justification for this is based on the fact that much of it is microstructurally insensitive and that the post-continuum phase, which is affected by microstructure, usually constitutes only a small proportion of the endurance.

Also excluded are the effects of environment (Chapter 6 in the earlier volume), except so far as the differences between air and vacuum are concerned. Consideration of these is likely to assist the understanding of the fundamental processes involved in laboratory testing.

With a rise in temperature, three time-dependent effects assume increasingly greater significance. These thermally activated processes may be categorised as follows:

(a) *Mechanical effects.* Enhanced dislocation cross slip and climb, relative weakening of grain boundaries, cavitation and creep.

(b) *Environmental effects.* More rapid diffusion leads to a faster and more extensive influence of oxidation, or any deleterious chemical reactions.

(c) *Alloy instability.* Since most engineering alloys are used in a metastable condition derived for optimum monotonic strength, exposure at elevated temperature accelerates their transformation to equilibrium, and usually inferior properties.

Clearly, the real situation will be much more complex than this rather simplistic overview, with exceptions and interactions in abundance. As a general rule, fatigue enhances the above effects, and in turn, cyclically induced changes are accentuated by high temperatures. However, the above description does provide a useful framework for the more detailed examination of the metallography of high temperature fatigue which follows.

While the contribution of Metallography will be gauged from the outside only by its ability to describe the behaviour of real engineering alloys, it must first be acknowledged that a basic understanding is usually derived from studies on pure metals, simple solid solutions and, frequently, single crystals. As a consequence, the format of the chapter will involve an initial examination of these fundamental studies prior to an assessment of more complex commercial alloys. In cases where general patterns of behaviour have been reasonably established, mention of particular alloys will not be made. Where, however, firm trends do not exist, or in situations for which only fragmented information is available, specific alloys will be cited. Since the effort devoted to studying fatigue at ambient temperature greatly exceeds that at high temperatures, it will usually be necessary to describe the former as the baseline.

4.3. DEFORMATION PROCESSES DURING FATIGUE

All fatigue involves irreversible plasticity which occurs on a micro- or a macroscale, with the role of dislocations paramount. This is now considered, commencing with ideal high purity materials and progressing through to commercial alloys. Several aspects of the subject have recently been reviewed in depth [3, 4, 5] and these publications represent the best sources of specific detail. The present overview aims to present a broader picture of the current state of the art. Some assessment of progress in the past 15 years may be obtained by comparison with an earlier review based at a similar level [6].

At this point, it is appropriate to cite the importance of slip mode in determining the type of dislocation arrangement induced by fatigue. Planar slip is restricted to a single-slip system, and is promoted by a low-stacking fault energy, large grain sizes, shearable precipitates and the presence within the matrix of obstacles to dislocation motion. In contrast, wavy slip is favoured by a high-stacking fault energy, small grain sizes, non-shearable particles and the availability of several slip systems.

4.3.1. Pure Metals

4.3.1.1. Ambient Temperature

In single crystals of metals exhibiting planar slip at low strain amplitudes, the principal products of dislocation–dislocation interactions are loops which agglomerate into bundles or 'veins' within an essentially dislocation free matrix (Fig. 1). With increasing concentration of dislocations in the

FIG. 1. Dislocation bundles or veins produced by low stress, room temperature, cycling. Original magnification 85 000 approx.

veins (and hardening) they become locally unstable, resulting in the formation of persistent slip bands (PSBs) into which subsequent deformation is concentrated. Within the PSBs dislocations are arranged in a wall- or ladder-like formation (Fig. 2), the actual walls being normal to the primary slip direction. During a period of cyclic stability the number of PSBs increases, until they fill the entire volume of the specimen (Fig. 3). Secondary slip then occurs, associated with further hardening and the

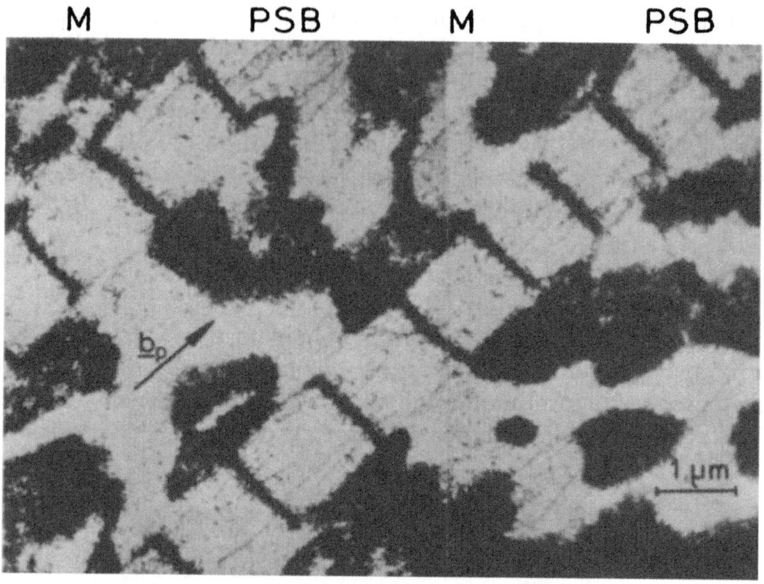

FIG. 2. Persistent slip bands in fatigued copper. M, Matrix; PSB, persistent slip band [3]. (Reproduced by courtesy of Professor H. Mughrabi.)

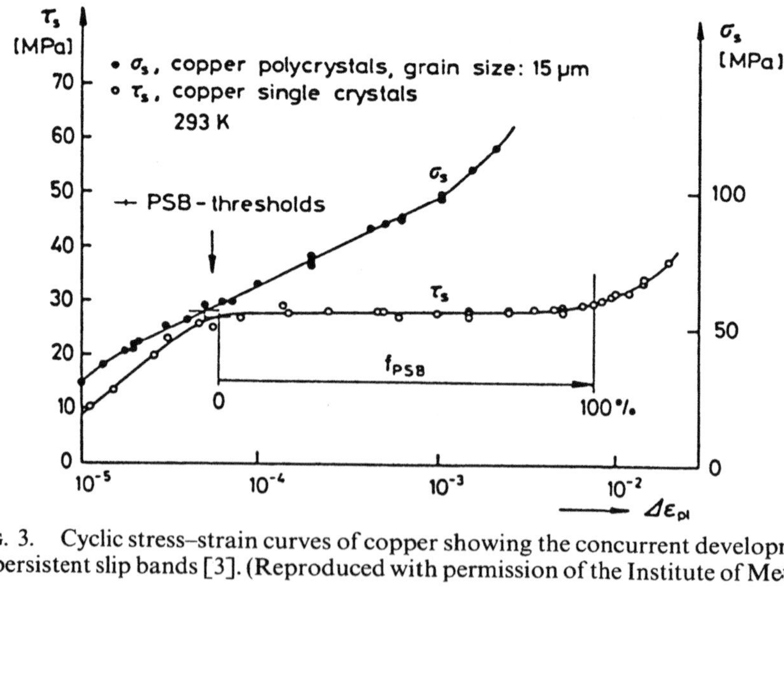

FIG. 3. Cyclic stress–strain curves of copper showing the concurrent development of persistent slip bands [3]. (Reproduced with permission of the Institute of Metals.)

formation of subcells and subgrains as the overall dislocation content and stored energy are reduced (Fig. 4).

Following stress saturation, the mechanisms of deformation in the PSB and the matrix are quite different. In the latter, strain accommodation is via 'quasi-reversible', to and fro, bowing of screw dislocations in the channels between the veins [7]. In the softer PSBs, which experience the majority of the strain, dislocation multiplication occurs by edge dislocation segments

FIG. 4. Dislocation subcells formed after secondary slip during fatigue of stainless steel at 600°C. Original magnification 10 000 approx.

bowing from the walls and traversing the channels (Fig. 5). In addition, the PSB walls yield to a limited extent. To maintain the saturated condition, an equivalent amount of dislocation annihilation also takes place.

In polycrystals, the onset of secondary slip is at much smaller strain levels and the saturation plateaux are less apparent in the cyclic stress–strain curve [8] (Fig. 3). PSBs are generally found only in suitably oriented-surface grains [9].

These are the basic descriptions for the dislocation arrangement caused by fatigue. Other terms exist, often for more specific features. For example, a 'labyrinth structure' is formed in grains, with their (100) direction close to the deformation axis [10, 11]. It consists of dislocation walls lying on intersecting (100) and (010) planes. Alternative deformation mechanisms, such as microtwinning, may also occur if dislocation mobility is restricted.

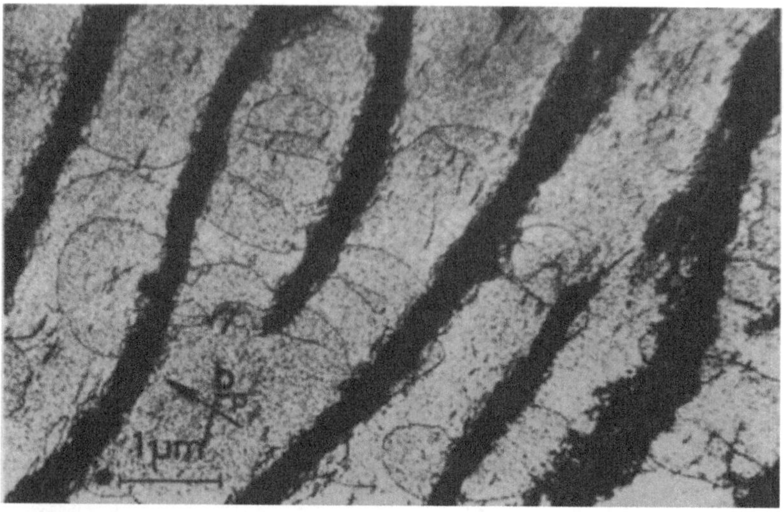

FIG. 5. Dislocation arrangement in PSB in fatigued copper showing bowing from walls. Section parallel to primary glide plane [15]. (Reproduced by courtesy of Professor H. Mughrabi.)

4.3.1.2. Elevated Temperatures

The gradual evolution of the dislocation structures from the veins through to the subgrains may be regarded as a thermodynamic description. However, the kinetics of the processes are probably more relevant and these are influenced by several factors. The onset of wavy slip and the development of lower-energy cell structures is promoted by increases in temperature (climb becomes more significant), the number of cycles (primary slip becomes more difficult), and plastic strain range (a greater driving force for cross slip). For example, pure copper subjected to 10^4 cycles at $\Delta\varepsilon_t = 0{\cdot}032\%$ exhibits a vein structure at temperatures up to 400°C, but doubling the (total) strain range leads to subgrain formation above 300°C [12] (see Fig. 6). The subcell size, d, at saturation, diminishes with increasing plastic strain range and saturation stress level, in line with an expression of the form $d \propto \Delta\varepsilon_p^{-m}$ or $\Delta\sigma^{-n}$. Because of the variation of stacking fault energy with temperature, attempts to produce a 'universal' equation, even for one material, have been unsuccessful, and Mughrabi [13] has argued that the only unique relationship possible is that between cell size and the athermal component of the saturation stress. Nevertheless, there is a well established trend for subcell sizes to become larger with increasing temperature. Further, at $T > 0{\cdot}5T_m$, ordering of the cell wall

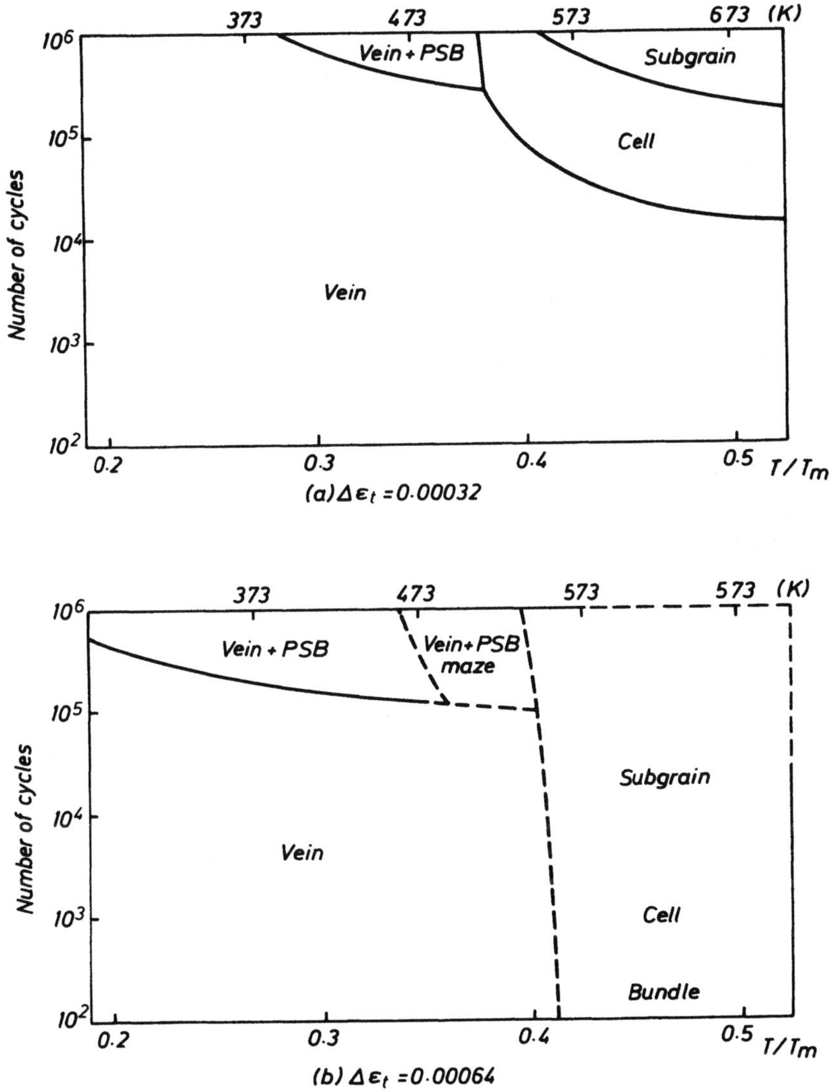

FIG. 6. Dislocation structure as a function of number of cycles and temperature: polycrystalline copper. (a) $\Delta\varepsilon_t = 0{\cdot}032\%$, (b) $\Delta\varepsilon_t = 0{\cdot}064\%$. (Based on Ref. 12, reproduced with permission of Pergamon Press.)

structure commences producing regular dislocation arrays and well defined subgrains. Fatigue damage maps are extremely useful in summarising the resultant dislocation structures (Fig. 6).

In situations with few slip possibilities, notably single crystals, orientation has a strong influence on cyclic strength. For copper, these differences become greater with increasing temperature [14].

The temperature and strain rate sensitivity of dislocation flow in BCC metals is much greater than that experienced by FCC structures. Around ambient temperatures ($\leq 0{\cdot}2T_m$ K) and at moderate strain rates ($\simeq 10^{-4}\,s^{-1}$), screw dislocations possess limited mobility and the transition to multiple slip occurs at intermediate strain amplitudes. PSBs are not usually observed in pure BCC metals near room temperature, since the strain can be accommodated within the open cell structure produced [3, 15]. With only small strain ranges ($\Delta\varepsilon_p \simeq 10^{-4}$), a phase of cyclic stability appears in the cyclic stress–strain curve as edge dislocations oscillate to and fro and induce no microstructural change. Near the surface, however, a dislocation cell structure may form when the loss of edge segments necessitates multiple slip [16] (Fig. 7).

At higher temperatures ($T > 0{\cdot}2T_m$), when mechanical behaviour is less temperature sensitive, fatigue produces a dislocation arrangement similar to that observed in FCC metals. The PSBs are less clearly defined within a vein-bearing matrix.

4.3.2. Solid-solution Alloys

The addition of solute atoms results in a reduction in the stacking fault energy with respect to the pure solvent, although a third element may either

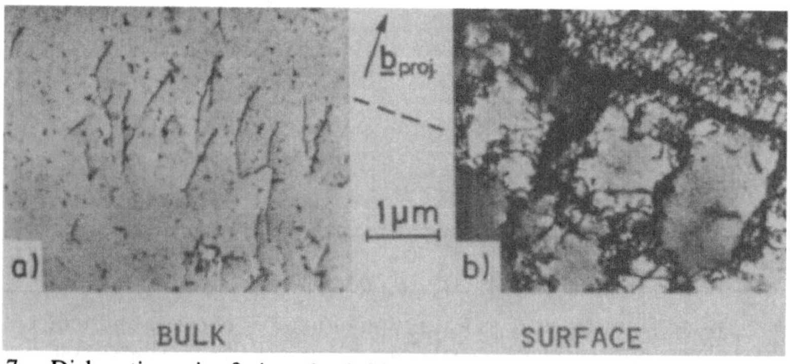

FIG. 7. Dislocations in fatigued niobium single crystal indicating differences between surface and bulk regions at low plastic strain range ($\Delta\varepsilon_p = 3 \times 10^{-4}$) [16]. (Reproduced with permission of Professor H. Mughrabi.)

increase or decrease the modified value. There is thus a tendency to promote planar slip and its consequences but, for dilute alloys, the microstructural response to fatigue generally follows the same pattern as the base material. The development of a cell structure is retarded, requiring either a greater strain amplitude or more cycles than in the pure solvent. Once formed, the cell size is usually smaller. In low-carbon steel, PSBs are formed both in surface grains and to a lesser extent within the body of the material [9, 17]. Their form is more akin to the 'ladder-like' structure than the lamellar formation seen in pure iron [18]. These differences are manifested in the stress–response curve by more rapid softening and more pronounced subsequent hardening in the alloy.

At elevated temperatures, the reduced tendency for cross slip in solid-solution alloys is counterbalanced by the temperature dependent increase in SFE (stacking fault energy). The montage shown in Fig. 8 provides a good overview of the effects of temperature and stress range on the dislocation arrangement and cyclic response of a low-carbon steel. Interstitial solid solutions are susceptible to dynamic strain ageing which is associated with an increasing strength over a critical temperature range. A marked change in dislocation structures induced by fatigue may result as the effects of an increased 'effective' solute concentration are experienced. For example, in type 316 stainless steel, the dislocation structure reverts to a planar form at temperatures between 200 and 550°C, as distinct from the cellular configuration at other temperatures [19] (Fig. 9). Under non-strain-ageing conditions, an increasing nitrogen content leads to a planar, rather than a cellular arrangement [20] and continuous cyclic hardening rather than stress saturation. In the low-carbon steel mentioned above, raising the temperature to around 150°C results in cyclic hardening at most strain ranges, unlike the significant softening observed at room temperature. This is attributed to dynamic strain ageing and is associated with the formation of a further variant of PSB—narrow, dislocation-free channels within the bulk of the material [18]. The interface with the matrix contains a high dislocation density, and the dislocation arrangement in the latter is cellular since the total alloy content is insufficient to lower the SFE to a level where planar slip predominates.

4.3.3. Two-phase Alloys

4.3.3.1. The Particle–Dislocation Interaction

The principal criterion governing microstructural changes induced by fatigue and the stress response in two-phase alloys is whether the particles

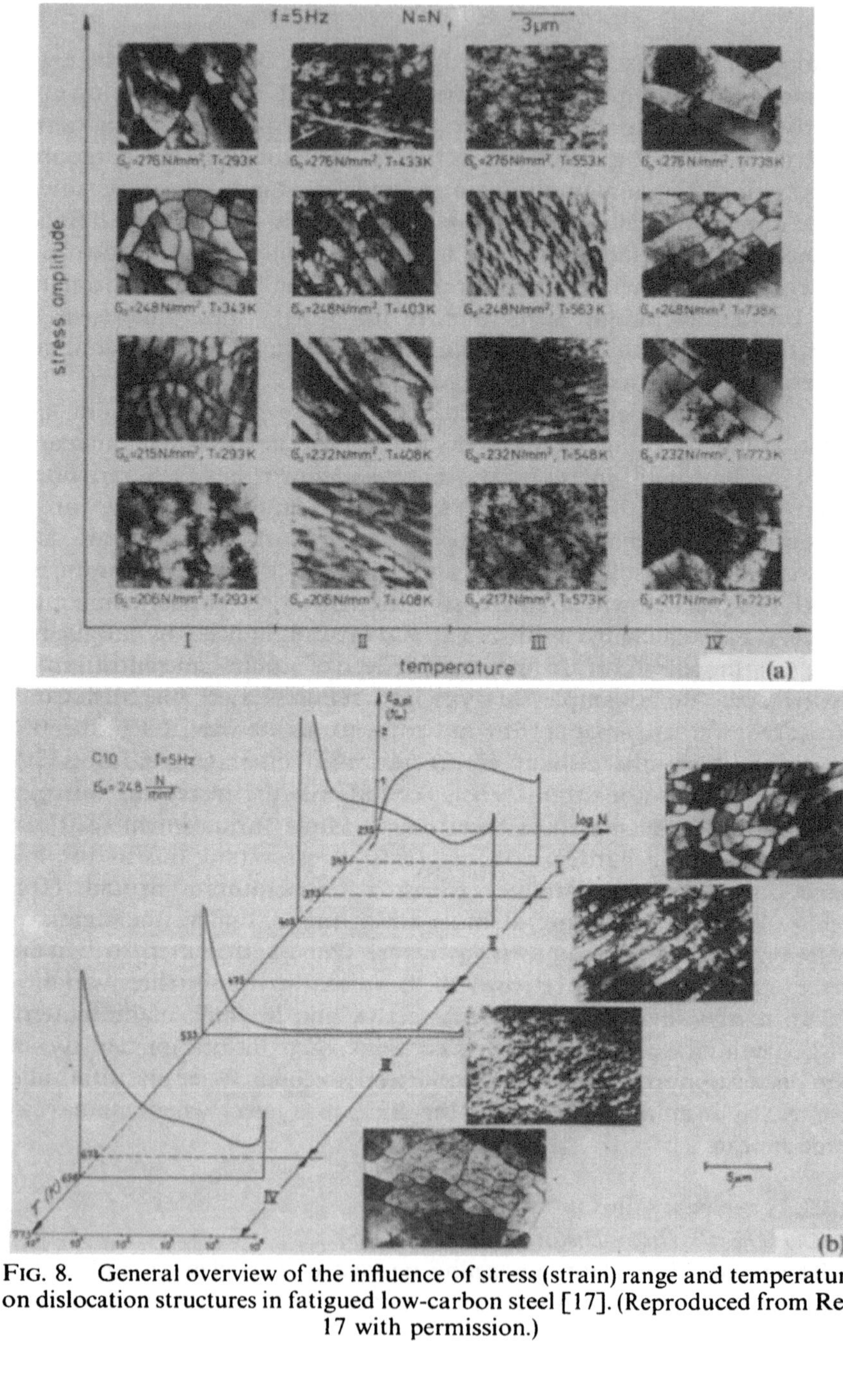

FIG. 8. General overview of the influence of stress (strain) range and temperature on dislocation structures in fatigued low-carbon steel [17]. (Reproduced from Ref. 17 with permission.)

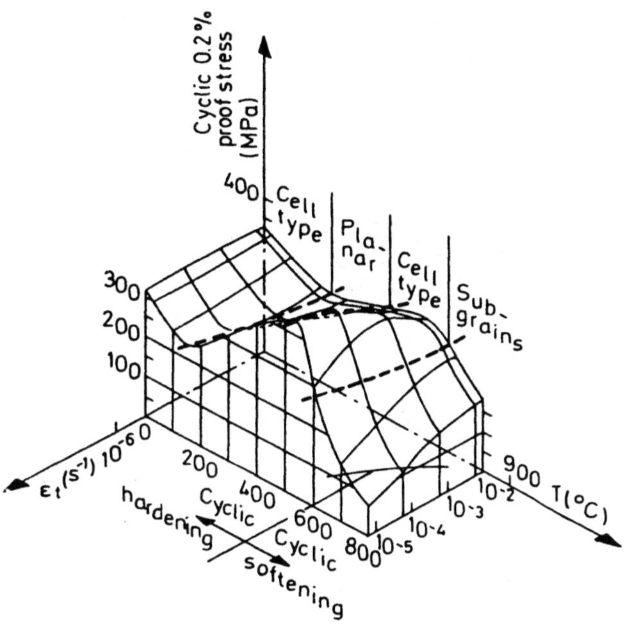

FIG. 9. Fatigue failure map for type 310 stainless steel showing suppression of subcell formation due to strain ageing [19]. (Reproduced by courtesy of Dr. K. Kanazawa.)

are sheared by the oscillating dislocations [21]. The consequence of such an event is the formation of thin PSBs in which the local precipitate structure is degraded by dissolution, a loss of order hardening or over-ageing [22]. When planar slip prevails, the effective particle size is diminished and some disordering may occur [23, 24], but under wavy slip conditions the precipitates in the PSB are often completely destroyed [24] (Fig. 10). A moderate increase in temperature enhances the mobility of the vacancies produced during fatigue and accelerates these effects (Fig. 11). These changes induce substantial softening which may be preceded by a brief period of hardening. Hardening up to a critical stress level has been shown to be necessary for under-aged Cu–Co and Al–Zn–Mg single crystals, which in turn is associated with a limiting plastic strain range below which PSBs do not form [25].

The presence of non-shearable second-phase particles acts to homogenise the dislocation arrangement (Fig. 12) caused by fatigue, minimising the deleterious strain localisation into PSBs and the cyclic instability. An enormous variety of microstructural change is possible according to the

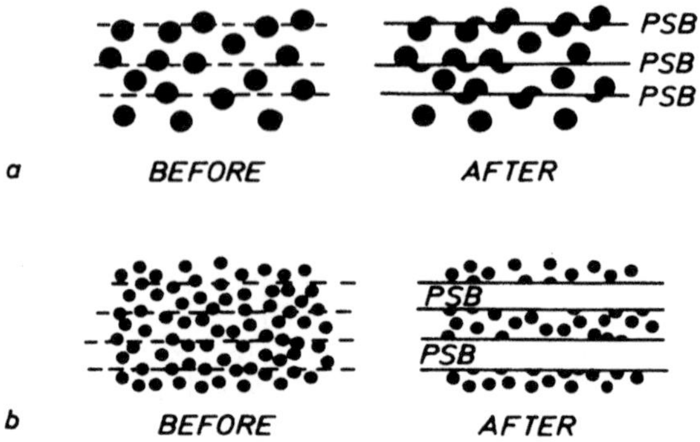

FIG. 10. Effect of shearable particle size and slip mode on precipitate–dislocation interaction during fatigue. (After Ref. 3.) (a) Planar slip, large particles; (b) wavy slip, small particles.

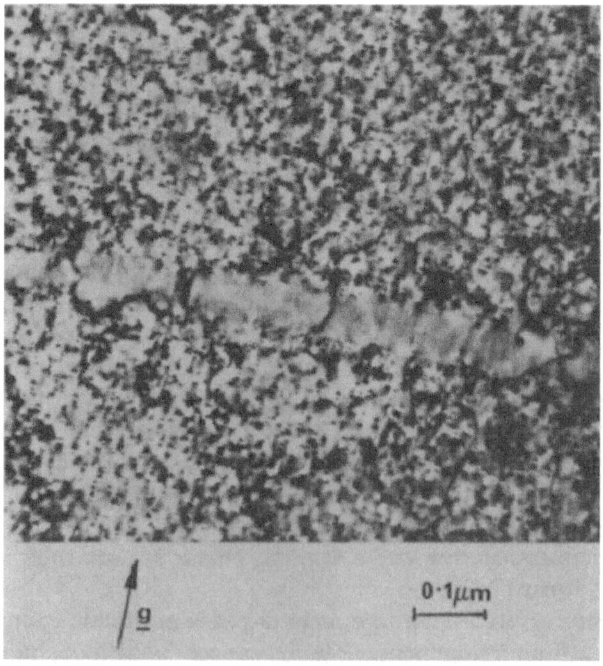

FIG. 11. Persistent slip band in Cu–2at%C alloy containing coherent cobalt particles [3]. (Reproduced by courtesy of Professor H. Mughrabi.)

FIG. 12. Homogeneous deformation around non-shearable particles during fatigue of a fully aged Al–Cu–Mg alloy. Original magnification 28 000.

alloy, the test conditions and, in many cases, the particular grain involved. Whether dislocations pile up against particles or loop around them is a key factor. Some common responses include alignment and redistribution of particles [26], cyclically induced precipitation [27] or coarsening of existing precipitates and cell formation, which may be controlled either conventionally by the plastic strain range or by the particle distribution [27]. Examples will be cited when particular alloy groups are considered, but it is probably useful to quote some general trends at this stage.

Most engineering alloys appear in service in a metastable condition and their properties deteriorate on prolonged exposure to high temperatures as equilibrium is approached. It is well known that both monotonic and cyclic strain will enhance either homogeneous or heterogeneous precipitation processes. Recently, this effect has been quantified for the coarsening of NiAl-type particles in an Fe-base alloy [28]. At 700°C, cycling at 0·05 Hz produced an increase in particle growth rate of 2·7 times, due, it was suggested, to fatigue-generated vacancies accelerating volume diffusion. An alternative mechanism of enhanced coarsening has been cited by

Antolovich *et al.* [29] for an Ni-base superalloy, René 80. Here, fatigue-induced dislocations link the γ' particles, providing easy diffusion paths.

4.3.3.2. Dislocation Structures Produced by Fatigue

The growing emphasis on commercial alloys has produced further guidelines for microstructural changes induced by fatigue. For example, studies on Alloy 800 suggest that a critical dislocation density is required for subcell formation, and that its value is inversely proportional to the absolute temperature [30]. Subcell sizes in Alloy 800 and type 304 stainless steel have been related to both saturation shear stress τ, and equivalent plastic shear strain γ_p, through the relationship [30, 31]

$$\tau/G = B(b\gamma_p/\lambda)^m \tag{1}$$

where λ is the cell diameter, b is the Burgers vector, and G is the shear modulus; B and m are constants (Fig. 13). An interesting correlation is then possible between cell diameter and the cyclic-strain hardening exponent, n ($\tau = A(\gamma_p)^n$), i.e. $n = -\mathrm{d}\ln(\lambda)/\mathrm{d}\ln(\gamma_p)$. Examination of interrupted tests on pre-aged type 304 stainless steel demonstrates that, once formed, the cell size decreases as saturation is approached, but for continuous cycling it remains constant after half-life [32]. In comparison to the solution-treated condition, prior ageing results in a lower dislocation density and a larger cell size [33].

The transition from veins to subcells to subgrains with increasing temperature, strain range and test duration at temperatures above $0{\cdot}5T_m$ is confirmed. When microtwinning is a preferred deformation mode, as in Alloy 718 at 427°C and below, raising the temperature leads to deformation by slip, but lowering the frequency (strain rate) opposes that trend [34]. In the same alloy, bilinearity in the Coffin–Manson plot is attributed to a mechanism change.

4.3.3.3. Influence of Microstructure on Fatigue Response

Recent work by Nilsson [35] demonstrates well how minor changes in heat treatment may produce microstructures which respond quite differently to high strain cycling. Ageing Alloy 800 at 600°C produced γ' precipitates of approximately 200 Å diameter, whereas ageing at 550°C led to finer precipitates, undetectable in the microscope, and probably of the order of 50 Å. During low-cycle fatigue at 600°C the former were non-shearable, resulting in a homogeneous dislocation density together with many Orowan loops, while the small particles were sheared and were associated with dislocations in planar arrays. Surprisingly, no softening

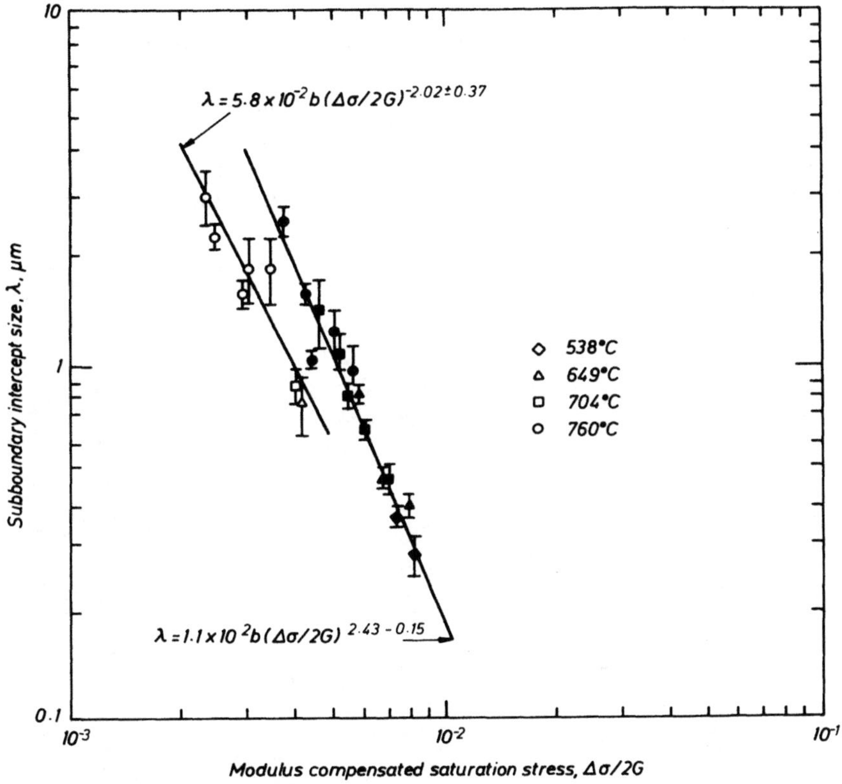

FIG. 13. Subcell size as a function of normalised saturation stress, $\Delta\sigma/G$, for Incoloy 800. (After Ref. 30.)

was observed in this latter case due to the concomitant precipitation of $M_{23}C_6$. Studies on a similar alloy have demonstrated an increased endurance and a reduced amount and rate of cyclic hardening as a consequence of prior ageing [36, 37].

In addition to second-phase particles appearing from thermodynamic considerations, commercial alloys frequently contain dispersoids or inclusions which may also influence fatigue behaviour. When these are relatively small (≤ 1 μm), improved high- and low-cycle properties result [38, 39] due to the homogenisation of slip. Larger inclusions may act as initiation sites for stage II cracks [40] which may spread into the matrix, although their ability to do this depends upon the nature of the particle–matrix interface [41].

So far, it has been implicitly assumed that the initial dislocation density has been low and typical of the annealed condition. However, well developed dislocation arrangements may exist in, for example, cold-worked or quenched and tempered alloys. In such cases, redistribution occurs and the configuration typical of fatigue is adopted, provided sufficient driving force in the form of strain range, temperature and time is available. Elongated subgrains in cold-worked stainless steel become

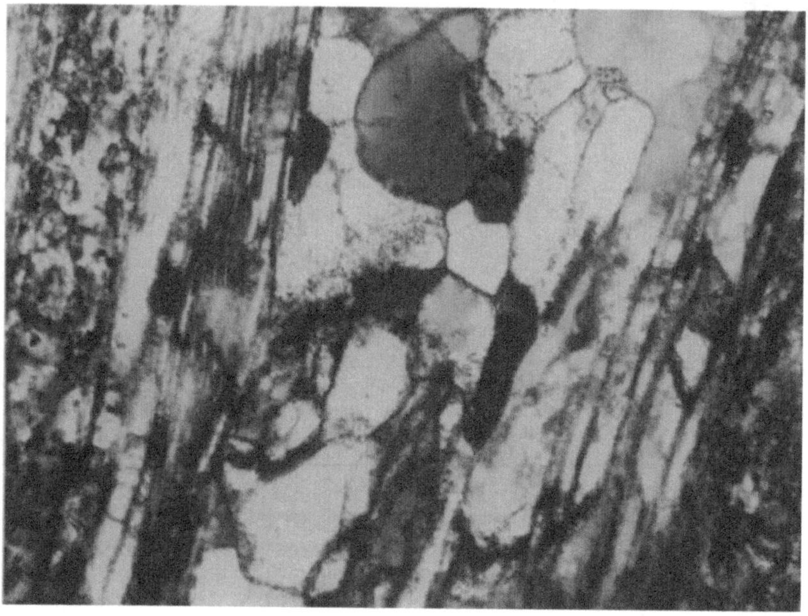

FIG. 14. Transition between initial cold-worked substructure (elongated subcells) and fatigued rearranged equiaxed cells due to fatigue at 600°C in type 316 stainless steel. A similar rearrangement occurs after room-temperature cycling [42].

equiaxed cells (Fig. 14) [42], the size of which is determined by the strain range and temperature. Subcells in martensitic and bainitic steels undergo a similar transition [26, 43]. In a cyclically softening, normalised and tempered, 9Cr–1Mo alloy steel, well developed subgrains and appreciable carbide coarsening occur during fatigue at 700°C, but cycling at 593°C is sufficient for both processes just to commence [43]. Prior ageing (10^4 h at 625 or 675°C), produces a small decrease in the endurance of type 316 stainless steel [36].

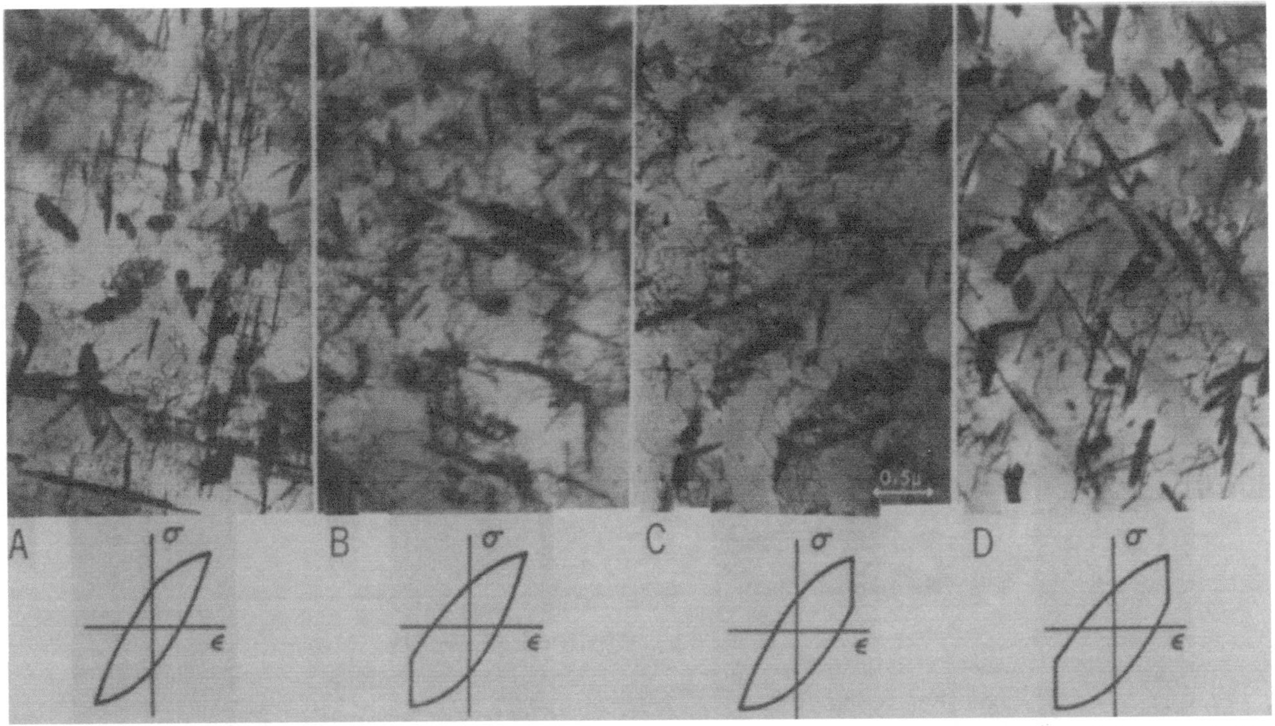

FIG. 15. Dislocation arrangements around carbide particles as a function of wave shape. Annealed $2\frac{1}{4}$Cr–1Mo, $\Delta\varepsilon_t = 0{\cdot}5\%$ at 593°C [27]. (Reproduced with permission of Professor K. D. Challenger.)

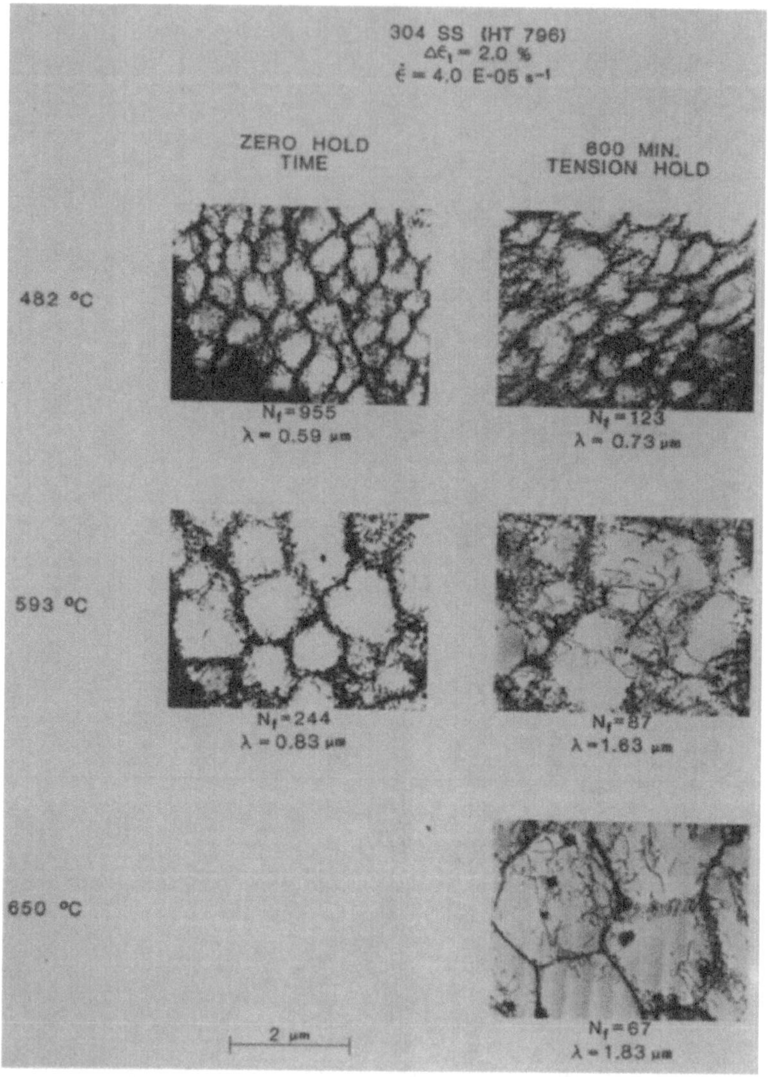

FIG. 16. Fatigue substructures in type 304 stainless steel for various temperatures and hold times: $\Delta\varepsilon_t = 2{\cdot}0\%$ [31]. (Reproduced with permission of Dr J. Moteff and ASM.)

4.3.3.4. Effects of Dwell Periods

It has further been implicitly assumed so far that cycling has been continuous and at a constant strain rate. The insertion of dwell periods leads to the possibility of fatigue–creep interactions, which will be considered later. The present section is confined to examining the effect of hold periods on pre-initiation microstructural changes. Although there is by no means an adequate amount of data on this topic, the studies of Challenger and Vining [27] exemplify the key factors involved (see also Chapter 2 for an interpretation in terms of back-stresses). For an annealed $2\frac{1}{4}$Cr–1Mo steel at 593°C, continuous and symmetrical dwell cycles produced a uniform dislocation distribution with many loops but little tangling around particles (Fig. 15). Unbalanced dwell cycles resulted in a considerably higher dislocation density with much tangling (Fig. 15). It was suggested that, during continuous cycling, dislocations were able to oscillate to and fro between M_2C carbides creating jogs on intersecting other dislocations. However, during a dwell there was opportunity for the dislocations to loop around the particles and not participate in the deformation in the reverse direction. Other dislocations had to be created for this purpose, giving rise to a progressive increase in their density in an unbalanced cycle. In balanced cycles the trapped dislocations had time to untangle from the carbides in the second dwell, thus maintaining the smaller number of active dislocations observed. Clearly, the temperature, strain range, particle size and spacing are the parameters which govern the influence of this model. In type 316 stainless steel at 600 and 650°C, it has been shown that tensile dwells produce a gradual transformation of the dislocation subcells to fully polygonised subgrains typical of creep deformation [31, 44]. Massive carbide precipitation and grain-boundary migration also occur. Precipitation rates are increased over that produced by prior cold work, together with a greater tendency for matrix formation and the appearance of intermetallic phases [33]. Figure 16 illustrates the effect of a tensile dwell at various temperatures in a pre-aged type 304 steel. The cell size was found to be inversely related to the relaxed tensile saturation stress, with constants of proportionality varying with strain rate [31]. For 60 min tensile dwells, the cell size diminished progressively throughout life [32].

4.4. PERMANENT DAMAGE DURING FATIGUE

The deformation processes and their associated strength changes may usually be eliminated by annealing (the Frontispiece of this volume being

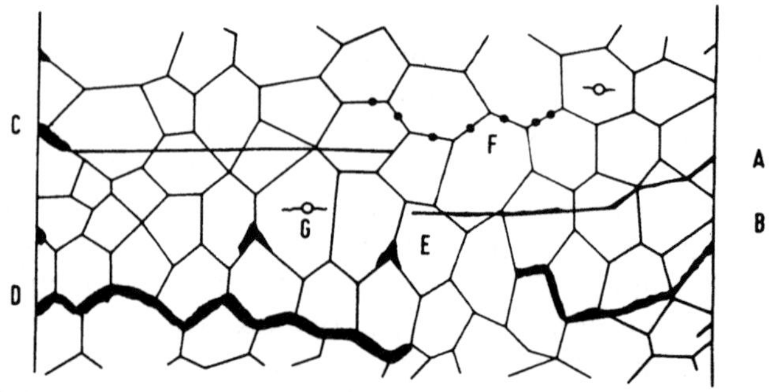

FIG. 17. Schematic of types of fatigue damage. (A) Transgranular crack initiation and growth; (B) transgranular initiation followed by intergranular growth; (C) intergranular initiation followed by transgranular growth; (D) intergranular initiation and growth; (E) triple point, wedge, cracks; (F) grain boundary voids; (G) initiation from matrix pore or inclusion.

an exceptional case), but following the localised saturation of dislocation effects, permanent damage begins to accrue in the form of cracks and cavities. Figure 17 shows various examples of these defects which originate either at the surface or internally:

(1) *Surface cracks.* These may be inter- or transgranular and may (a) extend at a rate independent of cyclic frequency (or strain rate), (b) exhibit a time or strain-rate dependence as determined by mechanical parameters, or (c) display a time dependence determined by environmental effects.
(2) *Internal cracks and voids.* These may grow in similar ways except, because of their location, no environmental sensitivity is apparent. Under certain conditions, however, intergranular damage may become surface connected.

It is important to appreciate that several damaging processes may take place concurrently, and the dominant failure mode is that favoured by the balance of the prevailing conditions (temperature, microstructure, environment and waveform). Fracture maps, despite their imprecision, have been a useful innovation, enabling large amounts of information to be displayed in a convenient visual form.

4.4.1. Transgranular Damage

4.4.1.1. Mechanisms for Initiation

The classic models for cyclic crack initiation based on surface observations of extrusions and intrusions (for a review, see Ref. 45) have been supplanted in recent years by a dislocation–PSB approach founded on information from TEM studies. Typical of these are those papers emanating from the groups of Mughrabi [46, 47] and of Brown [48, 49]. For FCC metals exhibiting wavy slip [46, 47], the former consider the PSB near saturation to have, at its interface with the matrix, unstable edge dislocations of opposite sign (Fig. 18). Under stress, these are driven out from the surface, resulting in the rapid production of extrusions, the outer surface of which subsequently roughens with continued cycling (Fig. 18(c)).

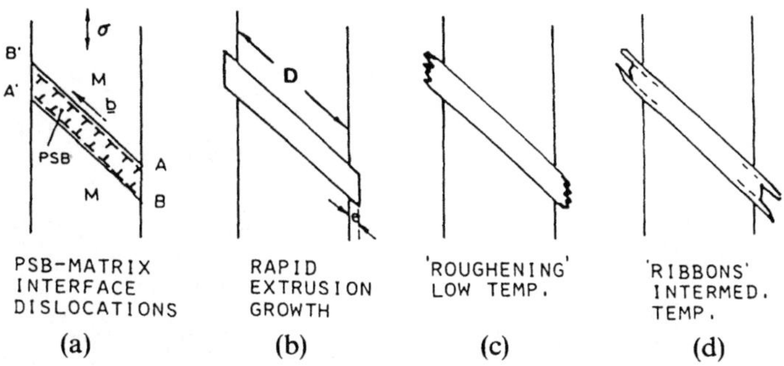

FIG. 18. Mechanism for the evolution of an emergent PSB in an FCC single crystal [46]. (Reproduced by courtesy of Professor H. Mughrabi.)

Crack initiation may then occur either at the slip steps between the emergent extrusion and matrix or at the micronotches at its outer edge. It should be noted that the simultaneous production of an intrusion is not required by this model, which defines an intrusion simply as the notch associated with the extrusion–surface interface. Initiation and growth of stage I cracks begin at this location, but much later than the formation of the extrusion.

At higher temperatures, extrusions are no longer static since vacancy mobility permits their continuous growth if they are thin but, when their overall thickness exceeds 1 μm, the formation of 'ribbons' occurs (Fig. 18(d)). PSBs are no longer produced when the temperature is greater than about $0{\cdot}4T_m$ [12]. Essmann *et al.* [46] have estimated the vacancy concentration produced by the annihilation of unlike edge dislocations in the walls of a PSB at saturation, and hence, the initial step height which is proportional to the size of the grain housing the PSB. In line with observation [47], the height of extrusions arising from single crystals is usually much greater than that from polycrystals.

The model of Brown *et al.* [48, 49] is broadly similar, except that the interface dislocations are stable. Large surface stresses are invoked which cause decohesion at the PSB–matrix interface. A gradual surface roughening occurs due to random irreversible slip, which is also the mode of initiation in planar-slip materials in general. Surface observations, especially those of low temperature fatigue tests, furnish good support for these models [3] (Fig. 19). At higher temperatures, the ribbons may often be associated with grooves occurring adjacent to or within the extrusions. These may be regarded as equivalent to intrusions.

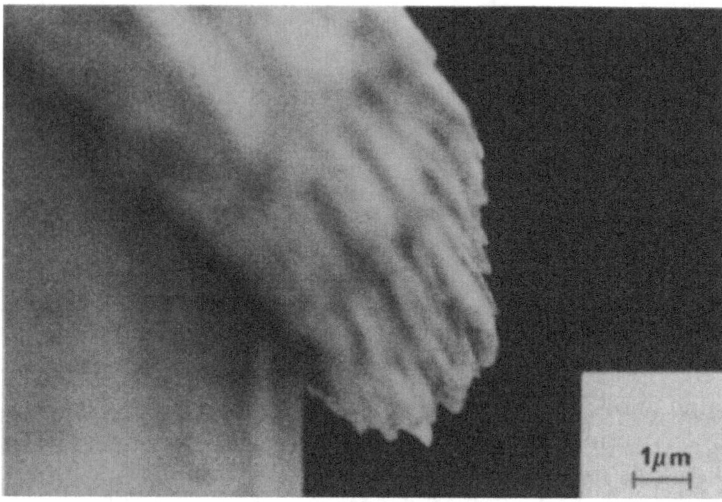

FIG. 19. A well developed **PSB** in single crystal copper after 4×10^4 cycles [3]. (Reproduced by courtesy of Professor H. Mughrabi.)

As mentioned earlier, extrusions in polycrystals are generally much smaller than those in single crystals and their significance in crack initiation is less. Surface roughening at emerging PSBs produces a 'hill and valley' profile on a much finer scale than the coarser extrusions in single crystals. If, because of slip restrictions, the PSB–matrix interface dislocations do not emerge from the surface, they are likely to impinge on grain boundaries and cause intercrystalline crack initiation. This can occur at ambient temperatures and below, and with small ($\simeq 10^{-4}$) plastic strain ranges [15]. Intergranular damage is considered later.

The transition from a grooved-profile to a fully developed stage I shear crack is sensitive to environment, but in a manner which depends on the definition of 'initiation' employed. Using data of Neumann [50] and Hunsche [51], and considering the similarity in PSB surface profiles produced in air and under vacuum, Mughrabi [4] ascribes the order of magnitude differences in plastic strain accumulation and length after identical cycles to the beneficial effects of vacuum on crack growth. However, if 'initiation' was taken as a larger crack (say 1 μm), then vacuum could be interpreted as affecting both initiation and growth.

4.4.1.2. Microstructure and Initiation

In commercial alloys, many examples of the idealised initiation models just discussed are to be found. At medium to high strain levels, slip steps play an important role in planar-slip alloys, in which a fine homogeneous distribution confers optimum resistance to initiation [52]. To prevent shearing, a coarse precipitate structure is desirable. The susceptibility to a fall in monotonic strength in this method may be avoided by employing a bimodal precipitate distribution, an intermetallic dispersoid [53] or oxide particles [54] to ensure homogeneous deformation. In Alloy 800H it has recently been demonstrated that initiation is retarded and the number of cracks reduced by raising the ageing temperature to produce homogeneous deformation [55]. Similar microstructural improvement may be achieved when the slip step height at the free surface is reduced by grain refinement, and initiation is subsequently retarded due to the smaller stress concentration [56].

With reducing strain ranges, initiation at internal discontinuities becomes significant. Examples include pores, inclusions and particles added intentionally to provide strength or homogeneous deformation. In aluminium alloys, particles greater than 10 μm in diameter greatly enhance cracking [57] which arises due to debonding or fracture of the particle under impingement of slip bands [58, 59]. At high temperatures, environment becomes increasingly important in crack initiation with surface oxide rumpling, penetration of oxide spikes (Fig. 20) and oxidation of emergent grain boundaries occurring [60, 61]. The last event may lead to complete intergranular failure, which is now considered.

4.4.2. Intergranular Damage

To a first approximation, the hallmark of high temperature fatigue is the preponderance of intergranular cracking in the form of grain-boundary voids and wedge cracks. These features, of course, are sometimes produced during ambient temperature fatigue, but to a much lesser extent. Elevated

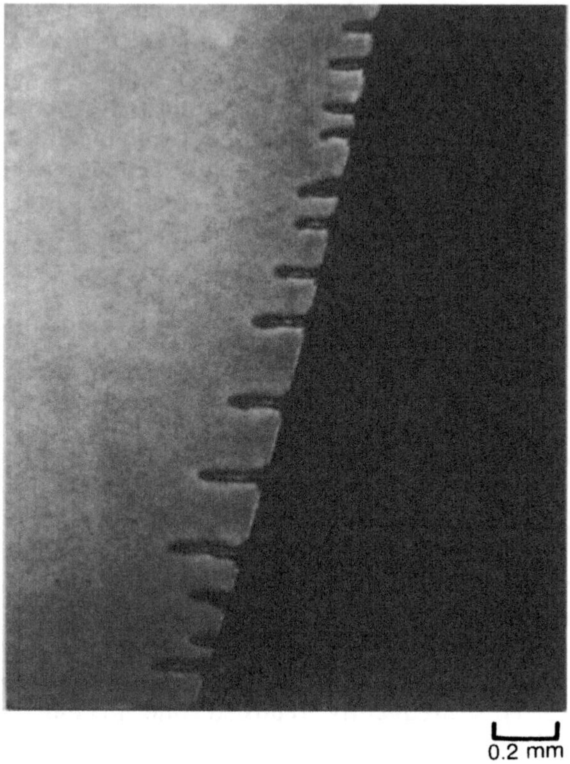

FIG. 20. Oxide penetration producing rapid initiation in a Ni-based superalloy cycled at 980°C.

temperatures provide greater scope for time-dependent mechanical and environmental effects to be exerted and, with air as the environment, it is only recently [62] that attempts have been made to separate the two. In contrast to creep, intergranular cavitation and cracking produced during fatigue is an under-investigated topic and one which generally is subjected to a 'creep-modified' approach. The review by Argon [63] of that area serves as a good foundation for the appraisal of fatigue intergranular failure which now follows.

4.4.2.1. Models for Cavity Nucleation and Growth

With regard to the nature of the mechanisms for cavity nucleation and growth, the former is associated with stress concentrations and decohesion at grain-boundary particles and ledges. These may be induced by either grain-boundary sliding or matrix-slip impingement [64, 65] and sustained by stress-driven vacancy diffusion or further sliding. Grain boundaries may become serrated in pure metals and solid-solution alloys due to the absence of any pinning effects [66, 67]. In the absence of sliding there is a good correlation exhibited between slip-band spacing and cavity spacing [68] (Fig. 21). Clearly, this type of cavity nucleation will be most significant under conditions in which planar, heterogeneous, slip prevails. The nature of the grain boundary itself is also influential in determining the ease of cavitation during slip impingement. Boundaries separating highly misoriented grains favour nucleation since the misfit strains are accommodated more easily [69]. Grain-boundary sliding is an essential

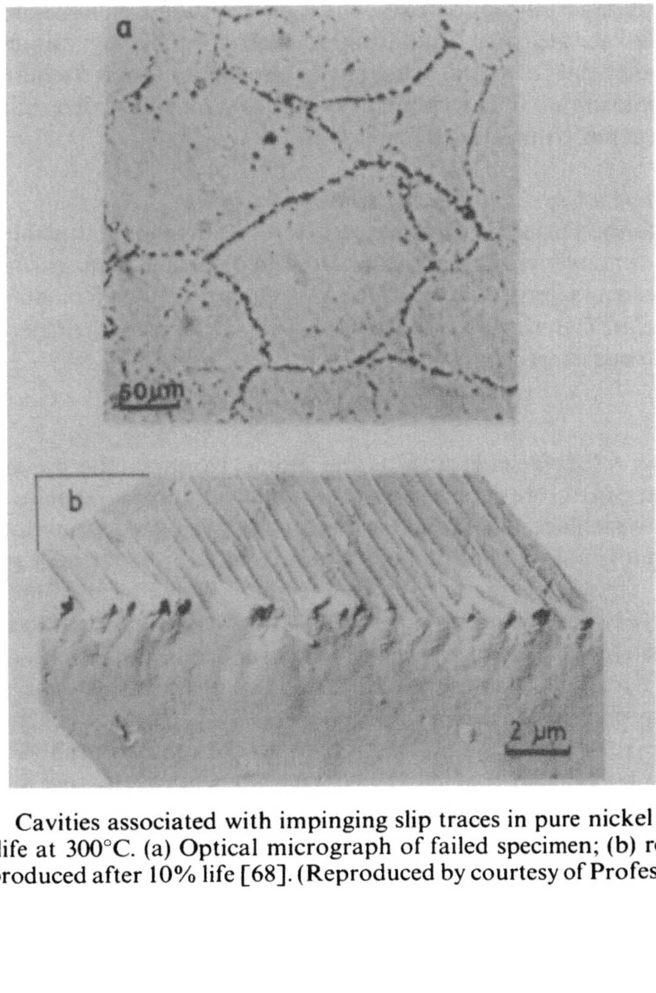

FIG. 21. Cavities associated with impinging slip traces in pure nickel fatigued to 10% of life at 300°C. (a) Optical micrograph of failed specimen; (b) replica from surface produced after 10% life [68]. (Reproduced by courtesy of Professor R. Raj.)

prerequisite for wedge cracking. Once nucleated, stability may be assisted by various factors, such as sliding of the adjacent grain over a particle–matrix interface [5], migration of fatigue-generated vacancies, segregation of impurities [70, 71] and, most commonly, the diffusion and development of an internal gas pressure [72]. In this final circumstance, the critical cavity radius of most engineering alloys is reduced to about $10^{-2}\,\mu$m at $0{\cdot}5T_m$ [73].

The growth of stable cavities and their eventual interlinkage occurs by broadly similar mechanisms to those reported for creep failure (for general reviews, see Refs 63 and 74), although vacancies produced during fatigue enhance growth by diffusion [72, 75]. Evidence from metallography and density measurements suggests that nucleation is relatively rapid, and endurance is generally controlled by the growth stage, which proceeds at a diminishing rate [67, 76]. Small-angle neutron scattering studies possess excellent potential for monitoring cavity growth but, as yet, there seems a significant mismatch in determined cavity size with values determined from high-voltage electron microscopy [77].

4.4.2.2. Analyses of Cavity Nucleation and Growth

Although the variables, temperature, stress, environment, strain rate (or frequency), and waveform are interrelated in a complex manner, it appears that there is an intermediate regime of frequency in which cavitation does not occur. The criterion for cavity nucleation [75] is that a critical radius, r_c, be attained, at which

$$r_c = 4KT\gamma/\sigma_0^2\Omega \tag{2}$$

where γ is the surface energy, Ω is the atomic volume, and σ_0 is the peak tensile stress in the fatigue cycle. The value of r_c may be reduced via a parameter which accounts for the roughness of the grain boundary. When sliding results in alignment of grain boundaries along planes of maximum shear, the frequency must exceed a critical value (the 'upper critical frequency') for cavitation to occur [75]. If the stress relaxation time, τ, is greater than t_1 (the cavity nucleation time) (or alternatively, the cyclic frequency $f > 1/2t_1$), cavitation can take place at the stress concentrations lying on grain boundaries at 45° to the stress axis [73]. At lower frequencies ($\tau < t_1$), relaxation may eliminate local stress concentrations and the stress levels may become insufficient for cavity nucleation. In terms of stress, estimates of the critical values, σ_{cw}, for wedge crack formation may be obtained from [78]

$$\sigma_{cw} = 2\gamma^* G/(1 - \nu)L \tag{3}$$

where γ^* is the effective fracture energy, v is Poisson's ratio, G is the shear modulus, and L is the length of the sliding boundary segment. Wedge cracking requires higher stress than void formation.

A second criterion must be invoked to explain the requirement for the frequency to be *below* a further critical level (the 'lower critical frequency') in the low-frequency regime. There should be sufficient time for the cavity embryo to acquire a stable size by vacancy diffusion or grain-boundary sliding and not sinter out during the compressive half-cycle. The critical size may be determined from eqn (2). In this domain, cavities grow mainly on grain boundaries with maximum principal stress (i.e. perpendicular to the stress axis). Data on most engineering alloys fall into this latter category since test frequencies are usually maintained at low levels for compatibility with service conditions. In any event, such alloys generally possess grain-boundary precipitates which inherently resist sliding. The expression for this lower critical frequency, f_c, is [73]

$$f_c = \frac{\sigma_0^7 \Omega^4 C_0 \delta D_b^0 \exp(-Q_b/RT)}{16\pi^2 k^4 T^4 \gamma^3 F_v} \tag{4}$$

where C_0, δ, D_b^0 and Q_b are, respectively, the equilibrium vacancy concentration, the effective width, the pre-exponential term in the diffusion coefficient, and the activation energy for diffusion of grain boundaries; F_v is a geometrical factor for cavity nucleation.

For most engineering alloys at $0{\cdot}5T_m$ and for a plastic strain range of 1%, f_c falls between 10^{-2} and 10^{-5} Hz [62]. Thus, overall, the microstructural requirement for resistance to cavitation is that grain-boundary particles should neither be too large and create high stress concentrations, nor should they be too small and permit grain-boundary sliding. Austenitic stainless steel, with and without $M_{23}C_6$ particles along grain boundaries, clearly illustrates this point [79]. The absence of such precipitates resulted in an intergranular failure, while their presence inhibited grain-boundary sliding and was associated with a transgranular fracture.

For intergranular crack growth to occur, two criteria must be satisfied [80]: (a) cavities must nucleate ahead of the crack, and (b) the fatigue crack growth rate must be of the same order as the intercavity spacing, $(\lambda - a)$, where λ is the spacing of grain boundary particles having a diameter a. Assuming that the stress level within the crack-tip plastic zone is equal to the yield value, eqn (4), for the lower critical frequency for void nucleation on grain boundaries perpendicular to the stress axis, may be modified to

$$f_c = \frac{\sigma_y^7 \Omega^4 \delta D_b}{16\pi^2 k^4 T^4 \gamma^3 F_v} \times 10^{-4} \tag{5}$$

The second requirement is met when the crack-tip opening displacement exceeds $(\lambda - a)$, or alternatively, when a critical stress intensity range, ΔK_{crit}, is achieved, i.e.

$$\Delta K_{crit} \geq \sqrt{2E\sigma_y(\lambda - a)} \quad (6)$$

For Inconel X750 at about 600°C, ΔK_{crit} equals $14\,\mathrm{MPa}\sqrt{m}$. Good experimental agreement has been found for the domain of intergranular crack growth in Inconel 718. Frequency–temperature diagrams for other alloys are shown in Fig. 22. It is appropriate to note that eqn (5) indicates that the highest-strength alloys are most vulnerable to intergranular cracking.

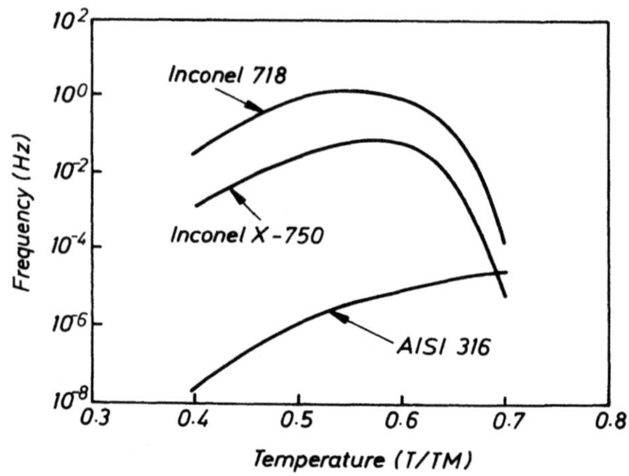

FIG. 122. Lower critical frequency as a function of temperature. (After Ref. 80.)

Cavity growth rates have recently been computed under unconstrained grain-boundary diffusion-controlled conditions as a function of stress waveform [81]. These calculations predict the observed beneficial effects of a compressive stress dwell, the deleterious consequences of a tensile stress dwell, and the diminishing growth rate with increasing cavity size (Fig. 23). Also demonstrated was the sensitivity of the critical cavity radius to its initiation site and hence its shape. Surprisingly, fast–slow stress cycles were indicated to be more damaging than slow–fast cycles, which is contrary to most findings (see Chapter 3). The operation of additional growth mechanisms such as grain boundary diffusion, surface diffusion, power law creep and plastic deformation was cited as an explanation.

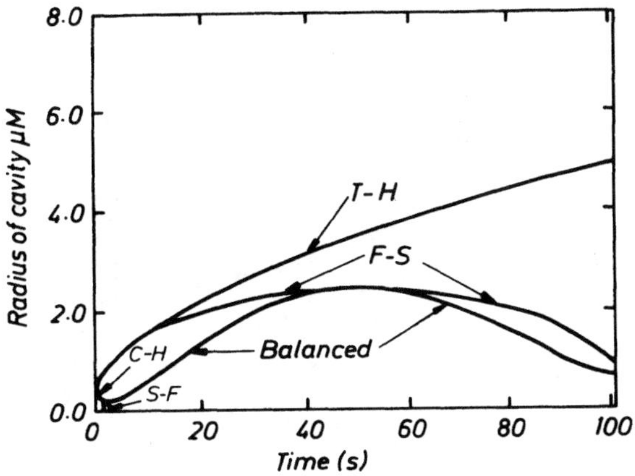

FIG. 23. Influence of wave shape on cavity growth (initial radius 3 μm). S–F, Slow–fast; F–S, fast–slow; C–H, compressive hold; T–H, tensile hold. (After Ref. 81.)

4.4.2.3. Environmental Effects

The value of the lower critical frequency, mentioned earlier, below which voids form on grain boundaries perpendicular to the stress axis, was found to be a minimum for tests in vacuum and to increase in air [62]. The model predicts that, under typical conditions for continuous cycling tests ($T = 0{\cdot}5T_m$ K and $\Delta\varepsilon_t = 1{\cdot}0\%$), this critical frequency was of the order of 10^{-5} Hz ($\simeq$1 cycle per day). Confirmatory evidence from experiments on AISI 304 and A286 was cited. Since this frequency was substantially lower than that usually employed in the laboratory, it was argued that life reductions should be attributed to environmental rather than mechanical (creep) effects. However, with cycles containing hold periods at maximum tensile stress (or strain), the time mean stress over a cycle increased and the critical cavity radius was diminished. Provided that a threshold stress level [82] was exceeded, true creep damage was incurred. Reductions in endurance could then be attributed to both mechanical and environmental factors. For brief dwells at low stress levels, environmental effects would predominate and a compressive dwell inserted in the cycle would have little effect [83]. Conversely, for prolonged dwells at high stress, creep processes would prevail and the inclusion of a dwell in compression would be beneficial in sintering out creep damage produced during the tensile half-cycle. Suggested mechanisms for the promotion of cavitation by air included absorption of oxygen into cavities, oxide-wedging at grain boundaries, oxide-induced porosity and reactions between oxygen and active species within the material, leading to pressurised gas bubbles. The

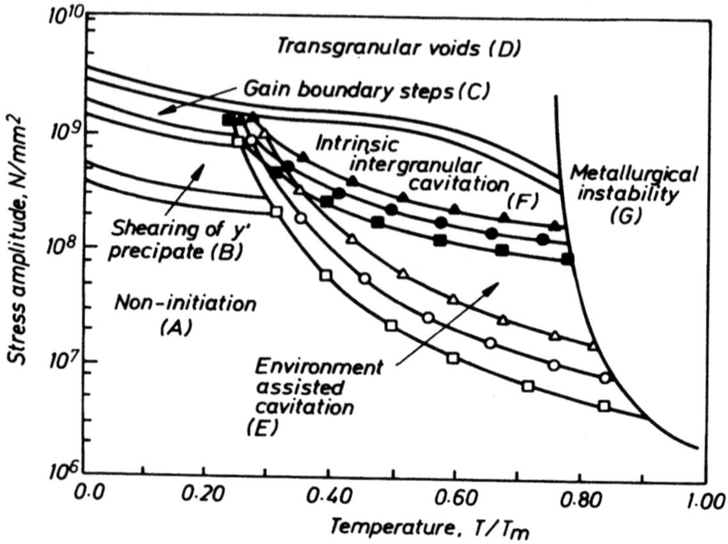

FIG. 24. A fatigue mechanism map for Inconel 718. (After Ref. 62.)

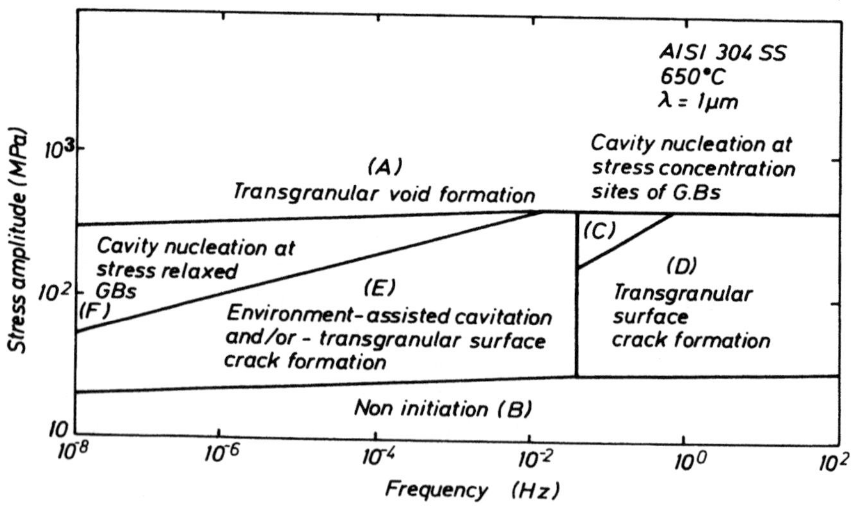

FIG. 25. A cavity nucleation map for fatigue of AISI 304 stainless steel at 650°C. (After Ref. 62.)

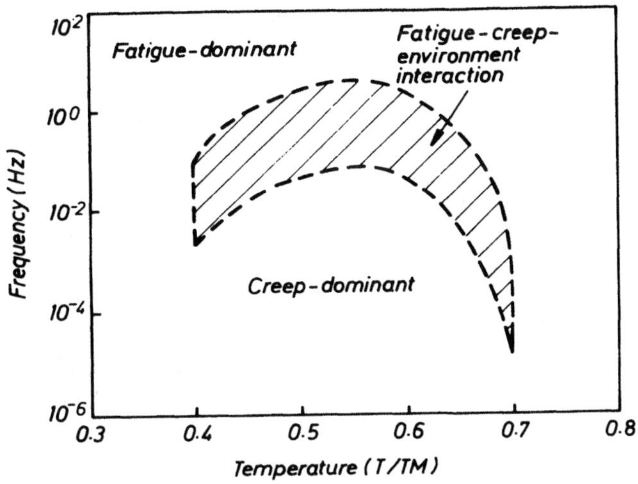

FIG. 26. A frequency–temperature map indicating the fatigue–creep–environment interaction regime. (After Ref. 62.)

sensitivity of the lower critical frequency to stress levels had two significant implications: (a) cyclically hardening alloys under strain control might not cavitate until hardening had enabled the appropriate stress level to be obtained (similarly, softening materials might cease cavitating); (b) during stress relaxation in a dwell, void formation might terminate giving rise to an apparent saturation hold time effect. A fracture map (Fig. 24) provides a good overview of the necessary conditions for various modes of failure. Such maps can describe intergranular failure alone (Fig. 25) or identify areas of a strong environmental influence (Fig. 26).

4.5. FATIGUE–CREEP INTERACTIONS

This heading is taken to describe a type of test rather than a failure mechanism, since, as will be seen later, a true interaction between the two failure modes is an uncommon event. Unlike much of the work described so far, metallography has generally been performed in this area as a useful adjunct to tests primarily aimed at obtaining endurance data on commercial alloys for life-prediction purposes. Some detailed observations of the internal deformation processes involved have been described earlier. In this section, attention will be focussed on the contribution of metallography to the interpretation and subsequent utilisation of the 'engineering' data.

Broadly, there are two types of fatigue–creep interaction, which have been termed 'sequential' and 'simultaneous' [84]. In the former category, the specimen experiences a complete period of fatigue or creep damage which is then followed by the other mode. In a simultaneous interaction, an element of fatigue (cycle dependent), and an element of creep (time dependent), occur in each cycle. Such tests are typified by strain cycles which contain hold periods usually at peak strain levels. As mentioned earlier, it is only very recently [62] that attempts have been made to discriminate between mechanical and environmental effects in the so-called 'creep' regime. For convenience, in the following paragraphs 'creep' will imply time-dependent damage which, for tests in air, will include both mechanical and environmental factors.

A good example of the contribution of metallography may be found in the interpretation of the effects of prior fatigue on subsequent creep. High strain cycling of 1Cr–Mo–V alloy steel significantly reduced the subsequent high-stress creep life, whereas creep at low stresses was unaffected [85, 86]. Identification of the various locations of damage revealed that both high-strain fatigue and high-stress creep occurred principally in the grain matrix [87]. Cyclic softening produced an essentially weaker material which then crept faster and failed earlier. In this case there was a strong interaction between deformation mechanisms. Under low-stress creep, damage (cavitation) occurred on grain boundaries and there was little consequence of prior high-strain fatigue. The limited data available so far suggest that prior creep is less influential on subsequent fatigue [86, 88]. Oxide blocking and possible blunting of the crack has been put forward as an explanation of this. However, in the case of a non-ductile alloy, internal creep cavitation would be expected to have a highly deleterious effect as the crack initiation phase, which constitutes a large proportion of life in high-cycle fatigue, would be complete.

In cyclically hardening alloys, fatigue has the reverse effect on subsequent creep, in that deformation rates are retarded [33, 89, 90]. At high creep stress levels this benefit is negated by a reduction in creep ductility, with little overall effect on life resulting [90]. It has been suggested that both prior fatigue and pre-ageing effects on creep may be accommodated via a friction stress approach [90].

The most notable success of metallography, however, has been in understanding the tensile-dwell behaviour of different alloys and the demonstration of the need to identify the dominant failure mode in simultaneous fatigue–creep interactions. Tension-only dwells were known to produce substantial reductions in the endurance of low-alloy steels and austenitic stainless steels [86, 91] (Fig. 27). It was thus tempting to analyse and extrapolate both sets of results in a similar manner, but simple optical metallography demonstrated that this would be totally unrealistic. Failure in the 1Cr–Mo–V steel, as revealed by examination of interrupted test samples, involved nucleation, growth and interlinkage of grain-boundary cavities (i.e. it was creep dominated). In the stainless steel, a fatigue crack progressed intergranularly through the specimen, accelerating when it encountered cavities (i.e. it was fatigue dominated) (Fig. 27). In this latter

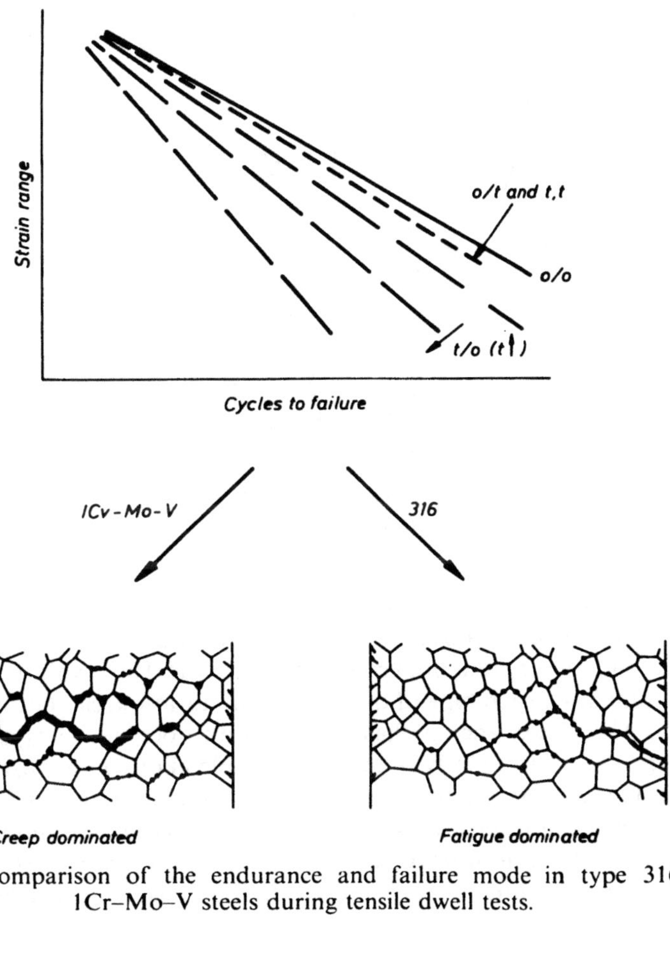

FIG. 27. Comparison of the endurance and failure mode in type 316 and 1Cr–Mo–V steels during tensile dwell tests.

case, a true mechanistic interaction took place between the two fracture processes. Obviously, for life-prediction purposes, a fatigue-type expression was required for the stainless steel and a creep-based equation for the 1Cr–Mo–V alloy. Current models are able to take into account the various mechanisms of cavity growth which may occur during a tensile dwell, in a strain-controlled cycle [92] (Fig. 28). These subsequently appear as separate domains on the 'creep–fatigue fracture map' [93] (Fig. 29). Attempts have also been made to compute the separate elements of fatigue and creep damage as they accumulated in a dwell test [94].

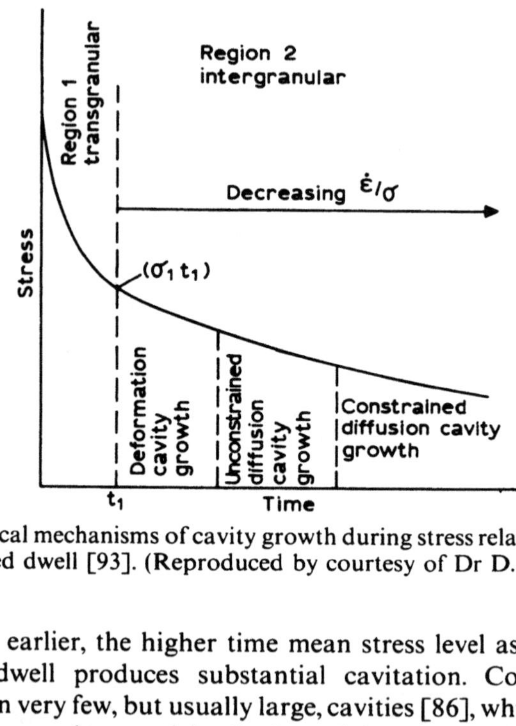

FIG. 28. Typical mechanisms of cavity growth during stress relaxation in a strain-controlled dwell [93]. (Reproduced by courtesy of Dr D. A. Miller.)

As implied earlier, the higher time mean stress level associated with a tensile-only dwell produces substantial cavitation. Compression-only dwells result in very few, but usually large, cavities [86], while in a balanced cycle containing each type of dwell of equal duration, cavitation is usually absent. The appearance of heavy cavitation does not always indicate a shortened life, particularly in cyclically stable alloys with a low transition fatigue life (e.g. some nickel- and titanium-based alloys). Here, substantial mean stresses of opposite sign to the dwell develop in asymmetric cycles and minimum endurance is often associated with compressive dwells due to enhancement of fatigue-crack growth by the tensile mean stress [95, 96].

Although grain boundaries may be heavily cavitated in the equivalent tensile dwell case (Fig. 30), failure may be retarded by the concurrent compressive mean stress. An alternative explanation has been suggested by Nazmy [97, 98], analysing IN 738 behaviour. He argues that a load-shedding effect occurs in the tensile dwell case because of the multiplicity of cracks, and that the fatal defect grows under a smaller driving force (stress intensity range), than would an isolated crack, as essentially exists in a compressive dwell specimen.

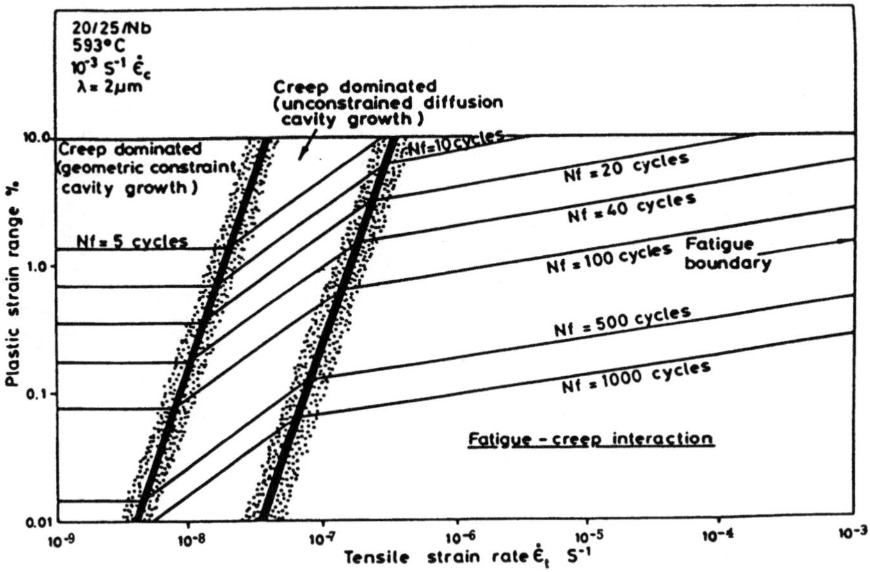

FIG. 29. Fatigue–creep fracture map for 20–25–1Nb stainless steel at 593°C, taking into account the different modes of cavity growth [93]. (Reproduced by courtesy of Dr D. A. Miller.)

In their response to simultaneous interactions, cast-to-cast variations in stainless steels present a notoriously difficult task to the designer. A type 316 alloy, resistant to tensile dwells, has been shown to age more rapidly than a susceptible cast [99]. Prior ageing (up to 10^4 h at 625 and 675°C) has little effect on the continuous cycling behaviour of each, but causes a fourfold improvement in the life of the susceptible alloy in a tensile dwell test [100]. The resistant alloy is unaffected by prior ageing. It is thought that the changes in creep ductility may be the controlling factor involved.

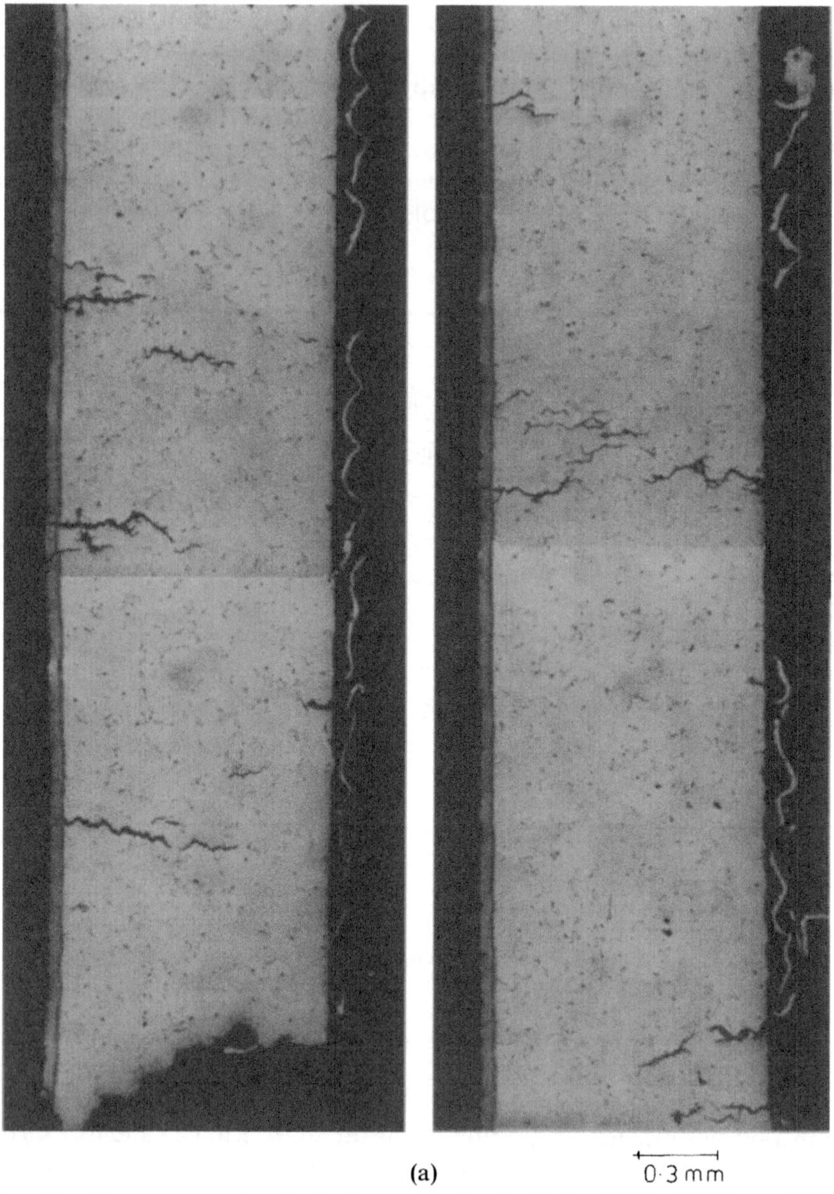

FIG. 30. Damage produced during a 5 min dwell in MAR M002 at 850°C ($\Delta\varepsilon_t = 0{\cdot}9\%$). (a) Tensile dwell; (b) compressive dwell. One wall of a coated cylindrical specimen shown [95].

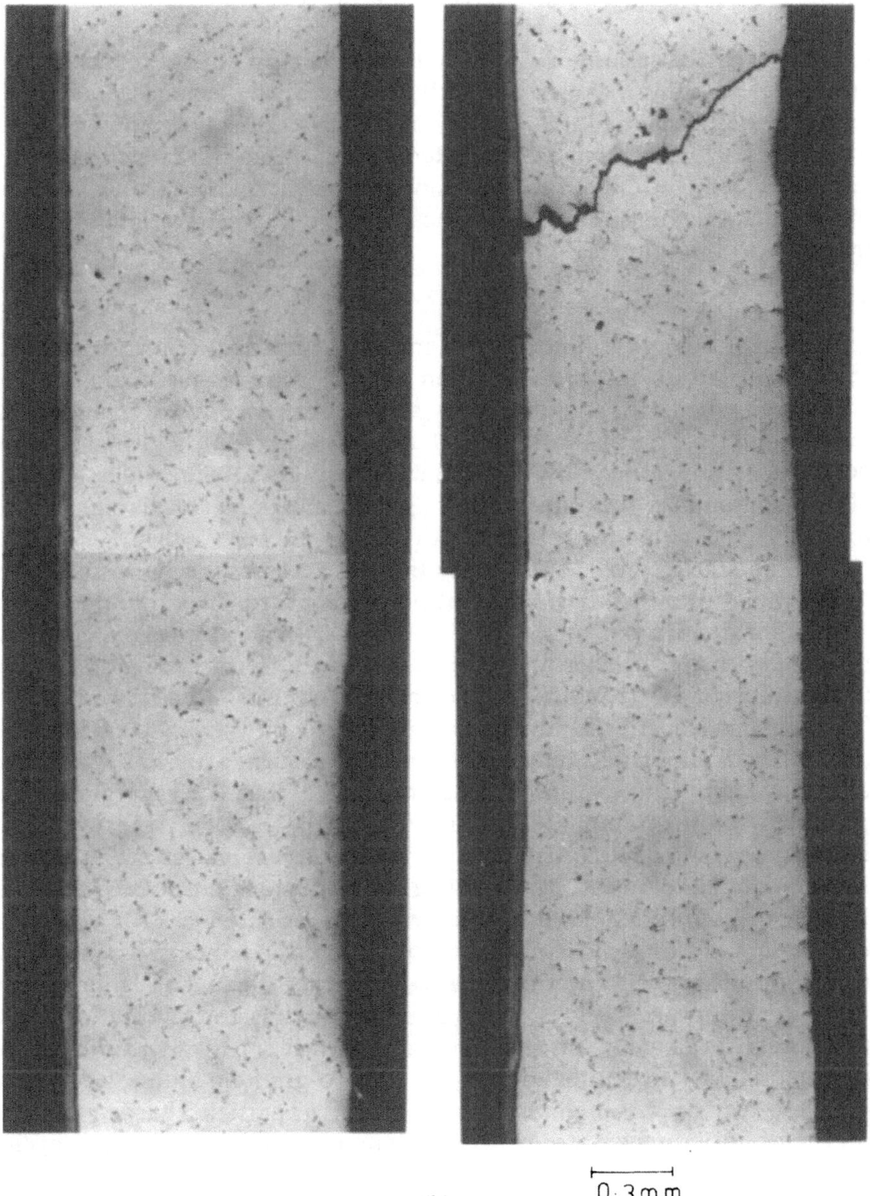

(b)

FIG. 30—*contd.*

4.6. ALUMINIUM AND TITANIUM ALLOYS

A great deal of attention was devoted in the earlier volume to steels and nickel-based alloys, which correctly reflected the emphasis of high temperature fatigue studies. In the following section two other groups, which are likely to be of increasing interest, are considered. In each case, the level of understanding of the high temperature fatigue behaviour is often fragmentary, but it is useful to link these specific examples with the general trends already outlined.

4.6.1. Aluminium Alloys

Wrought aluminium alloys are commonly precipitation-strengthened and, due to strain localisation, often exhibit poor fatigue properties in comparison to steels, as indicated by a low endurance to UTS ratio. The interactions between dislocations and the various intermediate phases have been extensively studied by Laird and co-workers [101]. Model Al–4Cu alloys containing fine coherent θ'' particles exhibit cyclic hardening followed by softening at temperatures up to 160°C ($0{\cdot}46T_m$) [102]. Softening is attributed to the formation of deformation bands, slightly misoriented from the matrix but not precipitate free. They contain a high dislocation density and are regarded as PSBs. With increasing temperatures the bands become broader, and after fatigue at 160°C transformation to the θ' phase occurs. The matrix of the material consists of a dispersion of fragmented dislocations and dislocation debris in all grains. No evidence of either over-ageing or reversion was found. Subsequent work on an identical alloy at room temperature has demonstrated that a larger grain size produces a more pronounced saturation stress plateau of a higher level, but is associated with a reduced endurance [103]. Recent studies at 250°C on a large grain size, annealed, Al–2·5Mg alloy demonstrated the range of fatigue damage possible according to local orientation [104]. Grains oriented for multiple slip exhibit intense surface-slip bands and banding of dislocations in their interior. Impingement of these bands on grain boundaries produces ledges which ultimately act as crack initiation sites. Favourably oriented grains show only fine surface slip lines and an interior dislocation wall structure, and do not lead to grain-boundary ledges.

Microstructures containing semi-coherent θ' particles harden rapidly and saturate permanently during room-temperature fatigue, whereas cyclic straining between 150 and 250°C results in softening after saturation [105]. The θ' plates are plastically deformed about an axis in the plane of the plate, and, additionally at high temperatures, plate slicing occurs. Nucleation of the equilibrium precipitate, θ, is accelerated.

As mentioned earlier, fatigue performance may be improved by dispersing slip, for example, via deployment of $MnAl_6$ particles. Alternatively, the use of a higher ageing temperature yields more stable precipitates, as illustrated by the superior fatigue behaviour of Al–Cu–Mg alloys over Al–Zn–Mg alloys [106]. The potential benefits of over-ageing are demonstrated in an Al–Mg–Ag alloy, in which large precipitate particles prevent localisation of slip and produce a 50% increase in fatigue limit [107]. The amelioration in all cases is much less for notched specimens than for smooth.

Unlike the wrought forms, cast aluminium alloys (most commonly Al–Si alloys) may be used at temperatures well in excess of 200°C. If they are to make further inroads into the automotive industry, more efficient design will require all the comprehensive knowledge and understanding equivalent to that which presently exists for steels and nickel-based alloys. For example, in LM-30 (Al–17Si–4Cu–1·1Fe–0·5Mg), fracture of primary silicon particles ahead of the crack accentuates growth at both room temperature and 250°C [108]. Temperature itself has a significant effect on early stage growth at R values around zero but virtually none when $R = 0{\cdot}5$.

4.6.2. Titanium Alloys

Titanium is an allotropic element which possesses a BCC structure at temperatures above 882°C and is HCP below. Three forms of alloy exist, α (HCP), β (BCC), and the combination ($\alpha + \beta$), in which the α is found in either a Widmanstatten or an equiaxed form. The fatigue properties of α-containing alloys are consequently more sensitive to microstructural and textural effects than alloys with isotropic FCC or BCC crystal structures. Common alloy additions include aluminium and oxygen (α-phase stabilisers), molybdenum and vanadium (β stabilisers), and zirconium, tin and silicon which have a neutral effect.

Very little detailed metallography of high temperature fatigue has been performed and much of the following, which principally relates to ambient temperature cycling, should be considered in conjunction with the general effects of increasing temperature, previously outlined.

4.6.2.1. Deformation

The phase usually cyclically hardens due to either dipoles hindering the subsequent motion of screw dislocations (low strain ranges), or monotonic hardening (high strain ranges) [109]. Twinning occurs at all strain levels and results in an upswing of the cyclic stress–strain curve. In the pure metal, slip takes place on the $\{10\bar{1}0\}$, prismatic planes, producing bands equivalent to PSBs, and this deformation mode predominates over twinning at high strain ranges [110]. In comparison to steels, the size of cyclically induced changes in strength in titanium alloys is usually small, and some alloys exhibit almost perfect stability. IMI 829 (Ti–5·5Al–3·5Sn–3Zr–0·25Mo–INb–0·3Si) is a 'near-α' alloy containing a very small amount ($<2\%$) of β. Fatigue at room temperature and 600°C with plastic strain ranges up to 2·5% is unable to induce strength changes in excess of 5% [111].

Beta alloys have a high flow stress and low work-hardening capacity, and exhibit initial cyclic softening due to an increase in mobile dislocation density [112, 113]. With increasing number of cycles, hardening may occur and restore the strength to above its original level [109]. Suggestions that metallurgical instability also contributes to softening remain, as yet, unsubstantiated. The alloys tend to be stronger than the α or $\alpha + \beta$ variety, and are more fatigue-resistant when assessed in terms of total strain per cycle, but not when the plastic strain range is considered. In an all-β, Ti–Mn alloy, halving the grain size produces about a 50% increase in endurance, an effect attributed to the reduction in slip distance and hence, surface offsets [114].

Two-phase ($\alpha + \beta$) alloys may exhibit cyclic hardening or softening, with the latter predominating at low strain amplitudes [115, 116]. Although α accommodates a greater proportion of the strain than β, the cyclic response cannot be explained by a simple law of mixtures [109]. Colony size, rather than the dimensions of individual α platelets, governs maximum slip distance.

Microstructural extremes in the form of equiaxed α (Eα), and Widmanstatten α plus grain boundary α (W + GBα), have been studied in a Ti–6Al–2Sn–4Zr–6Mo alloy [112, 117]. Softening occurs in both structures during fully reversed cycling ($R = -1$), when the initial quenching temperature is 927°C, but in fluctuating tension ($R = 0$), hardening is exhibited. Redistribution of the high quenched-in dislocation density is thought to be responsible for softening, and it is suggested that tension–tension fatigue produces a dislocation arrangement more resistant to subsequent dislocation motion, and hence, hardening. When the original quenching temperature is lowered to 871°C, cyclic hardening is observed which is attributed to transformation hardening.

4.6.2.2. Cracking

Crack initiation in the pure metal occurs at slip bands, twins and grain boundaries, particularly when they separate widely misoriented grains [110]. In the (W + GBα) type two-phase microstructures, few slip bands are produced by small strain range cycling. Extrusions form at the α–β interfaces and these, together with the GBα particles, act as the crack nucleation sites. Increasing the strain levels results in the formation of pores which subsequently link up to create a microcrack [117]. In the Eα configuration, cracking is associated with slip bands in the equiaxed α particles, in addition to the α–β interfaces. With higher strain ranges the former become predominant. Large colonies of aligned α platelets in β-processed alloys accelerate initiation during low-cycle fatigue. Incompletely healed casting porosity may have a similar effect when 'cleavage rosettes' subsequently appear on the fracture surface [118]. Transforming the β-phase of W$\alpha + \beta$ structures to martensite does not retard fatigue-crack initiation [119].

During high temperature low-cycle fatigue, cracking is again prevalent along α–β interfaces, but the suggested mechanism is that of embrittlement of the interface by interstitial elements [120]. In general, the longer the α platelet, the larger the initial crack size and the shorter the life. Consequently, β-processed alloys tend to have inferior LCF properties to (α–β)-processed alloys, since they are more susceptible to crack initiation. In terms of crack growth, however, β-processed alloys are superior, but in titanium alloys the initiation stage usually predominates.

The role of environment is clearly demonstrated when comparing tests in air and vacuum. Surface crack initiation occurs in the former, whereas subsurface formation is prevalent under vacuum [121]. The significant effects of frequency on endurance in air disappear under vacuum, substantiating the environmental, as opposed to the mechanical, time dependency.

4.6.2.3. *Dwell Effects*

Titanium alloys are unusual since they are susceptible to dwell effects at ambient temperatures, ($T \sim 0{\cdot}15 T_m$ K), at which traditional creep processes would not be expected [118, 122]. Substantial reductions (50 ×) in low-cycle fatigue endurance or accelerations in crack growth rate may be induced by dwells inserted at maximum load in those microstructures containing α colonies aligned roughly perpendicular to the applied stress. In some instances, crack growth rate appears unaffected and the life debit is solely associated with initiation effects [118]. The mechanisms of strain accumulation are similar in the continuous and dwell-containing cycles [123] which, to date, have all been performed with R values greater than zero (i.e. in fluctuating tension). Near-basal-plane facetting and predominantly subsurface initiation typified the dwell tests, while surface initiation is more representative of uniform cycling [124]. Suggested mechanisms include the formation of hydrides [118], strain-ageing due to hydrogen [122] and the presence of stacking faults obstructing slip on the basal planes [123]. This dwell phenomenon tends to disappear at temperatures above 150°C.

More conventional, strain-controlled dwell tests on a Ti–5Al–5Sn–2Zr–2Mo–0·25Si at 482°C demonstrate the greater severity of a balanced dwell compared to continuous cycling at the same overall frequency [125]. This was attributed to environmental effects and cracking occurred at the α–β interface in each case. At 600°C, IMI 829 falls into the less common category, mentioned in Section 4·5, in that it is compression-dwell sensitive [96]. The high tensile mean stresses developed during such a cycle result in a fatigue-dominated failure requiring fewer cycles than the equivalent tensile dwell test, despite the heavy creep damage sustained in the latter. Balanced dwells are also more deleterious than tensile dwells due to environmental effects. Again it should be emphasised that small changes in temperature might completely reverse the controlling failure mechanism.

4.7. CONCLUDING REMARKS

In his support role to the Engineer, the Materials Specialist has two prime functions; (1) to provide improved materials, and (2) to ensure that the behaviour of existing materials can be reliably predicted. Both objectives require a thorough understanding of the deformation and fracture processes which occur under realistic service conditions.

Sophisticated metallography, in particular transmission electron microscopy, has contributed substantially to this understanding, and better materials have resulted. Particle pinning of grain boundaries and the use of a bimodal precipitate distribution to disperse slip, while retaining strength, are just two examples mentioned. It is expected that this evolutionary process in the provision of better materials will continue.

Much less progress seems to have been made towards reliable life prediction; although notable exceptions such as the interpretation of fatigue–creep test data have been cited. A great deal more quantitative metallographic information is required. While it is conceivable that assessment of 'macrodamage', such as cracking and cavitation, will be of immediate use, the quantitative link between fine microstructural features, such as dislocations and precipitates and service performance, has yet to be established. The welcome growing emphasis on the quantitative aspects of fatigue mechanisms is well illustrated by comparison of the proceedings of two conferences [126, 127].

This chapter has demonstrated the heavy bias adopted towards the purist approach, of single crystals, simple alloys and uniform test conditions. Commercial nickel alloys and stainless steels have also received much metallographic attention, but generally with regard to their behaviour under idealised (isothermal, uniform, cycling) conditions. Much work remains to be done on other high temperature alloys in these areas, while in the potentially most fruitful fields of complex stress/strain cycling and non-isothermal testing, only token inroads have been made and the Engineer has yet to be convinced of the valuable contribution that Metallography can make.

REFERENCES

1. Forsyth, P. J. E., 1962, A two-stage process of fatigue crack growth, *Proc. Crack Propagation Symp., 1961, Cranfield*, pp. 76–94.
2. Miller, K. J., 1984, Initiation and growth rates of short fatigue cracks, *Eshelby Memorial Conf.*, Sheffield, (IUTAM), pp. 477–500.
3. Mughrabi, H., 1985, Dislocations in fatigue, *Dislocations and properties of real materials*, London, The Institute of Metals, pp. 244–62.
4. Mughrabi, H., 1985, Cyclic deformation and fatigue—some current problems, *ICSMA 7, Montreal*, pp. 1–26.
5. Bressers, J., 1985, Fatigue and microstructure, *Proc. Int. Conf. on High Temperature Alloys—Their Exploitable Potential, Petten*, pp. 1–26.
6. Plumbridge, W. J. and Ryder, D. A., 1969, Metallography of fatigue, *Metall. Reviews No. 136, Metals and Materials*, pp. 119–42.
7. Mughrabi, H., 1980, Microscopic mechanisms of metal fatigue, *Proc. Int. Conf. on Strength of Metals and Alloys, Vol. 3*, eds P. Haasen, V. Gerold and G. Kostorz, Pergamon Press, pp. 1615–38.
8. Mughrabi, H. and Wang, R., 1980, Cyclic strain localisation and fatigue crack initiation slip bands in face centred cubic metals and single phase alloys, *Int. Symp., Defects and Fracture, Tuczno, Poland*, eds G. C. Sih and H. Zorski, The Hague, M. Nijhoff, pp. 15–28.
9. Pohl, K., Mayr, P. and Macherauch, E., 1980, Persistent slip bands in the interior of a fatigued low carbon steel, *Scripta Metall.*, **14**, 1167–9.
10. Winter, A. T., Pederson, O. B. and Rasmussen, K. V., 1981, Dislocation microstructures in fatigued copper polycrystals, *Acta Metall.*, **20**, 738–48.
11. Boulanger, L., Bission, A. and Tavassoli, A. N., 1985, Labyrinth structures and persistent slip bands in fatigued 316 stainless steel, *Phil. Mag. A*, **51**, L5–L11.
12. Shirai, H. and Weertman, J. R., 1983, Fatigue dislocation structures at elevated temperatures, *Scripta Metall.*, **17**, 1253–8.
13. Mughrabi, H., 1979, Discussion at Conf. on Fatigue Mechanisms, Kansas City, STP 675, Philadelphia, ASTM, pp. 126–7.

14. Lisiecki, L. L. and Weertman, J. R., 1986, Orientation effects in copper single crystals fatigued at elevated temperatures, *Scripta Metall.*, **20**, 249–52.
15. Mughrabi, H., Ackermann, F. and Herz, K., 1979, Persistent slip bands in fatigued face centred and body central cubic metals, *Fatigue mechanisms*, ed. J. T. Fong, Philadelphia, ASTM 675, pp. 69–97.
16. Ackermann, F., 1981, Doctorate Thesis, Univ. Stuttgart.
17. Pohl, K., Mayr, P. and Macherauch, E., 1981, Cyclic deformation behaviour of a low carbon steel in the temperature range between room temperature and 850 K, *Int. J. Fracture*, **17**, 221–33.
18. Pohl, K., Mayr, P. and Macherauch, E., 1983, Shape and structure of persistent slip bands in iron carbon alloys, *Defects, fracture and fatigue*, eds G. C. Sih and J. W. Provan, The Hague, M. Nijhoff, pp. 147–59.
19. Kanazawa, K., Yamaguchi, K. and Nishijima, S., 1985, Mapping of low cycle fatigue mechanisms at elevated temperatures for an austentic stainless steel, *Low Cycle Fatigue: Future Directions*, Lake George, ASTM, Philadelphia, STP 942, ed. H. Solomon (*to be published*).
20. Nilsson, J. O., 1984, Influence of nitrogen on the high temperature low cycle fatigue behaviour of austenitic stainless steels, *Fatigue Engng Mat. Struc.*, **7**, 55–64.
21. Mughrabi, H., 1983, Cyclic deformation and fatigue of multiphase materials, *Deformation of multiphase and particle containing materials*, eds J. S. Bilde-Sorenson *et al.*, Riso National Laboratory, Roskilde, Denmark, pp. 65–82.
22. Laird, C., 1975, Cyclic deformation of metals and alloys, *Treatise on materials science and technology*, *Vol. 6*, *Plastic deformation of materials*, ed. R. J. Arsenault, San Francisco, Academic Press, pp. 101–62.
23. Stolz, R. E. and Pineau, A. G., 1978, Dislocation–precipitate interaction and cyclic stress--strain behaviour of a γ' strengthened superalloy, *Mat. Sci. Engng*, **34**, 275–84.
24. Gerold, V., Lerch, B. A. and Steiner, D., 1984, Fatigue behaviour of age hardened alloy single crystals, *Z. metall.*, **75**, 546–53.
25. Wilhelm, M., 1981, The cyclic stress–strain behaviour of age hardened Cu–Co and Al–Zn–Mg alloy single crystals, *Mat. Sci. Engng*, **48**, 91–106.
26. Plumbridge, W. J. and Bartlett, R. A., 1982, Cyclic response of a 1Cr–Mo–V low alloy steel, *Int. J. Fatigue*, **4**, 209–16.
27. Challenger, K. D. and Vining, P. G., 1983, Substructure and back stress changes resulting from the cyclic loading of $2\frac{1}{4}$Cr–1Mo steel at 755 K, *Mat. Sci. Engng*, **58**, 257–67.
28. Calderon, H. A., Weertman, J. R. and Fine, M. E., 1984, Effect of cyclic plastic deformation on elevated temperature Ostwald Ripening of an alloy with coherent precipitates, *Scripta Metall.*, **18**, 587–92.
29. Antolovich, S. D., Liu, S. and Baur, R., 1981, Low cycle fatigue behaviour of René 80 at elevated temperature, *Metall. Trans.*, **12A**, 473–81.
30. Nahm, A. and Moteff, J., 1981, Characterisation of fatigue substructure in Incoloy Alloy 800 at elevated temperature, *Metall. Trans.*, **12A**, 1011–25.
31. Ermi, A. M. and Moteff, J., 1982, Correlation of substructure with time dependent fatigue properties of AISI 304 stainless steel, *Metall. Trans.*, **13A**, 1577–88.

32. Ermi, A. M. and Moteff, J., 1983, Microstructural development and cracking behaviour of AISI 304 stainless steel tested in time dependent fatigue modes, *J. Engng Mat. Technol.*, **105**, 21–30.
33. Horton, C. A. P., Lai, J. K. L. and Skelton, R. P., 1983, The relationship between microstructure, fatigue and creep behaviour in a type 316 stainless steel, Mechanical properties of structural materials including environmental effects, IAEA Report IWGFR-49, Vol. 1, pp. 411–33.
34. Sanders, T. H., Frishmuth, R. E. and Embley, G. T., 1981, Temperature dependent deformation mechanisms of Alloy 718 in low cycle fatigue, *Metall. Trans.*, **12A**, 1003–10.
35. Nilsson, J. O., 1984, Influence of dislocation–precipitate reaction on low cycle fatigue resistance of Alloy 800 at 600°C, *Metal Sci.*, **18**, 351–5.
36. Dean, M. S. and Plumbridge, W. J., 1982, Ageing of Type 316 stainless steel and its effect on creep and fatigue behaviour, *Conf. on Mechanical Behaviour and Nuclear Applications of Stainless Steel at Elevated Temperatures, Varese*, Book No. 280. Institute of Metals, pp. 73–80.
37. Nilsson, J. O. and Thorvaldsson, T., 1985, Low cycle fatigue behaviour of alloy 800 H at 600°C—effect of grain size and γ' precipitate dispersion, *Fatigue Fract. Engng Mat. Struct.*, **8**, 373–84.
38. Lutjering, G., Daker, H. and Munz, D., 1973, Microstructure and fatigue behaviour of aluminium alloys, *Proc. Third Int. Conf. on 'The Strength of Metals and Alloys', Cambridge, Vol. 1*, pp. 427–31.
39. Edwards, L. and Martin, J. W., 1982, The influence of dispersoids on the low cycle fatigue properties of Al–Mg–Si alloys, *Proc Sixth Int. Conf. on 'The Strength of Metals and Alloys'*, pp. 873–8.
40. Morris, W. L., Buck, O. and Marcus, H. L., 1976, Fatigue crack initiation and early propagation in Al 2219-T851, *Metall. Trans.*, **7**, 1161–5.
41. Kung, C. Y. and Fine, M. E., 1979, Fatigue crack initiation and microcrack growth in 2024-T4 and 2124-T4 aluminium alloys, *Metall. Trans.*, **10**, 603–10.
42. Plumbridge, W. J. and Dalski, M. E., 1977, Unpubl. M.Sc. thesis, University of Bristol.
43. Matsuoka, S., Kim, S. and Weertman, J. R., 1983, Mechanical and microstructural behaviour of a ferritic stainless steel under high temperature cycling, *Topical Conf. on Ferritic Alloys for Use in Nuclear Energy Technologies* (Utah), pp. 507–16.
44. Rezgui, B., 1979, Interaction fatigue–fluage, effect d'un temps de mainten sur la resistance a la fatigue d'un acier (type 316) a 600°C, Rapport CEM-R-4982.
45. Laird, C. and Duquette, D. J., 1972, Fatigue crack initiation, *Corrosion fatigue—chemistry, mechanics and microstructure*, ed. O. J. Devereux, A. J. McEvily and R. W. Staehle, Nat. Assoc. Corrosion Engrs—2, Houston, pp. 88–117.
46. Essmann, U., Gaesele, U. and Mughrabi, H., 1981, A model of extrusions and intrusions in fatigued metals, Part 1, *Phil. Mag.*, **44**, 405–26.
47. Mughrabi, H., Wang, R., Diffort, K. and Essmann, U., 1983, Fatigue crack initiation by cyclic slip irreversibilities in high cycle fatigue, *Fatigue mechanisms—advances in quantitative measurement of physical damage*, eds J. Lankford *et al.*, STP 811, Philadelphia, ASTM, pp. 5–43.

48. Brown, L. M., 1981, Dislocations and the fatigue strength of metals, *Dislocation modelling in physical systems*, eds M. F. Ashby *et al.*, Pergamon Press, pp. 51–68.
49. Brown, L. M. and Ogin, S. L., 1985, Role of internal stresses in the nucleation of fatigue cracks, *Eshelby Memorial Lecture* (*Sheffield*), Cambridge University Press, pp. 501–28.
50. Neumann, P., 1983, Fatigue, *Physical metallurgy* eds R. W. Cahn and P. Raason, Amsterdam, Elsevier, pp. 1553–94.
51. Hunsche, A., 1982, Doctorate thesis, Rheinisch–Westfalisch Technische Hochschule, Aachen.
52. Graf, M. and Hornbogen, E., 1978, The effect of inhomogeneity of cyclic strain on initiation of cracks, *Scripta Metall.*, **12**, 147–50.
53. Pelloux, R. M. N. and Stoltz, R. E., 1976, Optimisation of microstructures for fatigue resistant engineering alloys, *Proc. ICSMA 4, Nancy*, pp. 1023–36.
54. Kim, Y. M. and Merrick, H. F., 1980, Fatigue properties of MA-6000E, a γ' strengthened ODS alloy, *Superalloys 1980*, Metals Park, Ohio, ASM, pp. 551–61.
55. Bressers, J., Schusser, U., Fenske, E. and De Cat, R., 1986, Time dependent processes in the low cycle fatigue of nickel base superalloys, COST 501, to be published.
56. Fine, M. E., 1980, Fatigue resistance of metals, *Metall. Trans.*, **11A**, 365–79.
57. James, M. R. and Morris, W. C., 1982, The fracture of constituent particles during fatigue, *Mat. Sci. Engng*, **56**, 63–71.
58. Lankford, J., 1977, Effect of oxide inclusions on fatigue failure, *Int. Metall. Rev.*, **11**, 221–8.
59. Tanaka, K. and Mura, T., 1982, A theory of crack initiation at inclusions, *Metall. Trans.*, **13A**, 117–23.
60. McMahon, C. J. and Coffin, L. F., 1970, Mechanisms of damage and fracture in high temperature, low cycle fatigue of a cast nickel based superalloy, *Metall. Trans.*, **1**, 3443–50.
61. Antolovich, S. D., Bomas, P. and Strudel, J. L., 1979, Low cycle fatigue of René 80 as affected by prior exposure, *Metall. Trans.*, **10A**, 1859–68.
62. Taplin, D. M. P., Tang, N. Y. and Leipholz, H. H. E., 1984, On fatigue-creep–environment interaction and feasibility of fatigue maps, *Sixth Int. Conf. on Fracture*, *New Delhi*, pp. 127–42.
63. Argon, A. S., 1982, Mechanisms and mechanics of fracture in creep alloys, *Recent advances in creep and fracture of engineering materials and structures*, eds B. Wilshire and D. R. J. Owen, Swansea, Pineridge Press, pp. 1–52.
64. Wells, C. H. and Sullivan, C. P., 1969, Interactions between creep and low cycle fatigue in Udimet 700 at 1400°F, *Fatigue at high temperature*, STP 459, Philadelphia, ASTM, pp. 59–74.
65. May, M. J. and Honeycombe, R. W. K., 1963/4, The effect of temperature on the fatigue behaviour of Mg and some Mg alloys, *J. Inst. Metals*, **92**, 41–9.
66. Saegusa, T. and Weertman, J. R., 1978, Influence of grain boundary morphology on void formation, *Scripta Metall.*, **12**, 187–91.
67. Snowden, K. U., 1966, Fatigue crack formation in bicrystals of lead, *Phil. Mag.*, **14**, 1019–29.
68. Lim, L. C. and Raj, R., 1984, On slip induced intergranular cavitation during

low cycle fatigue of nickel at intermediate temperature, *Acta Metall.*, **32**, 727–33.
69. Lim, L. C. and Raj, R., 1984, Effect of boundary structure on slip induced cavitation in polycrystalline nickel, *Acta Metall.*, **32**, 1183–90.
70. Scaife, E. C. and James, P. L., 1968, Effect of environment on intergranular cavitation, *Metal Sci.*, **2**, 217–20.
71. Bricknall, R. H. and Woodford, D. A., 1981, Cavitation in nickel during oxidation and creep, *Creep and fracture of engineering materials and structures*, eds B. Wilshire and D. R. J. Owen, Swansea, Pineridge Press, pp. 249–62.
72. Gittins, A., 1968, The stability of grain boundary cavities in copper, *Acta Metall.*, **16**, 517–22.
73. Tang, N. Y., Taplin, D. M. R. and Plumtree, A., 1985, Schema for depicting cavity nucleation during high temperature fatigue, *Mat. Sci. Technol.*, **1**, 145–51.
74. Beere, W., 1981, Theoretical treatment of cavity growth and nucleation, *Cavities and cracks in creep and fatigue*, ed. J. Gittus, London, Elsevier Applied Science, pp. 1–27.
75. Skelton, R. P., 1966, The growth of grain boundary cavities during high temperature fatigue, *Phil. Mag.*, **14**, 563–72.
76. Page, R., Weertman, J. R. and Roth, M., 1980, Investigation of fatigue-induced grain boundary cavitation by small angle neutron scattering, *Scripta Metall.*, **14**, 773–7.
77. Yoo, M. H., Ogle, J. C., Borie, B. S., Lee, E. H. and Hendricks, R. W., 1982, Small angle neutron scattering study of fatigue induced cavities in nickel, *Acta Metall.*, **30**, 1733–42.
78. Smith, E. and Barnby, T. J., 1967, Crack nucleation in crystalline solids, *Metal Sci.*, **1**, 56–64.
79. Driver, J. H., 1971, The effect of grain boundary precipitates on the high temperature fatigue strength of austenitic stainless steels, *Metal Sci.*, **5**, 47–50.
80. Tang, N. Y., Plumtree, A. and Taplin, D. M. R., 1984, A model for predicting intergranular propagation of fatigue cracks at elevated temperatures, *Second Irish Durability and Fracture Conference, Dublin*, pp. 1–13.
81. Tang, N. Y. and Plumtree, A., 1985, A note on grain boundary diffusion controlled cavity growth during elevated temperature fatigue, *Metall. Trans.*, **16A**, 300–2.
82. Fleck, R. G., Taplin, D. M. R. and Beevers, C. J., 1975, An investigation of the nucleation of creep cavities by 1Mv electron microscopy, *Acta Metall.*, **23**, 415–34.
83. Thorpe, T. W. and Smith, G. C., 1981, Elevated temperature low cycle fatigue of AISI 316 stainless steel, *Proc. Fifth Conf. on Fracture, Cannes, Advances in fracture research*, ed. D. François, pp. 2413–22.
84. Plumbridge, W. J., Priest, R. H. and Ellison, E. G., 1979, Damage formation during fatigue–creep interactions, *Third Int. Conf. on Mechanical Behaviour of Materials, Cambridge*, pp. 129–39.
85. Plumbridge, W. J. and Miller, K. J., 1972, The effect of prior fatigue deformation on creep, *Int. Conf. on Creep Strength in Steel and High Temperature Alloys, Sheffield*, pp. 50–3.

86. Ellison, E. G. and Paterson, A. J. F., 1976, Creep fatigue interactions in 1Cr–Mo–V steel, *Proc. Inst. Mech. Engrs*, **190**, 321–50.
87. Plumbridge, W. J., Bartlett, R. A., Chung, T. E. and Ellison, E. G., 1977, Mechanism of fatigue–creep interaction in a low alloy steel, *Fourth Int. Conf. on Fracture, Waterloo*, pp. 831–7.
88. Nazmy, M. Y., 1982, Sequential effect of creep and fatigue in a cast Ni-base alloy, *Scripta Metall.*, **16**, 823–6.
89. Wei, K. and Dyson, B. F., 1982, Creep–fatigue interactions in 316 stainless steel under torsional loading, *Conf. on Mechanical Behaviour and Nuclear Applications of Stainless Steel at Elevated Temperatures, Varese*, Institute of Metals, Book No. 280, pp. 136–44.
90. Neailey, K. and Plumbridge, W. J., unpubl. work.
91. Wareing, J., 1977, Creep–fatigue interaction in austenitic stainless steels, *Metall. Trans.*, **8A**, 711–21.
92. Gladwin, D. and Miller, D. A., 1982, The effect of cycle waveshape on the low cycle fatigue behaviour of 20%Cr–25%Ni–Nb stainless steel at 650°C, *Fatigue Engng Mat. Struct.*, **5**, 275–86.
93. Miller, D. A., 1983, Fatigue–creep mechanisms maps for 20%Cr 25%Ni/Nb stainless steel at 593°C and 650°C, *Int. Conf. on Advances in Life Prediction Methods*, New York, ASME.
94. Plumbridge, W. J., Dean, M. S. and Miller, D. A., 1982, The importance of failure mode in fatigue–creep interactions, *Fatigue Engng Mat. Struct.*, **5**, 101–14.
95. Plumbridge, W. J. and Ellison, E. G., 1986, Low cycle fatigue behaviour of superalloy blade materials at elevated temperature, *Int. Conf. on Mechanical Behaviour of Superalloys*, London, Institute of Metals.
96. Plumbridge, W. J. and Stanley, M., 1986, The effects of dwell periods on the high temperature fatigue behaviour of Titanium (829) alloy, *Int. Conf. on Fatigue of Engng Mat. and Structures, Sheffield*, pp. 377–86.
97. Nazmy, M. Y., 1983, High temperature low cycle fatigue of IN 738 and application of strain range partitioning, *Metall. Trans.*, **14**, 449–61.
98. Nazmy, M. Y., 1983, Effect of multiple crack propagation on the high temperature low cycle fatigue of a cast nickel-base alloy, *Scripta Metall.*, **17**, 491–4.
99. Neailey, K. and Plumbridge, W. J., 1986, Application of the friction stress approach to creep in Type 316 stainless steel, *Conf. on Designing with High Temperature Materials*, York, Institute of Metals.
100. Neailey, K. and Plumbridge, W. J., 1987, High temperature mechanical properties of two batches on Type 316 stainless steel, Paper submitted to Conf. on Stainless Steels, 1987, The Institute of Metals.
101. Laird, C., 1977, The general cyclic stress–strain response of aluminium alloys, Philadelphia, ASTM, *STP 637*, pp. 3–35.
102. Bhat, S. P. and Laird, C., 1979, High temperature cyclic deformation of precipitation hardened alloy—II. Fully coherent precipitates, *Acta Metall.*, **27**, 1873–83.
103. Horibe, A., Lee, J. K. and Laird, A., 1984, Cyclic deformation of Al–4%Cu polycrystals containing θ'' precipitate, *Fatigue Engng Mat. Struct.*, **7**, 145–54.
104. Mabuchi, H. and Nakayamo, Y., 1986, Microstructure and fracture of

fatigued polycrystalline Al–Mg alloys at elevated temperature, *Scripta Metall.*, **20**, 717–21.
105. Bhat, S. P. and Laird, C., 1979, High temperature cyclic deformation of precipitation hardened alloy—I. Partially coherent precipitates, *Acta Metall.*, **27**, 1861–71.
106. Polmear, I. J., 1981, *Light alloys—metallurgy of the light metals*, London, Edward Arnold, pp. 37–42.
107. Boyapati, K. and Polmear, I. J., 1979, Fatigue-microstructure variations in some aged aluminium alloys, *Fatigue Engng Mat. Struct.*, **2**, 23–33.
108. Culver, L. E., Balthazar, J. C. and Radon, J. C., 1983, Temperature effects on the fatigue behaviour of an aluminium–silicon alloy, *Trans. Can. Soc. Mech. Engrs*, **8**, 150–6.
109. Saleh, Y. and Margolin, H., 1980, Low cycle fatigue behaviour of Ti–Mn alloys: cyclic stress–strain response, *Metall. Trans.*, **11A**, 1295–302.
110. Suhua, A., Zhongguang, W. and Yuebo, X., 1985, The fatigue deformation and fracture characteristics of coarse grained polycrystalline titanium, *Scripta Metall.*, **19**, 1089–93.
111. Plumbridge, W. J. and Stanley, M., 1986, Low cycle fatigue of a titanium 829 alloy, *Int. J. Fatigue*, **8**, 209–16.
112. Mahajan, Y. and Margolin, H., 1982, Low cycle fatigue behaviour of Ti–6Al–2Sn–4Zr–6Mo: Part II. Cyclic deformation behaviour and low cycle fatigue, *Metall. Trans.*, **13A**, 269–74.
113. Koss, D. A. and Wojcik, C. C., 1976, Flow stress asymmetry and cyclic stress strain response in a BCC Ti–V alloy, *Metall. Trans.*, **7A**, 1243–4.
114. Saleh, Y. and Margolin, H., 1982, Low cycle fatigue of Ti–Mn alloys: fatigue life, *Metall. Trans.*, **13A**, 1275–81.
115. Wells, C. H. and Sullivan, C. P., 1969, Low cycle fatigue crack initiation in Ti–6Al–4V, *Trans. Am. Soc. Metals*, **62**, 263–70.
116. Steele, R. K. and McEvily, A. J., 1976, High cycle fatigue behaviour of Ti–4Al–4V alloy, *Engng Fracture Mech.*, **8**, 31–7.
117. Mahajan, Y. and Margolin, H., 1982, Low cycle fatigue behaviour of Ti–6Al–2Sn–4Zr06Mo: Part 1. The role of microstructure in low cycle crack nucleation and early crack growth. *Metall. Trans.*, **13A**, 257–68.
118. Eylon, D. and Hall, J. A., 1977, Fatigue behaviour of beta processed titanium alloy IMI 685, *Metall. Trans.*, **8A**, 981–90.
119. Shih, D. S., Lin, F. S. and Starke, E. A. Jr, 1985, The effect of microstructure and texture on the low cycle fatigue properties of Ti–6Al–2Nb–1Ta–0·8Mo, *Titanium science and technology*, eds G. Lutjering, U. Zwicker and W. Bunk, Deutsche Gesellschaft für Metallkunde, pp. 2099–106.
120. Eylon, D., Bartel, T. L. and Rosenblum, M. E., 1980, High temperature low cycle fatigue in beta processed Ti–5Al–5Sn–2Zr–2Mo–0·25Si, *Metall. Trans.*, **11A**, 1361–7.
121. Hoffman, C., Eylon, D. and McEvily, A. J., 1982, Influence of microstructure on the elevated temperature fatigue resistance of a titanium alloy, *Low cycle fatigue and life prediction*, eds C. Amzallag, B. N. Leis and P. Rabbe, STP 770, Philadelphia, ASTM, pp. 5–22.
122. Evans, W. J. and Gostelow, C. R., 1979, The effect of hold time on the fatigue properties of a processed titanium alloy, *Metall. Trans.*, **10A**, 1837–46.

123. White, J., Loretto, M. H. and Smallman, R. E., 1984, The effect of dwells on low cycle fatigue on IMI 829, *Fifth Conf. on Titanium, Munich*, pp. 2297–304.
124. Davidson, D. L. and Eylon, D., 1980, Titanium alloy fatigue fracture facet investigation by selected area electron channeling, *Metall. Trans.*, **11A**, 837–43.
125. Eylon, D. and Rosenblum, M. E., 1982, Effects of dwell on high temperature low cycle fatigue of a titanium alloy, *Metall. Trans.*, **13A**, 322–4.
126. Fong, J. T., ed., 1979, *Fatigue mechanisms*, STP 675, Philadelphia, ASTM.
127. Lankford, J., Davidson, D., Morris, W. L. and Wei, R. P., eds, 1983, *Fatigue mechanisms: advance in quantitative measurement of physical damage*, STP 811, Philadelphia, ASTM.

5

The Physical Metallurgy of Failure Criteria

R. HALES

Berkeley Nuclear Laboratories, CEGB, Berkeley, Gloucestershire, UK

SUMMARY

The failure of engineering components by thermal fatigue was originally treated as an extension of low-cycle fatigue. Time-dependent deformation was regarded as causing a reduction of endurance in much the same way that notches reduce endurance of plain specimens under similar conditions. Early design and assessment procedures were based on codes originally intended to assess the likelihood of failure under fatigue loading at ambient temperatures only.

Recent developments in the investigation of fatigue at high temperatures have shown that failure can occur by different physical micro-mechanisms which degrade the material in different ways. An appreciation of the processes which cause these different types of degradation is fundamental to any attempt to predict the long-term creep–fatigue endurance of engineering materials. The most severe reductions of endurance compared with those measured under conditions of rapid symmetrical loading are observed when intergranular defects of the type associated with low-ductility creep failure are generated.

An understanding and quantitative description of the formation of such defects has allowed the formulation of an alternative assessment procedure which is briefly outlined. Such a procedure allows an estimate of the lower-bound behaviour of materials from limited data and, more importantly, gives a clear indication to designers how best to overcome unacceptable service conditions, whether it is better to improve material properties, or to modify the design thereby reducing the service demands on the component.

5.1. THE ORIGINS OF THERMAL FATIGUE

The high thermal efficiency and flexibility demanded of modern process plant subjects their materials of construction to extremely arduous service conditions. The high operating temperatures mean that most materials are taken into the creep regime and time-dependent deformation must be taken into account when assessing the durability of such structures. Furthermore, prolonged exposure to aggressive environments at elevated temperatures may lead to chemical and microstructural changes in the material which can affect their long-term mechanical properties.

The flexibility demanded of modern plant results in frequent start-ups and shut-downs, sometimes as part of planned cycle and sometimes responding to unscheduled demand. Such operating practices impose cyclic loads which can lead to fatigue-related failures. Steam generating plant and turbines used in the electricity supply industry exemplify these trends. As new plant is commissioned it generates electricity more efficiently than stations already in service, and therefore starts out by meeting base load demand. As time passes it is displaced in turn by newer and more efficient units, and is required to generate less power but under the potentially more demanding conditions of 'two shifting'. At the end of its days, it is called into service less frequently to meet peak demands. A typical history is illustrated in Fig. 1 [1].

As the station function changes, it is necessary that an accurate assessment be made of the ability to withstand the new service conditions to ensure continued availability. Large expensive components such as steam turbines are subjected to detail design studies in an attempt to ensure continuity of service and safe operation. Chemical process reactors, which often comprise pressure vessels connected by lengths of pipe, are not usually designed to such stringent criteria. Instead, items which are known to suffer the most arduous service conditions are either inspected or replaced regularly. Clearly, under either stratagem, an accurate assessment of the consequences of accumulated service experience together with a prediction of future demands is necessary to ensure the safe and economic operation of components.

Although there is no internationally agreed procedure for designing and assessing structures which operate in the creep regime, ASME Code Case N-47 [2] is widely used under these conditions. The materials properties and failure criteria defined in this code case are based on a limited number of short-term tests which were performed on a small number of casts. In the case of materials not covered by the code case, e.g. Cr–Mo–V steels used in

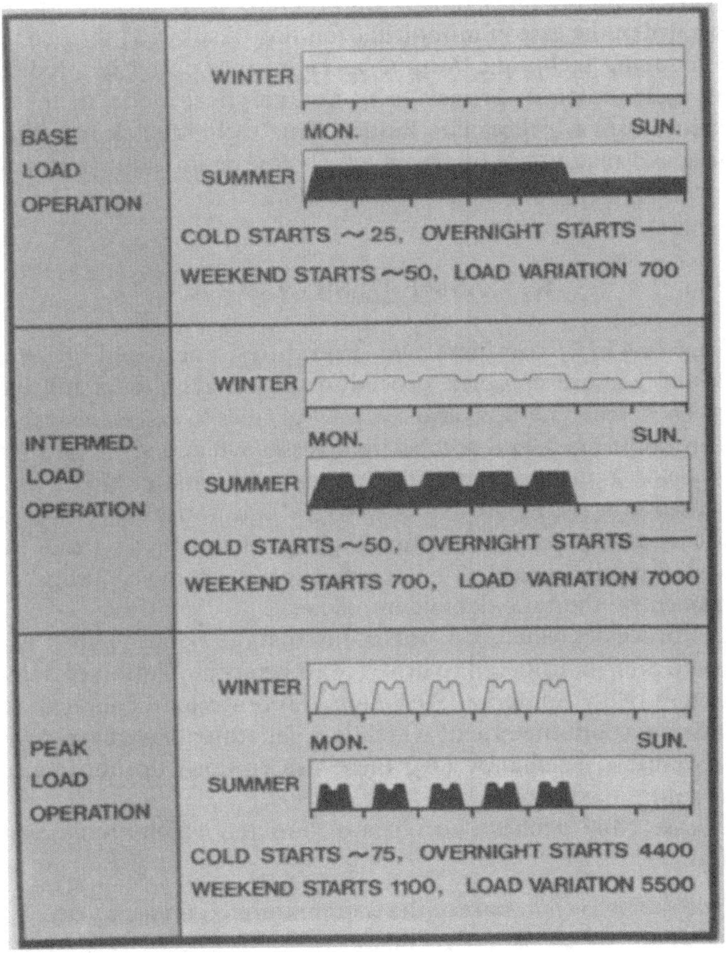

FIG. 1. An illustration of the changes in loading during 25 years' service of steam generating plant [1]. (Reproduced from Chapter 8 of the previous volume.)

the construction of turbine rotors and casings, designers have preferred to assess structures against long-term test data (see Chapter 7). Here again, the results may apply to a small number of casts within a given alloy specification.

Any code which is used to calculate the long-term performance of components should be based on an understanding of the processes which lead to failure. The importance of appreciating the conditions which can lead to various failure mechanisms (and thereby bounding them) has been demonstrated in the case of monotonic loading. Ashby [3] showed that, by using a mapping technique, long-term endurance could be predicted by applying basic materials properties to an analytical model of the correct micro-mechanism which causes failure. This technique clearly illustrates the source and magnitude of errors which can be introduced by a simple parametric extrapolation of short-term data.

5.2. ASME DESIGN CODE

Code Case N-47 [2] considers two procedures, one based on an elastic analysis and a more complex procedure which requires a full inelastic analysis. The former method employs a set of rules to assess the response of a structure to primary and secondary stresses which vary with time. The analysis yields a strain range through which the material is cycled and which is used to assess life from a 'slow-cycle' endurance curve. The use of a single life-assessment curve for slow cycling reflects the fact that thermal fatigue was initially regarded as an extension of low-cycle fatigue, which was modified by the time-dependent effects.

Failure processes which do not occur instantaneously, such as creep rupture, are often considered to arise from the accumulation of damage to the structure of the material. Although elaborate mathematical descriptions of damage and the rate at which it is generated have been developed (see, for example, Kachanov [4]) these descriptions do not identify the *physical* nature of damage.

Code Case N-47 defines two types of damage which are conveniently defined as:

> '*Fatigue damage* is that part of the total material damage caused by cyclic deformation which is independent of time effects. The damage is expressed in terms of a cyclic ratio'. $D_f = \sum N/N_f$ (Miner's law).

and

> '*Creep damage* is that part of the total material damage caused by time exposure to steady and transient stresses at elevated temperatures expressed as a time ratio (relaxation damage is a form of creep damage)'. $D_c = \int_0^t dt/t_r$ (Robinson's rule).

These definitions imply that the two deformation processes generate the same kind of damage. Indeed, its exact nature is not specified in the same way that the continuum damage invoked by Kachanov [4] is not specified, but rather it is regarded as a *state variable* which can lead to changes in behaviour, for example from steady-state to tertiary creep. This interpretation of damage allows the two processes to contribute equally to ultimate failure. In addition, the definition does not admit an initiation process. Low-cycle fatigue is considered to be controlled by cyclic crack growth, whereas it is recognised that a large fraction of the life in high-cycle fatigue is spent initiating a defect which then grows in a relatively small fraction of the life. The definition of fatigue damage and its contribution as determined by Miner's law takes no account of this transition.

The estimates of damage are combined to give a failure condition

$$N/N_{\mathrm{f}} + Nt/t_{\mathrm{r}} \leq D \tag{1}$$

where D is unity or less than one depending on the material under consideration. The definition of failure given in eqn (1) is used to determine the endurance of a point in a structure by non-linear analysis, or to construct a slow-cycle endurance curve for use in conjunction with an elastic design procedure.

The procedures adopted in the elastic analysis further emphasise the historic background to Code Case N-47 which is derived from a fatigue design code, ASME III. As in the fatigue code, the effects of discontinuities and notches are analysed using Neuber's rule (see Chapter 2). The presence of welds, multiaxial loading and other sources of uncertainty are taken into account by applying the same life-reduction factors to the slow-cycling curve as are applied to the low temperature fatigue endurance curve, viz. number of cycles divided by 20, or by twice the strain range. These procedures sometimes lead to very conservative estimates of life.

In addition to the shortcomings associated with the fatigue life fraction, the method adopted for assessing the creep life fraction also warrants closer examination. Wood *et al.* [5] have pointed out that the time fraction approach leads to the anomalous situation whereby a strong material which resists stress relaxation is attributed with a high creep damage fraction because it spends a long time under a large stress; even though in so doing it accumulates little or no creep strain. The converse is true for a weak material which rapidly sheds the applied load but accumulates significant creep strain in the process. The choice of time fraction rule also makes the calculation of creep damage very sensitive, since time to rupture is proportional to the applied stress raised to some power, often of the order of 5–15. Thus small errors in the description of stress relaxation are amplified and lead to significant variation in the estimated creep damage.

Despite these considerations, Code Case N-47 has been widely used in the design of high temperature plant. Numerous alternative life-assessment methods have been proposed [6, 7, 8] but until recently there has been little enthusiasm for change. Recent developments, however, have placed N-47 under pressure from opposing points of view. On the one hand, designers are finding increasing difficulty in obtaining the required endurance under certain operating conditions, particularly at lower strain ranges. This has led to a debate on how, in particular, the elastic design procedure should be carried out.* In seeking alleviation from the restrictions of N-47, designers have inevitably questioned the applicability of the relatively severe factors by which the endurance curve is reduced. On the other hand, there has been an increasing concern that certain materials will perform less well than anticipated as a result of long-term ageing or exposure to hostile environments such as carburising gases or irradiation.

5.3. CREEP–FATIGUE BEHAVIOUR

5.3.1. Types of Test

The behaviour of various materials under high temperature cyclic deformation has been widely reviewed elsewhere [6–11]. Any design code which is intended to assess the long-term behaviour of materials under service conditions should be based on an appreciation of the physical processes which lead to failure. Therefore it is worthwhile summarising the various types of laboratory test which have been carried out and the interpretations placed on the results.

5.3.1.1. Rapid Continuous Cycling

Cycling between strain limits at rates sufficiently fast to suppress time-dependent effects results in fatigue failures similar to those observed at ambient temperature. In general, increasing the temperature leads to a slight reduction in endurance which is interpreted as being due to increased plastic strain range, $\Delta\varepsilon_p$, for a given total strain range, $\Delta\varepsilon_t$. It should be noted that this is not universally the case. For example, type 316 stainless steel exhibits very complex cyclic behaviour, hardening at some temperatures but not at others (see Chapters 2 and 4).

* Recently published Codes RCC-MR and a Japanese Code have concentrated on modifying the stress analysis procedures while retaining essentially the same failure criteria (see Chapter 7).

5.3.1.2. Slow Continuous Cycling

As the strain rate is reduced, time-dependent deformation can occur, increasing still further the contribution of plastic strain. Provided that the loading pattern remains symmetrical, the reductions of endurance are small (less than 10% of life at ambient temperature) and, most importantly, the failure mode remains unaltered.

The increased time spent at high temperatures can result in structural changes, e.g. precipitation of a second phase, or oxidation of the surface. In the latter case, subsequent cycling can cause cracking of the surface layer, thereby enhancing the initiation stage of fatigue failure.

5.3.1.3. Asymmetric Loading

Asymmetric loading can be achieved either by imposing different strain rates in the tensile and compressive parts of the cycle or by introducing one or more periods at constant strain (see Chapter 3). In the latter case, elastic strain is converted to plastic strain at a variable decreasing rate by stress relaxation. The plastic strain rate is given by

$$\dot{\varepsilon}_p = -\frac{1}{E}\frac{d\sigma}{dt} \tag{2}$$

which is controlled by processes similar to those encountered in high temperature creep. Deformation histories of this kind have caused the greatest reduction in endurance compared with continuous cycling. The most striking reductions in endurance are observed when the slow strain rate deformation occurs under the influence of a tensile stress.

It is this last and most damaging type of deformation which must be assessed to determine the safe operating limits of materials in service.

5.3.2. Observation of Damage

5.3.2.1. Qualitative Observations

In general, symmetrical cyclic loading leads to a fatigue failure characterised by a transgranular crack, the faces of which are marked with striations. Each striation records the extent of crack advance by shearing off each cycle. Even when the strain rates are reduced and corresponding slight reductions of endurance are recorded, the failure mechanism remains unchanged.

Upon increasing the contribution of tensile strain at 'slow' rates a change in the fracture mode is observed. Increasing areas of intergranular fracture appear on the crack faces until, in the limit, it becomes impossible to find striations at all [12, 13]. This change is accompanied by the separation of a fraction of the grain boundaries throughout the material, either by the production of discrete cavities or by wedge cracks [13]. These types of defect are usually associated with creep failures.

A few materials, e.g. $2\frac{1}{4}$Cr–1Mo ferritic steel, also exhibit reductions in endurance when dwell periods under compression are imposed [14]. However in this case, these reductions did not result from a change in failure mechanism, which was transgranular fatigue crack growth, but by accelerated crack initiation from stress concentrations at the root of cracks in the surface scale.

5.3.2.2. The Origins of Intergranular Damage

The observation of intergranular damage similar to that generated by monotonic creep prompts the question of how such damage can accumulate in material which does not itself accumulate net strain. Unless compressive strains cause cavities to grow, which is contrary to any physical model for growth, the growth of cavities must arise as a consequence of the non-uniform deformation rates encountered around the hysteresis loop. Variations in cavity growth rates have been reported in a number of materials and have been placed on a firm theoretical footing by Beeré [15]. The effect of the changes in the rate-controlling step for cavity growth is manifested in the change in monotonic creep ductility and strain rate. Priest and Ellison [16] pointed out a sudden change from creep-ductile to creep-brittle behaviour with decreasing stress (and hence strain rate) of 1Cr–Mo ferritic steel [17, 18]. Many austenitic steels show a more gradual change from ductile to brittle behaviour, as shown in Fig. 2. The

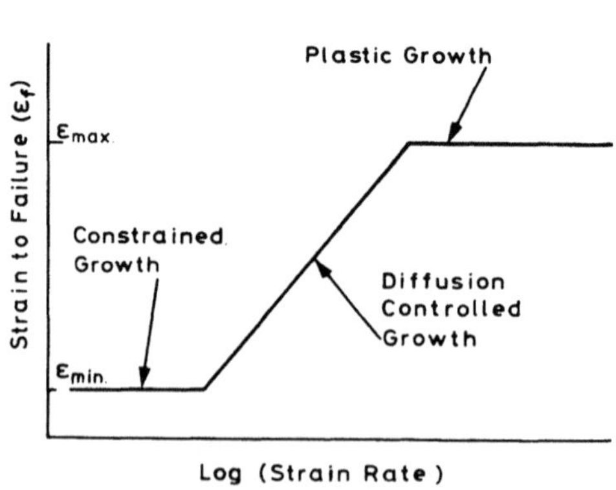

FIG. 2. The variation of creep ductility with strain rate due to changes in processes controlling failure.

three regimes correspond to three different rate-determining mechanisms of cavity growth. In the low-ductility regime, the growth rate is determined by creep rate, hence the constant value for the strain to failure.

To accumulate creep damage it is necessary that parts of the cycle are conducted at rates either side of the transition strain rate, and that more tensile strain should occur at low strain rates than compressive strain at the same rate. In typical laboratory test cycles, imposed strain rates are of the order 10^{-3} to $10^{-4}\,s^{-1}$, i.e. faster than the transition strain rates, which in the case of stainless steels are about $10^{-6}\,s^{-1}$. During stress relaxation, the strain rate given in eqn (2) rapidly declines from the imposed ramping rate to rates much less than the transitional value. It is these small fractions of the total cyclic strain range which give rise to the accumulation of creep damage. Thus not all strain contributes equally to the formation and growth of cavities, but the growth is determined by the instantaneous strain rate. In practical applications, variations in cavity growth rates arise from differentially applied loading rates, from stress relaxation and from non-isothermal deformation in which tensile strain takes place at high temperatures but the strain is reversed at low temperatures (see Chapter 3).

An important extrapolation of this interpretation of creep damage production and its role in creep–fatigue is that only materials which fail by intergranular cavitation will exhibit a strong creep–fatigue effect. Materials such as normalised and tempered 9Cr–1Mo which fail by mechanical instability and consequently show no variation of ductility with strain rate, do not suffer reductions of cyclic endurance due to creep–fatigue [10].

5.3.3. Quantitative Measurement of Damage

5.3.3.1. Under Tensile Stress

The observation of intergranular defects arising from slow plastic straining means that creep damage has a physical meaning which can be related to microstructural changes. The original definition only regarded damage as a state variable. The present author [12] and Levaillant and Pineau [19] have measured intergranular creep damage, which is defined as the length of a grain boundary separated per unit area of section examined. (For comparisons between different casts the measurements should also be normalised with respect to grain size.) It has been found from numerous tests, some interrupted before failure, that the observed damage can be correlated with accumulated strain normalised with respect to the strain-rate-sensitive ductility.

From a limited number of tests performed at low strain ranges and with long tensile hold times, a condition of failure was identified which corresponded well with a ductility exhaustion criterion of failure [17, 20, 21]:

$$\int_0^{t_h} \frac{\dot{\varepsilon}}{\varepsilon_f(\dot{\varepsilon})} dt = D \leq 1 \tag{3}$$

In some cases, especially with saw-tooth loading, large amounts of intergranular damage can accrue without the specimen parting. However, the above definition reflects the fact that when internal damage reaches a critical density the material must be considered to have failed because it is unfit for further service [22]. For example, a fatigue crack in a component made of such heavily damaged material would propagate at a catastrophically fast rate. This definition is analogous to the concept that fatigue failure of a small, smooth specimen corresponds to initiation of an unacceptable defect in a large component.

5.3.3.2. Under Compressive Stress

Models of creep failure at low ductilities rely upon the growth of intergranular cavities under the action of tensile stresses. It is attractive for the purposes of assessing engineering components to assume that compressive loads do not contribute to cavity growth (the existing code assumes that all periods of creep are equally damaging). Beeré and Roberts [23] have suggested that compressive holds will reverse the damaging processes which occur in the tensile hold period. Test cycles comprising both tensile and compressive holds are slow and very time-consuming if continued to failure and, therefore, the effect of such cycles has been inferred from interrupted tests.

Strain control tests have been carried out on the same batch of steel with hold periods at both the peak tensile and compressive strain. The combinations used were 30 min/2 min, 30 min/5 min and 15 min/5 min in tension and compression respectively. These combinations were applied for 2000–3000 cycles. The specimens were sectioned longitudinally, and the damage measured [24]. The damage was calculated by evaluating the integral in eqn (3) for both the compressive and tensile dwell periods, and the results are summarised in the Table 1 [24]. The term *b* will be defined in Section 5.7.1.3.

It can be seen that if tensile and compressive creep strains are regarded as equally damaging, as required by the existing code, the predicted endurances are of the order 500 to 650 cycles, whereas the specimens were subjected to between 2000 and 3000 cycles. The endurance based on the

TABLE 1

THE CALCULATED DAMAGE, D, FOR THE RESPECTIVE TENSILE AND COMPRESSIVE HOLD PERIODS

Tension			*Compression*			$N=\frac{1}{D_t+D_c}$	$N=\frac{1}{D_t-D_c}$
Time (min)	*b*	*D/cycle*	*Time (min)*	*b*	*D/cycle*		
15	3·8	$9{\cdot}8\times10^{-4}$	5	3·8	$8{\cdot}2\times10^{-4}$	555	6 250
30	2·9	$9{\cdot}10\times10^{-4}$	2	2·9	6×10^{-4}	667	3 333
30	3·7	$1{\cdot}11\times10^{-3}$	5	4·2	$8{\cdot}8\times10^{-4}$	502	4 347

difference between the forward and reversed damage lies between 3000 cycles and 6250 cycles. When the observed amount of intergranular damage is plotted as a function of the net damage then good agreement is obtained with the data previously obtained for tensile only dwell periods, as shown in Fig. 3. Hence uniaxial compressive creep strains may be regarded as not contributing to creep damage.

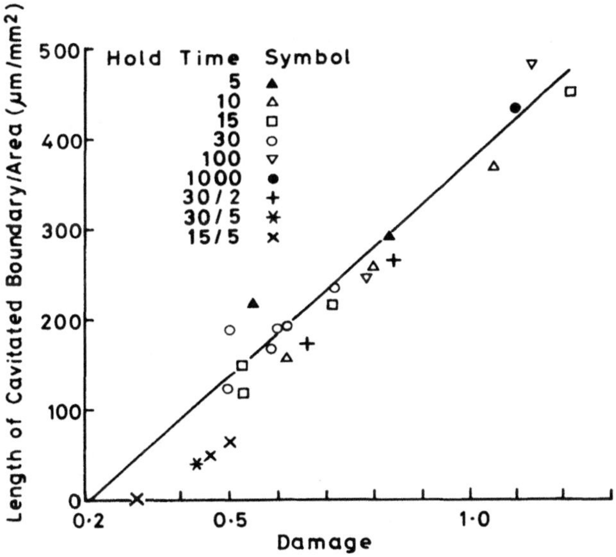

FIG. 3. A plot of measured intergranular damage versus calculated net damage for different combinations of hold time [24].

5.4. CREEP–FATIGUE FAILURE MAPS

The interpretation of time-dependent effects on fatigue endurance described here represents a broad concensus on the interpretation of experimental observations. However, this accepted interpretation has only been reached after prolonged debate which has arisen because, at varying times, the body of experimental evidence has been in conflict. Apparently inconsistent results have been obtained because of mechanistic differences between classes of materials, and because of variations in the detailed properties of casts within an alloy specification. Perhaps most noticeably, AISI type 316 stainless steel, which has been extensively investigated in many countries, exhibits wide variability in properties, in particular the strain to rupture and the mechanisms causing creep failure. This can result in fundamentally separate failure mechanisms when apparently similar test conditions are imposed. As already stated, such changes in mechanism are often accompanied by widely different endurances.

Similar differences have been observed in monotonic deformation and failure mechanisms. Ashby [3] used a mapping technique to bound the conditions which give rise to each process. A similar technique has been developed to bound creep–fatigue behaviour and to explain some of the variations in observation.

Edmunds and White [25] constructed failure curves for $2\frac{1}{4}$Cr–1Mo based on separate creep and fatigue failure mechanisms. They chose to connect the two very distinct curves with a smooth transitional curve to estimate the endurance they anticipated in the intermediate regime. The present author has extended this presentation by considering the initiation as well as failure criteria to produce a more detailed map [12], which is now considered.

5.4.1. Interaction Model

5.4.1.1. Fatigue Criteria

For the purpose of illustration, the behaviour of AISI 316 stainless steel is considered, but the principles can be equally applied to any material. Figure 4 shows a two-dimensional section at a given hold time through a three-dimensional construction (time is the third axis). First consider fatigue failure: most materials show limited variability in continuous cycling endurance with limited scatter, and a lower-bound fatigue failure line can be drawn of the form

$$N_f = C \,.\, \Delta\varepsilon_t \qquad (4)$$

(where C and α are constants) and is given by the line abc in Fig. 4.

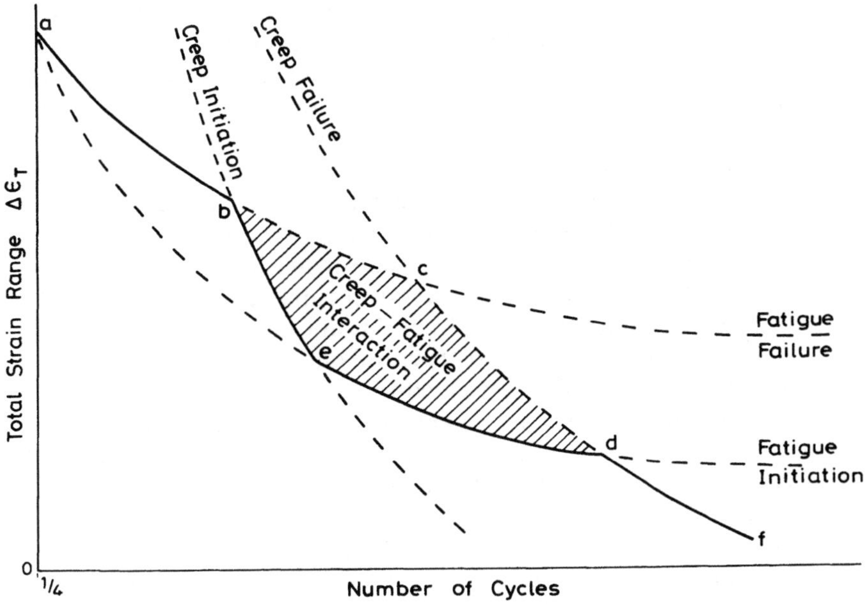

FIG. 4. A schematic creep–fatigue failure mechanism map [12].

Fatigue failure is usually considered to occur via three stages: stage I, an initiation stage controlled by dislocation processes which leads to a surface defect one or two grains deep; stage II, a growth process by bulk deformation which leads to a striated fracture surface; and finally stage III, failure by tearing off a small remaining ligament. Over the strain ranges of interest, the dominant process in determining endurance changes from stage II crack growth at high strain ranges to stage I initiation at lower strain ranges. Maiya [26] has determined the relative contributions of these two processes by counting the number of striations on fracture surfaces. Assuming that each striation corresponds to one cycle, the number of cycles spent propagating the crack (stage II) can be determined, and the number of cycles in stage I can be deduced by difference. Using this technique a relationship between strain range and the number of cycles to initiate stage II growth similar to eqn (4) has been deduced, although in this case the plastic strain range is generally found to afford a better correlation. The line aed in Fig. 4 describes fatigue-crack initiation.

5.4.1.2. Creep Criteria

Creep failure can be defined either on the basis of a time fraction rule or, as is preferred here, a strain fraction rule. The creep failure line, cd in Fig. 4, can be computed from a relatively few short-term tests using constitutive equations to calculate the accumulated strain during the dwell period [17]. The general shape of the curve arises from the stress–strain properties of materials; after yield, the stress rises only slowly with increasing strain. Hence, as the total strain range increases, the stress hardly changes and the accumulated strain per cycle changes by a correspondingly small amount. At strains near to yield the stress changes much more rapidly, with corresponding large increases in endurance for relatively small changes in strain range. These aspects are discussed further in Section 5.7.

The condition for the nucleation of creep damage is much more difficult to define. Some interpretations of continuum damage as formulated by Kachanov [4] suggest that microstructural damage is nucleated late in the creep life of a material and is associated with the onset of tertiary creep [27]. This cannot be universally true because, for example, in the case of type 316 stainless steel, identifiable intergranular defects have been observed at relatively low fractions of life without any related reduction of peak stresses or accelerated stress relaxation. The present author has extrapolated measured levels of microstructural damage and deduced that nucleation occurs at approximately 20% of the failure criterion (see Fig. 3). This definition of initiation has been used here to construct a line parallel to the creep-failure line at one fifth the number of cycles, i.e. the line be.

5.4.1.3. Discussion

In Fig. 4, two regimes are clearly identifiable: at high strain ranges fatigue failure occurs before creep damage can nucleate and the endurance is given by the continuous cycling curve ab. Similarly, at low strain ranges, creep failure defined by df can occur before a fatigue crack nucleates. It is important to remember that this failure does not necessarily correspond to the separation of a test specimen into two halves, but rather must be considered as being a level of damage which would render the material or component unfit for further duty [22].

Thus it can be seen that the question of whether a synergism exists between processes producing creep damage and fatigue damage is reduced in importance, because such an interaction can only occur over a narrow range of conditions. If there is no interaction between the two processes, endurance is given by the lines abc and cdf. The lowest endurance an interaction can cause is defined by lines abe and edf.

A failure map of this type is a useful aid in interpreting experimental observations. As hold times are increased, the creep-failure and initiation lines move to lower numbers of cycles. If the tests are conducted at high strains so that failure is always controlled by fatigue (line ab), it will appear that the effect of hold times is small and of little consequence. If, however, the strain range is lower and the increase in hold time causes a change in failure mechanism from fatigue to creep, it can be seen that a dramatic reduction in endurance will be recorded. Without the aid of such a diagram it would be easy for different workers to conclude that they were observing different and unrelated behaviour.

The effects of other service conditions can also be interpreted with the aid of the failure mechanisms map. Irradiation with thermal neutrons reduces the creep ductility of some steels, which again displaces the creep failure line to lower endurances but, in general, has little effect on the fatigue endurance.

5.4.2. Verification of Model

Support for the use of creep–fatigue failure mechanisms maps and the general form they should take is afforded by the measured endurance of the ferritic steel $\frac{1}{2}$Cr–Mo–V at 550°C [6, 28]. Two types of cycles have been studied: the normal laboratory cycle with a dwell period at peak tensile strain, and a cycle in which the dwell is introduced at zero strain in the tensile part of the cycle (designated 'Service cycle II' by the original authors) [28]. The results for cycles with 16 h hold periods are summarised in Fig. 5. The laboratory cycle at a strain range $\sim 1\%$ resulted in failure after 300 cycles, which corresponds to a test lasting nearly 5000 h. It would be tempting to regard this as a long-term test and conclude that life-reducing effects have saturated. Such an assumption would support the use of the application of a life-reduction factor to the continuous cycling curve. However, as the data obtained from tests at lower strain ranges show, such an assumption would lead to severely non-conservative life expectancies. The observed endurances at low strain ranges correlate well with a creep-based failure criterion.

Clearly, the construction of a failure mechanism map, even from a tentative database, is a valuable adjunct to ensure that laboratory tests will lead to the correct failure mechanism and thereby provide relevant endurance data for the life assessment of components in service.

5.4.3. Alternative Approaches

An approach based on failure mechanism maps has also been proposed by Sidey and co-workers [29, 30]. Initially, the effects of saw-tooth continuous cycles was represented by mapping fields of common failure mode as strain rate versus plastic strain range. The resulting failure map in

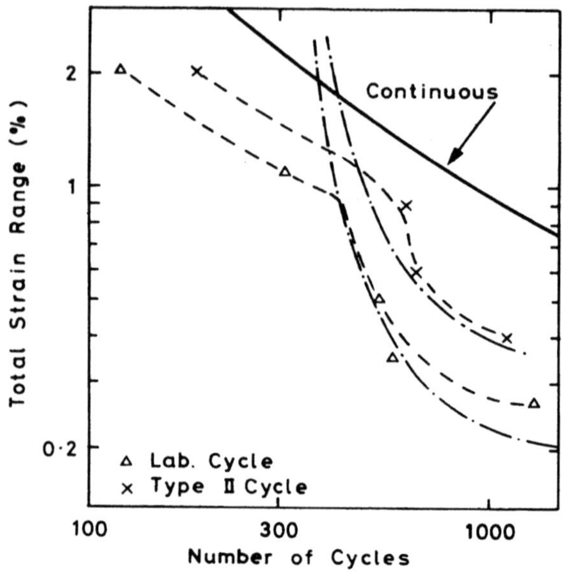

FIG. 5. The creep–fatigue endurance of a ferritic steel at 550°C compared with the calculated endurance based on ductility exhaustion [24].

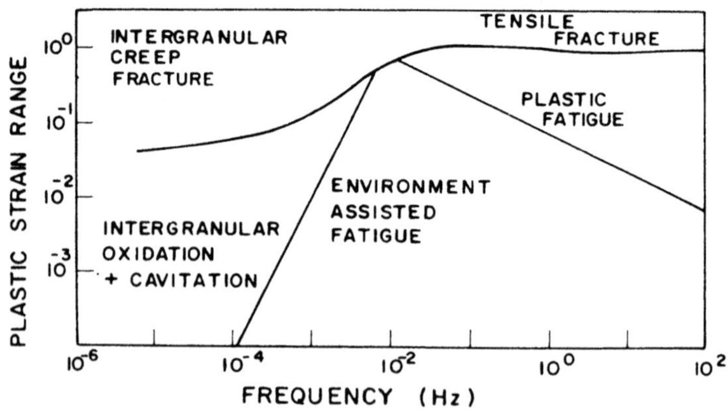

FIG. 6. A failure mechanism map illustrating the role of environmental effects under varying cyclic frequency. (Reproduced from Ref. 30 with permission of Pergamon Journals Ltd.)

Fig. 6 is limited to the saw-tooth cycle, and relies almost entirely upon experimental observation rather than being mechanistically based. It is interesting to notice that, for the low-alloy steel A286, the role of surface oxidation is identified, although again the exact nature of the environmental interaction is not modelled.

In an attempt to identify the role of strain rate asymmetry, this approach has been extended using Manson's 'Strain Range Partitioning' description of creep fatigue [31]. The waveform is specified by the ratio of the tensile and compressive strain rates which is plotted against the total cycle

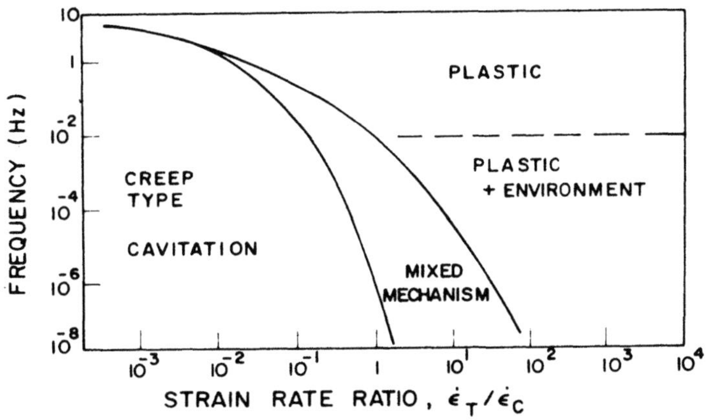

FIG. 7. A failure mechanism map showing the effect of unbalanced cyclic strain rates. (Reproduced from Ref. 30 with permission of Pergamon Journals Ltd.)

frequency. The maps of this type (Fig. 7) do not explicitly identify the effect of total strain range in determining the failure mode and must be considered to be a single slice through a three-dimensional diagram. The failure fields are again determined from metallographic examination of the failures.

Collins *et al.* [30] pointed out that experience shows that the effect of a rapid ramp plus hold period at constant strain does not necessarily produce the same type or amount of damage if the same strain is applied at a uniformly slow strain rate. However, this difference cannot be presented in either of these maps, since neither identifies nor qualifies the processes leading to creep cavitation and its contribution to failure. Further developments on frequency effects are discussed in Chapter 4.

5.5. FAILURE ASSESSMENT DIAGRAMS

Bestwick and Clayton [32] have converted the concepts of the failure mechanisms map shown in Fig. 4 into a failure assessment diagram (Fig. 8(a)). For any given set of conditions, failure occurs when the damage accumulated by either creep, D_c, or fatigue, D_f, exceed unity. This corresponds to the conditions given by the boundary ac′e. The lower safe bound to the area of the mechanism map which corresponds to a creep–fatigue interaction is defined by the boundary abcde. In this

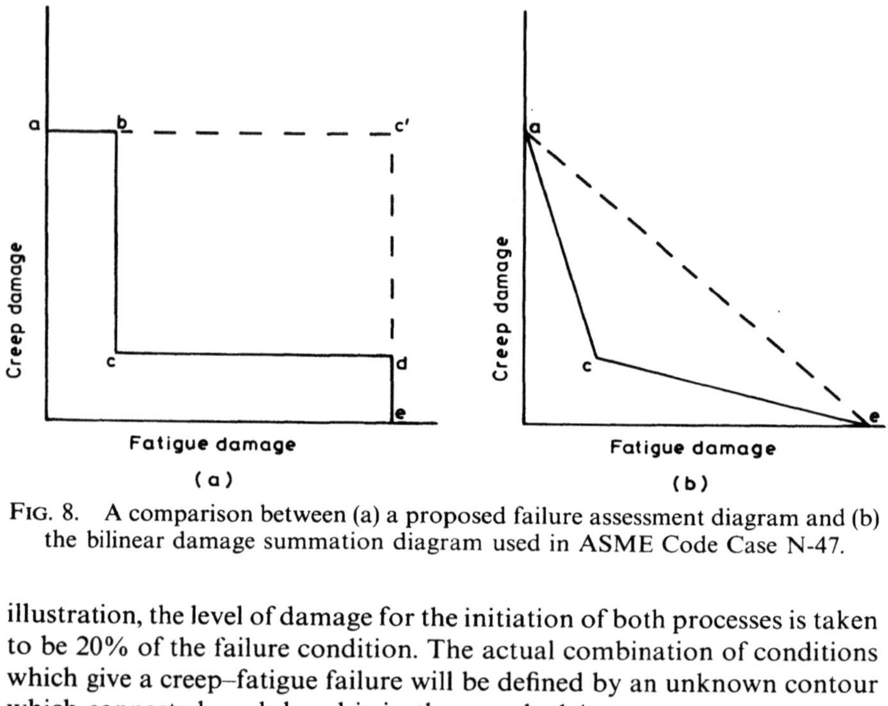

FIG. 8. A comparison between (a) a proposed failure assessment diagram and (b) the bilinear damage summation diagram used in ASME Code Case N-47.

illustration, the level of damage for the initiation of both processes is taken to be 20% of the failure condition. The actual combination of conditions which give a creep–fatigue failure will be defined by an unknown contour which connects b and d and is in the area bcdc′.

It is interesting to note that, in the original Code Case N-47, two failure diagrams were proposed, one a linear summation and the other a bilinear summation shown in Fig. 8(b). The path ace is remarkably similar to the boundary abcde, although at the time of its construction the underlying mechanisms which determine this shape were not appreciated.

As the total strain range is reduced and fatigue initiation takes a greater fraction of the endurance, the line bc will move to higher fractions of damage and a creep failure becomes more probable. Similarly, environmental effects such as irradiation by thermal neutrons can enhance creep initiation condition and the line cd will move towards zero.

5.6. REVISION OF CODES

Since the high temperature design Code Case N-47 was first published it has been applied to many problems. The cyclic strain range is calculated by an elastic analysis procedure which contains numerous conservatisms. The allowable number of cycles is obtained by reference to a design curve of number of cycles versus strain range for slow-cycling conditions, so that it embodies the effects of creep damage. This design curve, known as T-1430, is further reduced with respect to the estimated number of cycles to failure. In certain circumstances these conservatisms can combine to give an assessment which predicts failure of the component and necessitates either a more detailed non-linear analysis or a re-design. Two new design codes, one developed in France and one in Japan (see Chapter 7) have tried to eliminate a number of conservatisms in the stress analysis method, but they have largely retained the design curve against which failure is assessed.

Many research programmes have been conducted to investigate the creep–fatigue endurance of a wide range of alloys in recent years. As a result of these investigations, a number of different methods for predicting creep–fatigue endurance have been proposed. Miller *et al.* [7] have reviewed many of these, and although no one method stands out as being better able to predict experimental results for any individual cast of material, some are more amenable to being extended to define lower-bound behaviour for a whole range of materials within an alloy specification.

The observations and interpretations outlined in the previous sections underwrite the application of a strain-based method for assessing creep damage, and how it should be combined with fatigue damage to yield an assessment curve. In addition to re-examining the reasoning behind the construction of the assessment curve for use in conjunction with an elastic stress analysis method, two aspects of the currently accepted curve need scrutinising:

(1) Simplified analysis of the service conditions often indicates that a period at constant strain does not always occur at the extremes of the cyclic range. An approximate description of the strain history can be obtained from an elasto-plastic analysis of the component under the prevailing

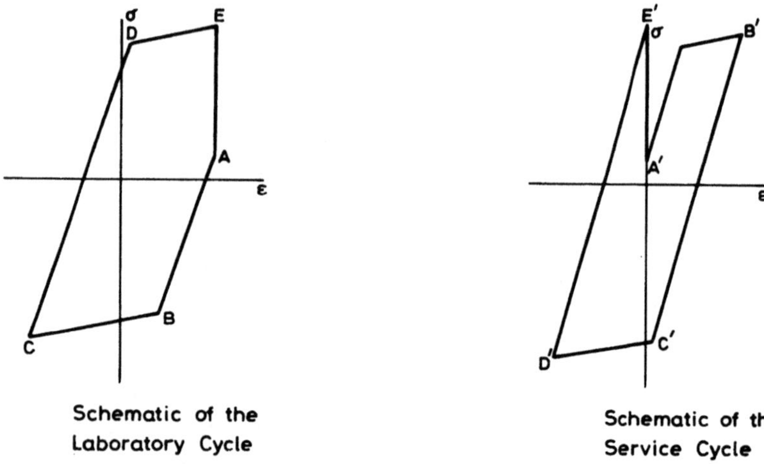

FIG. 9. Schematic hysteresis loops for the laboratory cycle and the type II service cycle.

operating conditions. For example, the thermomechanical loading of the inside surface of thick sectioned turbine casings leads to a dwell period at zero strain, but under a residual tensile stress. This cycle has been referred to previously as the 'Type II Service Cycle' [28] and is illustrated together with the laboratory cycle in Fig. 9. It should therefore be possible to determine endurance curves for different cycles both experimentally and by calculation, and to take advantage of the consequent improved endurance shown in Fig. 5.

(2) The presently accepted design curves are based on the assumption that the creep–fatigue effect saturates after a relatively short period, which is less than 24 h [33]. Therefore, a single design curve is regarded as being adequate to afford a lower bound to all circumstances. Analysis on a ductility exhaustion approach implies, however, that creep damage continues to accumulate as long as the remaining stresses continue to relax. Two cases in the operation of high temperature plant illustrate extremes of behaviour which must be accounted for: (i) where equipment 'two-shifts', i.e. when a cycle is applied at a regular frequency, typically after 8 or 16 h dwell; (ii) where a fixed operating life is interrupted infrequently, a different approach being necessary to determine the allowable number of interruptions for a given level of loadings.

The remainder of this chapter outlines a method of generating creep–fatigue endurance curves which is based on physical arguments.

5.7. PREDICTION OF ENDURANCE

A procedure whereby endurance may be calculated for a set of conditions of temperature, strain range and frequency is divided into a number of logical steps (some of which are similar to those used to construct T-1430 in Code Case N-47). Wherever possible, experimentally derived data should be used in such a way as to produce lower-bound endurances for the alloy specification under consideration. When this is not possible (for example in the case of very long endurances when even representative cyclic stress–strain data would take a prohibitively long time to acquire) parameters should be derived in such a way as to give lower-bound performance.

Design curves can be constructed for different temperatures as follows:

(1) For the temperature under consideration the stress at the beginning of the period at constant strain must be determined. This will depend on material, cycle type and cycle duration (see Chapter 2).
(2) The lower-bound ductility $\varepsilon_f(\dot{\varepsilon})$ must be described as a function of temperature and strain rate [17, 18] (see Fig. 2).
(3) The creep strain (ε_r) is calculated as a function of strain rate, using an appropriate function to describe the stress relaxation behaviour. Creep damage is given by strain divided by ductility at corresponding strain rates.
(4) The number of cycles at a particular temperature to cause failure by creep is calculated from the simple relationship given by eqn (3).
(5) The two damage processes, creep and fatigue, both require time to initiate microstructural damage features, and it is not possible for a synergistic interaction to occur unless both types of defect are present in the material [34]. Thus, over a wide range of conditions, failure is dominated by either fatigue or creep, and the creep–fatigue interaction accounts for only a very small range of conditions. The way in which this interaction can be assessed is described in detail by Bestwick and Clayton [32].
(6) Finally, an endurance curve for the cycle and endurance under consideration is prepared for use in conjunction with an appropriate structural analysis procedure.

5.7.1. Determination of Parameters

The ways in which these parameters may be obtained are outlined briefly.

5.7.1.1. *Stress–Strain Relationships*

In the case of two-shift, 16 h cycles, it is possible to obtain cyclic stress–strain data for all materials, whether they exhibit hardening or softening with increasing cycles, and it is recommended that data for the relevant cycle type should be obtained wherever possible (see for example, Chapter 2). However, for cycles of significantly longer duration, it is impractical to obtain such data experimentally (cycles may be up to 3000 h long). Under these circumstances, the peak stress must be estimated from short-term cyclic or monotonic data. In materials which cyclically soften it may be adequate to use the peak stress determined from short term tests, since this will produce a pessimistic lower-bound endurance curve.

Reference to Figs 9(a) and (b) shows how a commonsense argument can be applied to the problem of an austenitic steel which exhibits cyclic hardening. After a long hold period, the final stress will be very low, and much of the hardening will have been removed by recovery. Therefore, the first portion of the cycle can be described by monotonic data starting from zero stress (A–B–C in the case of the laboratory cycles and A′–B′ in the case of the service cycle). A worst case is obtained by assuming isotropic hardening; that is, subsequent yield occurs at the previous peak stress (see Fig. 23(a) of Chapter 2). Hence, on reverse loading, yield occurs at $D = -C$ or $C' = -B'$ in the two cycles illustrated, followed by plastic deformation to the opposite limit of the cycle. Point E in the laboratory cycle marks the start of another relaxation event. In the case of the service cycle, the strain returns to zero, the material obeying an isotropic hardening law, to restart relaxation. After relaxation, virgin properties are considered to be restored.

If later calculations show that the stress 'A' is not small then recovery of virgin properties cannot be assumed, and experimental data derived from the cyclic frequency closest to the case under consideration should be used to define the peak stress E.

5.7.1.2. *Creep Failure*

Ideally, the creep damage, D_c, should be evaluated (using forward creep data) from a detailed knowledge of the ductility, given by $\dot{\varepsilon}_{min}$, t_f, and the instantaneous deformation rate during the cycles (see for example eqn (2)). However, in most cases, detailed long-term creep data are not available and it is only possible to deduce an average deformation rate from ε_f/t_f [35]. A plot of ε_f versus ε_f/t_f usually exhibits decreasing strain rate until a lower ductility plateau is reached. Creep rates during stress relaxation often straddle the transitional behaviour and the low constant ductility regime shown schematically in Fig. 2. Increments of damage are calculated by dividing increments of relaxation strain by the ductility at the appropriate strain rate. When the damage totals one, failure by creep is considered to have occurred.

In some alloys, cast-to-cast variations in behaviour make the identification of a ductility transition curve difficult. In such circumstances the behaviour can be bounded by a 'step-function' which defines a minimum ductility below a critical strain rate ($\dot{\varepsilon}_{crit}$) [16]. The evidence presented in Section 5.3.3.2 strongly suggests that only creep strains under tensile stresses should be regarded as damaging and contributing to failure. Under multiaxial loading it is not clear how a combination of compressive and tensile loads should be treated, although a simple calculation of equivalent stress or strain based on a von Mises type criterion will very probably prove unduly pessimistic.

5.7.1.3. Stress Relaxation Behaviour

Numerous relaxation models of varying degrees of complexity have been derived. The more sophisticated models require substantially more data to evaluate the various constants than is often available from creep–fatigue tests; in many cases the published data consists only of the peak stress, σ_0, the relaxed stress σ, and the duration of the relaxation event, t. Simple relationships such as the Gittus equation [36]

$$\ln(\sigma_0/\sigma) = At^m$$

or the Feltham equation [37]

$$\sigma_0 - \sigma = b\ln(at + 1)$$

where A, m, a and b are material constants, have been fitted to the limited stress relaxation data available, and have been found to give an adequate description of the behaviour of several materials (e.g. type 304 and 316 stainless steels and $\frac{1}{2}$Cr–Mo–V ferritic steels).

The strain rate throughout the relaxation (eqn (2)) must be determined, and if the assumptions concerning the mechanical behaviour of long-duration cycles are to be met, the final relaxed stress must be small. Creep damage for the predicted relaxation history is calculated as follows:

(1) By subdividing the relaxation strain into small increments for which the mean strain rate is calculated. Damage is determined by dividing the fraction of strain by the measured ductility at the same strain rate.

(2) In the absence of detailed ductility behaviour, the relaxed strain is

divided into two components by the conditions of stress and time which correspond to the 'critical strain rate'. Strain generated at rates greater than $\dot{\varepsilon}_{crit}$ gives rise to damage proportional to $1/\varepsilon_{f\,max}$, while the remainder at the lower strain produces damage proportional to $1/\varepsilon_{f\,min}$.

5.7.2. Endurance Curve

From the calculations described above, the endurance curves for cycles containing 16 h hold periods are readily calculated. For the case of constant endurance (i.e. lifetime) the product of the number of cycles to failure and the time of each cycle is fixed:

$$N_f \times t_h = \text{endurance} \qquad (5)$$

Consequently, eqn (3) cannot be integrated over a specific time, and the cycles to failure at each strain range must be interpolated graphically, as shown in Fig. 10. The constant endurance in eqn (5) defines a straight line. A family of lines or curves at different strain ranges is determined by plotting the number of cycles to failure over a span of relevant cycle times. The intersections shown in Fig. 10 determine the failure condition for constant endurance at each strain range, and allow a failure curve, N_f vs. $\Delta\varepsilon_t$, to be plotted in the usual way.

5.7.3. Creep–Fatigue Interaction

At high strain ranges and short cycle times, failure will be dominated by fatigue and will be defined by the continuous cycling curve. Conversely, long cycle times and low strain ranges favour creep-dominated failures. There has always been concern that the two processes can interact synergistically to give very low endurances. Historically, the two damage parameters have been summed by a linear or bilinear law. More recently, Bestwick and Clayton [32] proposed the failure diagram illustrated in Fig. 8(a). To a first approximation, initiation of the two damage mechanisms is considered to occur at 20% of the creep and fatigue endurances, although this value may vary with material. There are insufficient data to determine which method of damage summation, linear, bilinear or interaction, best describes material behaviour, although the last method does have the virtue of being based on an appreciation of the physical processes involved.

5.7.4. Limiting Endurance

Before analysing the damaging mechanisms in detail, it is worthwhile considering the properties which bound endurance behaviour. As a starting

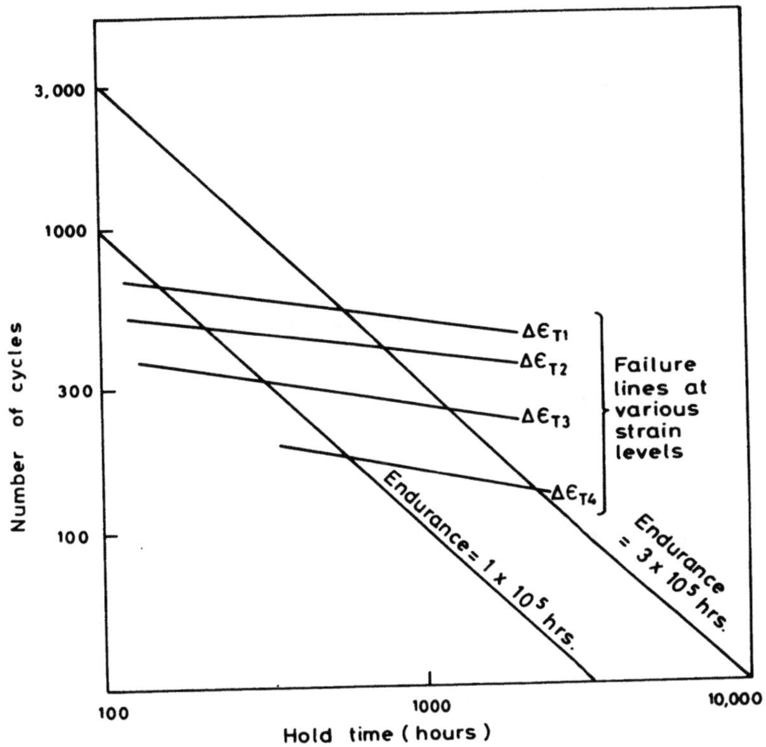

FIG. 10. The interpolation of creep-dominated failure conditions at constant total endurance.

point, the continuous cycling endurance can be expressed as a simple power relationship between the total strain range and the cycles to failure, the line AB in Fig. 11.

Depending upon the exact nature of the loading and the position of the hold period in the cycle, there is a limiting strain range below which relaxation strain cannot be regenerated. For example, in the case of the laboratory cycle with a hold period at peak tensile strain then, irrespective of the duration of the hold periods or number of cycles, if the total strain range is less than or equal to the elastic limit then creep strain is limited to the yield strain ε_y. This will be less than creep ductility, even in the most brittle materials, and hence sets a conservative lower limit below which creep failure cannot occur (line CD in Fig. 11). Similarly, in the case of a cycle with a hold period at zero strain, no creep strain will arise until $\Delta\varepsilon_t$ is greater than $2\varepsilon_y$. Finally, at strain ranges which cause repeated yielding, greater than $2\varepsilon_y$, the peak tensile stress is relatively insensitive to total strain range. Consequently, in a sufficiently long dwell period, the relaxation strain per cycle is approximately ε_y; thus the number of cycles to failure is given by $\varepsilon_f/\varepsilon_y$ and is plotted as line EF in Fig. 11.

The solid line represents a lower bound to endurance and may suffice to

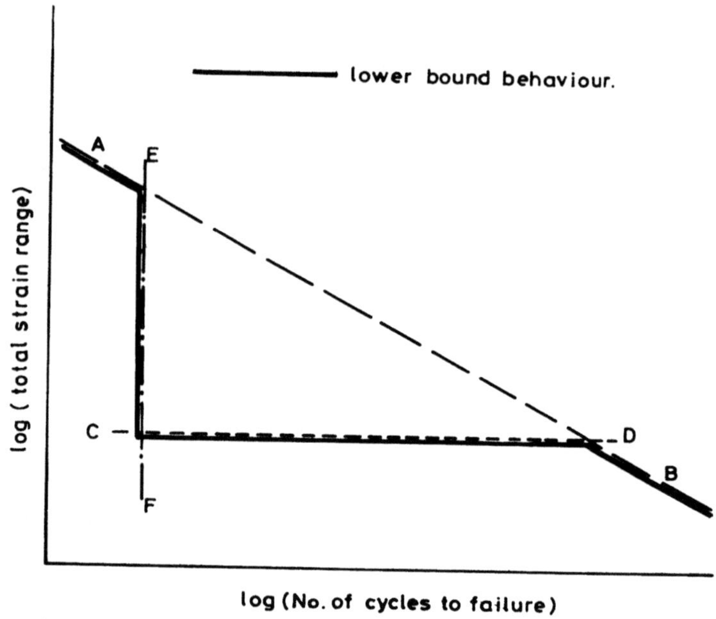

FIG. 11. An approximate lower-bound creep–fatigue endurance curve.

demonstrate the integrity of some components. More importantly, the three distinctly different contributing factors limit performance under different conditions of service. For example, the line CD is determined by cyclic yield properties and is not influenced by creep ductility, although it limits the area of a creep-type failure. On the other hand, service conditions which reduce minimum creep ductility by introducing intergranular cavities [38], for example during irradiation or thermal ageing, will not affect the level of the line CD, but they will reduce the number of cycles to failure given by the line EF.

5.8. CONCLUDING COMMENTS

A number of outstanding areas remain which give rise to uncertainty about the long-term creep–fatigue endurance of materials.

5.8.1. Back-stress

Most stress relaxation models, based on relatively short-term creep, assume that stresses eventually relax to zero. However numerous models of forward creep suggest the presence of a back-stress opposing the forward motion of dislocations [39]. The back-stress arises from the elastic stress fields of dislocation networks, which can equally well be produced by creep and cyclic deformation [40]. Nothing is known about the long-term stability of such networks: however, they could attenuate the long-term stress relaxation, thereby reducing the accumulated creep strain during each relaxation event.

In addition to affecting long-term stress relaxation, back-stresses can influence the cyclic stress–strain properties. They can affect the width of the stress–strain hysteresis loop and determine the cyclic yield strength of the material (see Chapter 2). Because of the two equally important ways in which these factors can influence cyclic endurance, the role of back-stresses and their long-term stability represents an important and potentially fruitful area for further work.

5.8.2. Multiaxial Effects

The factors leading to creep failure under multiaxial loading are poorly understood. The time fraction approach to creep damage used in Code Case N-47 uses a multiaxial equivalence based simply on von Mises equivalent stress. Several other equivalent values such as the Tresca equivalent stress are possible. However, low-ductility creep failure by the growth of intergranular cavities requires the action of a tensile stress. Furthermore, the growth of such cavities requires an accommodating tensile strain and therefore it seems reasonable to propose that only tensile strains contribute to creep failure. When all the stresses are tensile, the several methods of calculating equivalent uniaxial values all give approximately the same answer. When one or more stress is compressive the methods give widely different answers. For example, increased ductility has been recorded under torsion compared with the equivalent tensile conditions [41].

Once again, the development of a clear physical model together with the identification of the nature and disposition of structural damage could lead to alleviation of the failure criteria under multiaxial loading.

5.8.3. Thermal Ageing

Finally, the long times for which modern structural steels are expected to remain in service means that, in many cases, the microstructure will evolve by ageing processes. Wareing [42] showed that pre-ageing certain casts of AISI 316 stainless steel resulted in an improvement of the creep–fatigue endurance. More recently, Levaillant *et al.* [43] have shown a saturation of fatigue reduction factors with increasing hold time, even though stress relaxation is still generating creep strain. The interpretation of these observations is that the long exposure of the material at service temperatures results in an improvement in the creep ductility of the material.

The phenomena about which there remains uncertainty all tend to improve the creep–fatigue endurance with respect to the lower bounds calculated by the procedure outlined in the previous section. Further developments aimed at increasing confidence in the calculation of long-term endurances must be based on an appreciation of the underlying physical processes.

REFERENCES

1. Mayer, K.-H. and Tremmel, D., 1979, The thermal fatigue of components in steam power plant, *Proc. DVM Int. Symp. on Low Cycle Fatigue Strength and Elasto-plastic Behaviour of Materials, Stuttgart*, pp. 105–16.
2. Anon., 1977, *Boiler and pressure vessel code section III*, Code Case N-47-12, New York, ASME.
3. Ashby, M. F., 1978, Progress in the development of fracture mechanism maps, *Fracture-ICF4*, University of Waterloo, pp. 1–14.
4. Kachanov, L. M., 1958, Time of the fracture process under creep conditions, *Izv. Akad. SSSR OTN Teckh. Nauk*, **8**, 26.
5. Wood, D. S., Wynn, J., Baldwin, A. B. and O'Riordan, P., 1980, Some creep-fatigue properties of type 316 steel at 625°C, *Fatigue Engng Mat. Struct.*, **3**, 39–57.
6. Batte, A. D., 1983, Creep–fatigue life predictions, *Fatigue at high temperatures*, ed. R. P. Skelton, London, Elsevier Applied Science, pp. 365–401.
7. Miller, D. A., Priest, R. H. and Ellison, E. G., 1984, A review of material response and life prediction techniques under fatigue–creep loading conditions, *High-temp. Mat. Proc.*, **6**, 155–94.

8. Cailletaud, G., Nouailhas, D., Grattier, J., Levaillant, C., Mottot, M., Tortel, J., Escavarage, C., Heliot, J. and Kang, S., 1984, A review of creep–fatigue life prediction methods; identification and extrapolation to long term and low strain ranges, *Nucl. Engng Design*, **83**, 267–78.
9. Wareing, J., Bretherton, I. and Hales, R., 1985, Failure criteria during elevated temperature creep–fatigue cycling of austenitic stainless steel, *5th Int. Seminar on Inelastic Analysis and Life Prediction in High Temperature Environment, Paris*, pp. 31–54.
10. Wareing, J., 1983, Mechanisms of high temperature fatigue and creep–fatigue failure in engineering materials, *Fatigue at High Temperatures*, ed. R. P. Skelton, London, Elsevier Applied Science, pp. 135–86.
11. Hales, R. and Tomkins, R., 1982, Creep–fatigue failure in austenitic stainless steels relevant to structural performance, *Proc. ASME Conf. Pressure Vessels and Piping*, Orlando, 82-PVP-70.
12. Hales, R., 1980, A quantitative metallographic assessment of structural degradation of type 316 stainless steel during creep–fatigue, *Fatigue Engng. Mat. Struct.*, **3**, 339.
13. Pineau, A., 1983, High temperature fatigue behaviour of engineering materials in relation to microstructure, *Fatigue at high temperatures*, ed. R. P. Skelton, London, Elsevier Applied Science, pp. 305–65.
14. Challenger, K. D., Miller, A. K. and Brinkman, C. R., 1981, An explanation for the effects of hold periods on the elevated temperature fatigue behaviour of $2\frac{1}{4}$Cr–1Mo steel, *J. Engng Mat. Technol.*, **103**, 7–14.
15. Beeré, W., 1981, Theoretical treatment of creep cavity growth and nucleation, *Cavities and cracks in creep and fatigue*, ed. J. Gittus, London, Elsevier Applied Science, pp. 1–58.
16. Priest, R. H. and Ellison, E. G., 1981, A combined deformation map–ductility exhaustion approach to creep–fatigue analysis, *Mat. Sci. Engng*, **49**, 7–17.
17. Hales, R., 1983, A method of creep damage summation based on accumulated strain for the assessment of creep–fatigue endurance, *Fatigue Engng Mat. Struct.*, **6**, 121–35.
18. Sikka, V. K., 1978, Elevated temperature ductility of type 304 and 316 steel, Oak Ridge National Lab. Rep. ORNL/TM-6608.
19. Levaillant, C. and Pineau, A., 1980, Assessment of high temperature low cycle fatigue life of austenitic stainless steel by using intergranular damage as a correlating parameter, *Low cycle fatigue and life predictions*, STP 770, Philadelphia, ASTM, pp. 169–93.
20. Miller, D. A., Hamm, C. D. and Philips, J. L., 1982, A mechanistic approach to the prediction of creep dominated failure during simultaneous creep–fatigue, *Mat. Sci. Engng*, **53**, 233–44.
21. Wareing, J., Bretherton, I. and Livesey, V. B., 1986, Life prediction for elevated temperature components subjected to cyclic deformation, *Proc. Inst. Mech. Engrs Conf. Engng Materials and Structures, Sheffield*, pp. 147–64.
22. Bocek, M., Armas, A. and Piel, D., 1983, Life time calculations for L.C.F. loading combined with tensional hold periods: application to Zircaloy-4 and AISI 304, *J. Nucl. Mat.*, **115**, 159–68.
23. Beeré, W. and Roberts, G., 1982, Fatigue failure by creep cavity growth at high temperatures, *Acta Metall.*, **30**, 571–80.

24. Hales, R., 1984, The application of creep–strain as a parameter for estimating creep–fatigue endurances, *Proc. Second Int. Conf. on Creep and Fracture of Engineering Materials and Structures, Part II*, ed. B. Wilshire and D. R. J. Owen, Swansea, Pineridge Press, pp. 1015–27.
25. Edmunds, H. G. and White, D. J., 1966, Observations on the effects of creep relaxation on high strain fatigue, *J. Mech. Engng Sci.*, **8**, 310–21.
26. Maiya, P. S., 1979, Effects of notches on crack initiation in low-cycle fatigue, *Mat. Sci. Engng*, **38**, 289–94.
27. Levy, A. J., 1985, Physically based constitutive equation for creep-damaging solids, *J. Appl. Mech.*, **52**, 615–20.
28. Batte, D., Murphy, M. C. and Stringer, M. B., 1978, High strain fatigue properties of a 0·5Cr–Mo–V steam turbine casing, *Metall. Technol.*, **5**, 405–13.
29. Sidey, D. and Coffin, L. F., 1978, Damage mechanisms at high temperature, G.E. Report 78CRD050.
30. Collins, A. L. W., Sidey, D. and Taplin, D. M. R., 1979, Fatigue maps and wave-shape effects on fatigue, *Can. Metall. Q.*, **18**, 231–8.
31. Manson, S. S., 1973, The challenge to unify treatment of high temperature fatigue—a partisan proposal based on strain range partitioning, *Fatigue at elevated temperature*, STP 520, Philadelphia, ASTM, pp. 744–82.
32. Bestwick, R. D. W. and Clayton, A., 1985, Design methodology for creep–fatigue assessment using creep ductility criteria, *5th Int. Seminar on Inelastic Analysis and Life Predictions in High Temperature Environment, Paris*, pp. 55–60.
33. Anon., ASME Boiler and Pressure Vessel Committee, 1974, *Criteria for design of elevated temperature class 1 components in Section III of the ASME B. and P. V. Code,* ASME, New York.
34. Goodall, I. W., Hales, R. and Walker, D. J., 1980, On constitutive relations and failure criteria of an austenitic steel under cyclic loading at elevated temperature, *Creep in Structures, 3rd IUTAM Symposium, Leicester*, pp. 103–27.
35. Hales, R., 1984, An appraisal creep ductility for use in the assessment of creep–fatigue failure. CEGB Report No. TPRD/B/0554/N84.
36. Gittus, J. H., 1964, Implications of some data on relaxation creep in Nimonic 80A, *Phil. Mag.*, **9**, 749–53.
37. Feltham, P., 1960, Stress relaxation in copper and alpha brasses at low temperatures, *J. Inst. Metals*, **89**, 210.
38. Swindeman, R. W., Farrell, K. and Conway, J. B., 1983, Time-dependent fatigue of type 304 stainless steel containing microvoids in the starting microstructure, *Thermal and environmental effects and fatigue: research/design interface*, ASME Portland, Oregon, pp. 121–37.
39. Bolton, C. J., 1974, The role of internal stresses in high temperature creep, Ph.D. Thesis, Berkeley Nuclear Laboratories.
40. Challenger, K. D. and Moteff, J., 1972, Characterization of the deformation substructure of AISI 316 stainless steel after high strain fatigue at elevated temperatures, *Metal Trans.*, **3**, 1675–8.
41. Manjoine, M. J., 1983, Damage and failure at elevated temperature, *J. Press. Vessel Technol.*, **105**, 58–62.

42. Wareing, J., 1981, Creep fatigue behaviour of four casts of type 316 stainless steel, *Fatigue Engng Mat. Struct.*, **4**, 131–45.
43. Levaillant, C., Grattier, J., Mottot, M. and Pineau, A., 1985, Creep and creep–fatigue intergranular damage in austenitic stainless steels: discussion of the creep dominated regime, *Symposium on low cycle fatigue, 1985*, STP 942, Philadelphia, ASTM, to be published.

6

The Case for Standards in High Temperature Fatigue

G. B. THOMAS

ERA Technology Ltd, Leatherhead, Surrey, UK

SUMMARY

It is generally accepted that relatively simple mechanical tests to measure a material property should be carried out in a standard way in order to provide a consistent basis for comparison of data for design or quality control purposes. However, more complex tests that may tend to represent an aspect (or aspects) of service conditions are generally devised in different ways by different laboratories, at least in their early stages of development.

In this chapter it is argued that although this was initially the case with low-cycle fatigue (LCF) testing at high temperatures, the subject has now developed to the point that standard test procedures are required. These would facilitate comparison of data and the assessment of methods for life prediction, reduce duplication of testing and generally assist our understanding of the processes of deformation and damage under cyclic conditions. The interdependence of increased knowledge of LCF phenomena and the improvement of test methods is stressed, and the principal factors affecting reliability of results are identified.

The importance of specimen shape and dimensions and the associated choice of the method of strain measurement are discussed with reference to the relative advantages and limitations of the main types currently in use. Other factors considered include the effect of surface finish, axiality of loading, waveform, strain rate and, of particular importance, the need for a standard definition of failure. The problems of defining and measuring one of the fundamental parameters, i.e. plastic strain in situations where stress–strain hysteresis loops do not show the theoretically ideal shape, are considered and possible solutions are suggested. The lack of sufficient systematic information to enable quantification of the effects of variations in procedures is noted.

The present situation with respect to activities aimed at producing LCF testing standards is summarised. It is recognised that not all LCF test programmes can be, or necessarily should be, carried out solely according to a standard method; however, it is concluded that the development of a standard procedure is feasible and that the potential benefits are significant.

6.1. INTRODUCTION

6.1.1. Purpose of Standards

The prime objective of standards is to ensure that the quality and consistency of products and procedures meet a specified level, irrespective of the source of the product or the organisation carrying out the procedure.

The materials engineer takes note of several types of standard including design codes, product specifications, material specifications and methods of test. The last of these generally underpins the others and is used to demonstrate that a material or product matches the specification. The *method of test* is a standard that defines mandatory procedures and conditions of measurement for establishing whether the level of properties or performance called for in the material or product specification are met.

A subgroup of the 'method of test' standard can be identified that specifies a *test procedure* or 'code of practice' for performance of a test to determine material behaviour that is not considered to be a simple and inherent material property such as tensile strength. Whereas the 'method of test' can be followed in a routine way to measure a standard material property for quality control or design purposes, the 'code of practice' tends to make *recommendations* rather than mandatory statements, and the approach is typically more explanatory and is often orientated towards research applications, although the information obtained from the test may also be used in design. However, it is probable that tests covered by a code of practice will eventually become part of the characterisation of a material (or component) and will become the subject of a 'method of test' standard.

6.1.2. Relevance to High Temperature Fatigue

The last case describes the situation for fatigue at high temperature. Consideration of the preceding chapters makes clear that fatigue at high temperatures includes a range of phenomena associated with cyclic operation of high temperature components and plant. In its broadest sense the term covers; high-cycle fatigue due to high-frequency vibration in components operating at elevated temperature, e.g. 'flutter' of gas turbine blades; cyclic system stresses on hot components due to an on/off or low-power/high-power cycle of operation, e.g. the typical low-cycle fatigue (LCF) exposure of turbine discs in jet engines; thermal fatigue of regions of components that are subjected to cyclic stresses over a range of temperatures due to transient temperature gradients combined with constraint to thermal expansion, e.g. in the leading or trailing edges of turbine blades, in thick-section power plant components and in piston crowns and valves in internal combustion engines; and creep–fatigue loading which arises in situations where cyclic stresses and strains of thermal or mechanical origin are combined with creep occurring during periods of steady running, e.g. in rotating components of power plant and aeroengines.

These classifications of high temperature fatigue are generalised and simplified and, in practice, a clear distinction between thermal fatigue and creep–fatigue, for example, may not be appropriate. The variety of loading conditions and the complex combinations of temperature, loading rate and frequency that occur in operation have been discussed in the preceding chapters, and it is thus clear that the concept of the ability of a material to resist fatigue at high temperature does not imply a unique property such as, for example, the 0·2% proof stress.

In practice, high temperature fatigue phenomena and, in particular, thermal fatigue and creep–fatigue, are life-limiting factors, the significance of which is a function of the physical and mechanical properties of the particular material. The relevant properties include thermal conductivity, coefficient of thermal expansion, stiffness, elastic limit, work-hardening behaviour, strength and ductility. It is these parameters that dictate the stresses and strains experienced by the material under given conditions of operation, the extent of elastic and plastic deformation, the damage sustained per cycle and, ultimately, the life of the component.

The significant differences between fatigue at low and high temperatures arise directly from the time-dependent effects of creep and oxidation, that become important in the latter case. The distinction therefore depends on the material under consideration and is generally dictated by the temperature at which the influence of either oxidation or creep become apparent. Clearly, in constructing standards, the need to take account of time-dependent effects is of paramount importance. In general, their significance becomes greater as the frequency of fatigue cycling is reduced, either as a consequence of a decreased rate of straining (or loading) or by the incorporation of hold periods into the fatigue cycle.

The more demanding high temperature test regimes are those of low-cycle fatigue, in which the fatigue cycle results in macroscopic plastic strain in the specimen, and creep–fatigue, in which creep strain develops during hold periods either at constant load or as a result of stress-relaxation at constant strain. It is these regimes in particular which should be addressed when considering the requirements for standards. The majority of high temperature fatigue tests are carried out isothermally for reasons of simplicity, although there is increasing interest in tests involving variable temperature since in many cases these more closely simulate service conditions. Such tests are usually referred to as thermomechanical fatigue or anisothermal fatigue tests (see Chapter 3).

6.1.3. Need for Standard Test Procedures

The benefits arising from the existence of standards for high temperature fatigue testing would include reduction in the duplication of testing and the removal of barriers to trade. At present, a company supplying material or components may need to carry out a different type of test for each customer on the same material, with all the tests having the common objective of assessing the quality of the material with respect to its high temperature fatigue properties. This duplication of effort is attributable, in many cases, directly to the absence of an accepted standard and the consequent specification of 'customised' test procedures by each client organisation. Although the existence of a standard will not necessarily completely eliminate individually specified tests, which may be intended to produce design information for a particular set of conditions, it would remove the need to carry out a range of quality control tests each differing in comparatively minor ways.

These minor differences in test requirements, coupled with the lack of a standard against which a laboratory's procedures can be accredited, also lead to barriers to trade. These can apply to a variety of activities including production of material, manufacture of components and contract testing. Barriers most frequently mean that the quality-control tests are carried out by a specified organisation in a particular country, thereby putting other organisations in other countries at a competitive disadvantage in spite of their competence or economic efficiency.

When testing is carried out as part of a quality-control procedure, e.g. to establish that a minimum level of performance is achieved, the existence of a standard simplifies matters for those requiring the data and provides guidance to those producing the data. Furthermore, there is an ever-increasing and necessary emphasis on quality in manufactured products, and a corollary is the recent worldwide growth of organisations that have responsibility for ensuring the traceability of standards and for accrediting testing laboratories. (In the UK the relevant organisation is NAMAS—the National Measurement Accreditation Service—based at the National Physical Laboratory). The process of accreditation of laboratories as competent to perform these tests is facilitated by the existence of suitable standards. The development of a standard for fatigue testing at high temperatures is therefore seen as an effective means of assisting the process of improving quality and maintaining competitiveness.

In addition to the need for standards defining test procedures, there may also be a need to standardise low-cycle fatigue as a *test type* and even to specify detailed conditions for obtaining low-cycle fatigue data, i.e. test temperatures, strain ranges and strain rates. This would facilitate direct comparison of data from a wide range of different sources, a procedure which is at present often difficult even when the test procedures have otherwise been well controlled and well characterised. Such a development appears to be timely in view of the growing activity in the compilation and use of materials property databanks. Consideration of the extent of the information and experience that have now been accumulated on low-cycle fatigue properties and test procedures suggests a reasonable chance of a successful outcome.

6.1.4. Interdependence of Improved Test Procedures and Understanding of Materials Behaviour

References in available standards-orientated literature to the response of a material to the loading conditions described above are sparse. They do not appear in general product specification standards but are sometimes given in specific design codes, e.g. ASME Code Case N47 [1] and also in particular specifications demanded by customers purchasing materials or components. This is because the complexity of the phenomena involved and the interactions between them are not fully understood, and it also reflects the embryonic state of development of design procedures that fully take account of high temperature fatigue. In view of these factors it is not surprising that the development of standards is also at a preliminary stage and such as there are, principally ASTM E606-80 [2], are mainly aimed at specifying a level of measurement performance rather than at defining the whole test in a strict and detailed way, i.e. they are more concerned with test procedures than test type.

A major benefit of the code of practice standard is in enabling data to be obtained under controlled and reproducible conditions. This permits subsequent analysis and comparison of results from different sources and advances our knowledge of the subject. The development of improved test procedures relies on greater insight into the phenomena involved and, in turn, increased knowledge of the parameters governing the response of a material to a given set of test conditions allows better, i.e. more relevant and more specific, standards to be written. This interdependent process of improved test procedures and increased understanding of materials behaviour may eventually lead to a test (or a series of tests) that will quantify the response of a material to high temperature fatigue as though it were a mechanical property.

6.1.5. Data Requirements

Data are required for different purposes, for example, quality control or design. A quality-control test determines cycles to failure so that the endurance at a given range of strain can be shown to exceed a minimum value. However, the design information required is generally of two types. One concerns the stress–strain behaviour of the material (an inherent material property) and whether it is affected by factors such as the magnitude of the applied strain or stress and the temperature. It is also necessary to know how this property changes throughout life, i.e. whether the material cyclically hardens, softens or exhibits substantially stable stress–strain properties, as discussed in Chapter 2. The second concerns the life to failure as a function of the test conditions. Life is not a material property in the strict sense since it depends on the specimen form and the definition of failure (discussed later). Once a crack has formed, the testpiece becomes a structure and can no longer be regarded as a specimen within the definition of a standard. The relative importance of stress–strain and endurance data to the design process depends upon the specific application, but in most cases both are needed.

There are several methods of predicting low-cycle fatigue life. These may involve a number of parameters, but the central relation is that between endurance and the magnitude of the strain per cycle. This is expressed as a combination of the Coffin–Manson and Basquin equations to give:

$$N_f = A\Delta\varepsilon_p^{\alpha} + B\Delta\varepsilon_e^{\beta} \qquad (1)$$

where N_f is the number of cycles to failure, $\Delta\varepsilon_p$ is the plastic strain range, $\Delta\varepsilon_e$ is the elastic strain range, and A, α and β are material constants. This equation leads to the typical relationships between endurance and elastic, plastic and total strain ranges exemplified graphically in Fig. 1.

However, because of time-dependent effects, the constants in eqn (1) are functions of frequency and temperature. To ensure comparability between results for quality control and to allow reliable interpretation of data for design and research purposes, the test conditions must be controlled and the parameters of interest measured with accuracy. A standard that defines such parameters and accuracy limits therefore makes a valuable contribution to achieving consistency and to the understanding of the results.

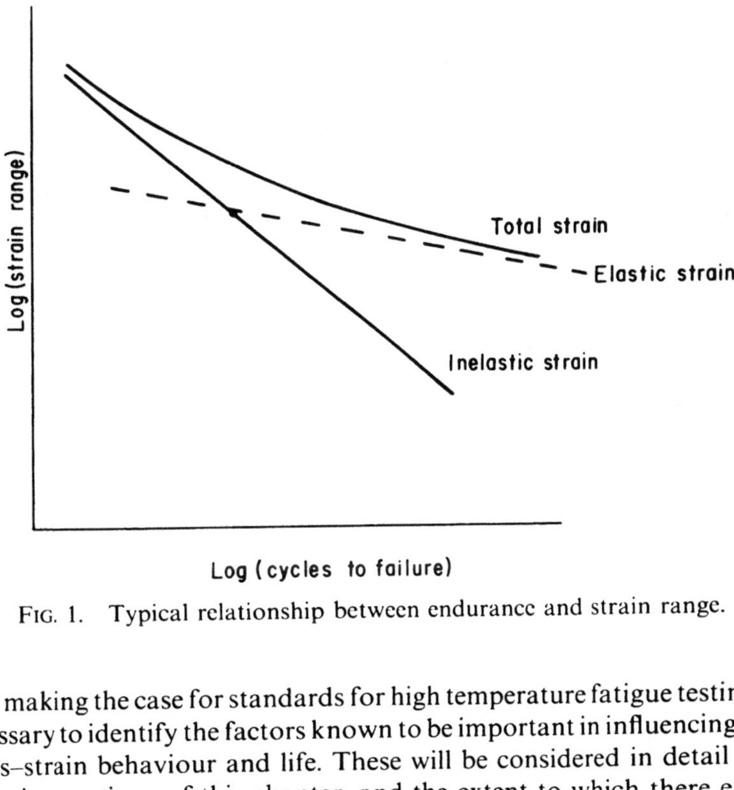

FIG. 1. Typical relationship between endurance and strain range.

In making the case for standards for high temperature fatigue testing, it is necessary to identify the factors known to be important in influencing cyclic stress–strain behaviour and life. These will be considered in detail in the following sections of this chapter, and the extent to which there exists a consensus view of their significance and of the need for specification will be discussed. The most commonly used technique involves axial, constant-strain amplitude under isothermal conditions, and the following remarks will be addressed principally to this method, although the relationship between this and other approaches will also be considered.

6.2. FACTORS AFFECTING RELIABILITY OF RESULTS

6.2.1. Specimen Design and Extensometry

6.2.1.1. Relation Between Specimen Form and Method of Strain Measurement

Strain-controlled fatigue testing generally involves applying strains that are great enough to produce measurable plastic strain in the specimen in each cycle. The corollary is the existence of a stress–strain hysteresis loop whereby the specimens experience both tensile and compressive loads. Consequently, they have to be designed so as to avoid buckling during the compressive part of the cycle, and this in turn dictates a relatively low ratio of length to diameter, e.g. certainly not greater than 4:1, and more commonly about 2:1 in the gauge section.

The other major constraints on the design of the gauge portion are (a) the need to measure or control strain with precision, and (b) the condition that the behaviour of the material should not be influenced by the technique used for strain measurement. These requirements are to some extent incompatible and various designs of specimen exist, each representing an attempt to resolve this conflict using the equipment available at the time and in the light of the contemporary knowledge of the processes involved in high temperature fatigue. The gauge portions generally fall into one of three categories; viz. hourglass profile with a round section, smooth parallel (usually cylindrical but can be rectangular section) and parallel with ridges (again usually cylindrical), as exemplified in Fig. 2. Typical diameters are in the range 6–12 mm.

Evidently, the specimen is not designed in isolation but is subject to the

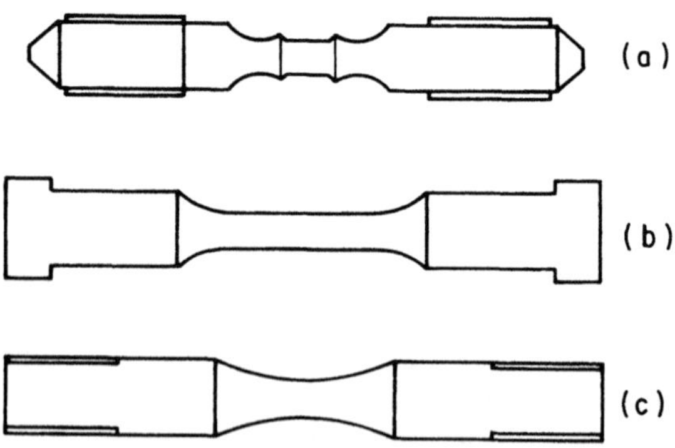

FIG. 2. Examples of commonly-used specimen forms.

limitations imposed by the type of extensometer to be used. Conversely, the choice of extensometer system depends on the type of specimen. Selection of the optimum extensometer/specimen combination is therefore a compromise (which is also influenced by the characteristics of the test material).

6.2.1.2. Considerations in the Selection of Specimen Gauge Shape and Type

(*a*) *Hourglass specimens.* Hourglass specimens were used extensively in the early work on high temperature fatigue. Clearly, this profile has the advantage of reducing the risk of specimen buckling to a minimum, and it is also significant that this shape was used in conventional fatigue testing at ambient temperature. However, for strain-controlled testing there are significant disadvantages associated with this type of specimen and these have become increasingly evident as techniques and understanding have developed. The drawbacks are particularly apparent when time-dependent strains are involved, and hourglass profiles are now generally reserved for specific applications only.

An inescapable accompaniment to the hourglass specimen when used for strain-controlled testing is the diametral extensometer, and this contributes to several of the disadvantages of this design. The limiting sensitivity of diametral strain measurement is inevitably lower than that of an axial method to an extent determined by Poisson's ratio and the specific dimensions of the specimen. In fact the sensitivity can be lower by a factor of up to five [3].

A consequent disadvantage of the need to use Poisson's ratio in converting from diametral to axial strains is encountered when the material to be tested is anisotropic or has a grain size that is large in comparison with the specimen gauge diameter. In these cases the measured value of Poisson's ratio at a given position of the extensometer probes on the specimen will be dependent upon the local orientation of the grains with respect to their crystal structure. Hence the measured diametral strain will, in general, not reflect the average applied axial strain. However, it has been demonstrated that, in the case of large-grained polycrystalline specimens, a diametral location can be selected, by trial and error, that will produce an average value for Poisson's ratio. Measurements of diametral strain at this position can be used to control the test to produce lifetime data that show reasonable agreement with those obtained from tests using axial-strain control [4].

When significant oxidation of the test material occurs, the tendency for the diameter of the specimen to increase results in considerable error in diametral strain control and consequently in lifetime determination.

The volume of material under test is very small in hourglass specimens, and this is a disadvantage when non-continuum damage processes, such as the development of intergranular cavitation, are significant causes of failure, since a representative number of grain boundaries may not be subject to the required test conditions. However, this confinement of the applied strain to a specific location can be an advantage when the volume of material to be tested is itself limited, e.g. in testing welded joints. Here, the ability to select which region shall occur at the minimum diameter can be useful, since it permits specific regions of the weldment, e.g. weld metal or heat-affected zone, to be selected for testing.

The hourglass shape also has advantages when testing at particularly high strain ranges, since it reduces the chances of buckling to a minimum. The stress-state is, of course, to some extent multiaxial which may be a disadvantage in some cases, e.g. notch-sensitive materials. This latter problem and that of the limited volume of material could be overcome by incorporating a small parallel portion at the minimum diameter, although the disadvantages of diametral strain measurement would still be present.

(*b*) *Cylindrical specimens.* The use of specimens with a relatively large parallel portion allows axial extensometers to be used with a consequent gain in sensitivity of strain measurement. The attachment of the extensometer limbs to the specimen can be achieved in several ways: (a) by the use of spring-loaded probes with knife edges or conical points located on the surface of the parallel portion of a smooth specimen; (b) by fixing the extensometer probes to ridges raised from the parallel length of the specimen; or (c) the extensometer may be attached to the specimen outside the parallel portion.

Attachment to ridges has been extensively used, as it allows the use of robust extensometry and provides a secure method of fixing. However, the system introduces a stress concentration into the specimen at the transition from parallel portion to ridge and gives rise to some uncertainty in definition of the gauge length because the whole of the portion over which displacement is measured is not parallel. The stress concentration is only of concern in tests to determine the endurance of material, as distinct from the stress–strain behaviour, and its influence depends upon the characteristics of the alloy. However, the region of non-uniform diameter in the vicinity of the ridge leads to a systematic error in the derivation of strain from the measured displacement between the ridges. Correction can be made by calculating an effective gauge length (for example, see Ref. 5), discussed in more detail below.

In notch-sensitive materials, premature crack initiation at the site of maximum stress concentration will give endurances that are considerably less than would be obtained from a specimen without ridges. The difficulty is compounded in materials which harden as a result of repeated cyclic straining, since the material within the parallel gauge length hardens, with the consequence that the first cracks tend to develop at the transition from gauge length to ridge. For this reason, the use of ridges presents a particular problem in testing, for example, austenitic stainless steels. Materials which exhibit cyclic strain softening or show essentially stable stress–strain behaviour under cyclic loading appear to be less affected by the presence of a ridge. Even though the location of the fracture may be close to the ridge, cracks are also seen at several locations along the parallel length, and current evidence suggests that the effect on life is not significant in materials of this type [6]. However, the evidence is limited and further information comparing endurance in specimens with and without ridges is required before definitive statements can be made.

To avoid problems with premature failure the ridges on the parallel portion can be dispensed with. Axial strain measurement can then be achieved in two ways; either by holding knife-edged or conical-pointed probes against the smooth parallel gauge portion using spring pressure, or by fixing the extensometer limbs to the specimen shoulders. In the latter case the calculation of effective gauge lengths is essential and, for a specimen with a relatively gentle transition radius, selected specifically to avoid difficulties with stress concentration in highly notch-sensitive materials, the correction required can be greater than the actual parallel length [5]. The effective gauge length depends upon the relative amounts of elastic and plastic strain in the specimen and will therefore change throughout a test if the material exhibits strain-hardening or strain-softening behaviour. It is further modified when creep strains are involved, and evaluation of an effective gauge length for creep strain involves complex calculation.

The use of a parallel specimen with an extensometer having spring-loaded, side-contacting probes (usually of quartz or other ceramic) avoids difficulties with stress concentration and effective gauge lengths and is increasingly gaining favour as a consequence. However, rather more care is required in setting up and the points or knife-edged contacts of the probes must be maintained in good condition to avoid slipping on the specimen surface giving rise to inaccuracies in strain measurement or control.

The pressure applied to maintain contact between the probes and the specimen must be great enough to ensure positive contact without slipping, and yet sufficiently light to avoid applying significant bending forces or producing a notch in the specimen that could lead to premature failure. At relatively low temperatures for a given material, the allowable spring force can be greater than at high temperatures, at which the material may be softer and more easily notched. However, at higher temperatures oxidation generally assists the probe to retain its location on the specimen, thus compensating for the lower spring pressure that the material can withstand. Nevertheless, in some highly oxidation-resistant materials, light spring contact may be needed because of creep considerations, but there may be insufficient oxidation to improve stability of fixing. There is thus a regime of temperature and material characteristics in which considerable care in setting up the test is required if reliable and consistent results are to be obtained.

6.2.1.3. Influence of Surface Finish of the Specimen

Another factor requiring standardisation is the condition of the surface of the specimen. It is well known that the quality of surface finish can have a considerable effect on life to crack initiation during fatigue testing at ambient temperatures. Consequently, when fatigue endurance is being characterised, careful polishing of specimen surfaces is required. This can be achieved using fine-grade abrasives with the polishing direction parallel to the specimen axis, combined with a final electropolish.

However, when testing at sufficiently high temperatures in air, oxidation would be expected to reduce or nullify the influence of surface finish and render such a high degree of polish unnecessary. Where there is sufficient information available to specify such temperatures, then the expenditure of time and effort on specimen preparation can be reduced by accepting a ground or machined finish to a specified level of surface roughness, e.g. $R_a = 1\ \mu m$. In the absence of such information, it may prove economical to carry out some preliminary testing to establish whether a better degree of surface finish is necessary at the test temperature. Of course, if it is required to establish the endurance of a specimen with a surface representative of a component, e.g. as-cast, then polishing is clearly inappropriate.

6.2.1.4. Requirements for Specimen Grip Ends

Thus far, specimen design has been considered only with respect to the gauge portion. While the recommendation of a single standard design of grip ends may not be appropriate, it is essential that certain features are incorporated in order to ensure that the loads are transferred from the loading bars to the specimen without significant bending and with no backlash. Consequently, the grip ends require surfaces through which tensile and compressive loads can be transmitted to the specimen and a means of ensuring that axiality of loading is achieved. The specimen head should be of sufficient size (cross-sectional area) to ensure that the stresses applied to it during testing are not great enough to result in plastic deformation or bending.

Several designs which satisfy these requirements are in use, each having their own particular combination of advantages and drawbacks. The specimens shown in Fig. 2 indicate some of the more usual types of grip end, and the subject has been extensively discussed [7]. A common method is the use of button ends clamped to the loading bar, using split collets and locking nuts. In this case the transverse surfaces of the button ends must be perpendicular to the specimen axis, (e.g., see Fig. 2(b)). This design is relatively insensitive to deviations from concentricity between the loading axis of the machine and the specimen axis. This is because the loading bars are, of necessity, substantially greater in cross section than the specimen gauge portion and thus resist the transference of bending forces to the specimen. Another frequently used design incorporates a cone seating between specimen head and loading bar (Fig. 2(a)). This has the advantage of automatically centring the specimen axis with the machine loading axis provided, of course, that the cone angle is correct, that the surfaces of the cone seatings are accurately made, and that the cones are concentric and true with respect to the specimen axis. A common alternative to button ends involves threads on the specimen head, upon which threaded collets are mounted to provide a bearing surface for the locking nut (Figs 2(a)(c)). This can often allow a specimen to be obtained from a smaller cross section of material than required for button-headed specimens, but the mutual concentricity of the axes of both thread and specimen is of paramount importance.

In general, the use of any of these designs requires that specimens are manufactured to a high level of accuracy in at least one aspect concerning concentricity or angular relationships. This implies that if a workshop can successfully manufacture specimens to any one of these designs then the potential accuracy is there to produce any of the others. The choice of design of grip end should therefore be dictated by the size of the available samples of material and the need for economy in specimen preparation. In any case, the demands on tolerances generally mean a relatively expensive specimen compared with, for example, a tensile testpiece.

6.2.2. Loading Procedures

6.2.2.1. Axiality of Loading

The important requirement for axiality of loading needs to be quantitatively expressed, i.e. standardised, to ensure comparability of results from one laboratory to another. The bending strains associated with significant non-axiality have the effect of subjecting some regions of the specimen to greater strain ranges than others. The consequences of this are twofold: (a) the stress–strain information obtained will, in general, be in error since for a given stress the measured strain will generally be greater or less than the average strain, depending on the location of the extensometer attachment position relative to the plane of bending; and (b) the endurance obtained will generally be related to the maximum strain range experienced by the most highly strained region of the specimen and, again, this will not necessarily coincide with the measured (or controlled) strain range.

The ASTM standard [2] specifies that the proportion of bending strain should be less than 5% of the applied axial strain. This figure is probably dictated by consideration of the best that can be achieved without excessive cost in terms of specimen manufacture and adjustment of alignment, rather than by a secure knowledge that no significant effect is evident from bending strain below this level. Indeed, it has been shown that bending strains of $\pm 5\%$ of applied axial strain can result in errors of up to 30% in life [8]. This suggests the need for a rather lower limit on bending strain but this in turn would require much tighter tolerances on specimen preparation and loading bar alignment. It would also require the existence of a sufficiently simple yet accurate means of identifying very low levels of bending strains in each specimen, and would possibly mean the strain-gauging of every specimen.

A standard procedure for the verification of specimen alignment under tensile loading has recently been published [9]. This provides useful guidance on alignment in general, although in fatigue testing the compressive phase of the loading cycle is the more demanding with respect to accuracy of alignment in order to avoid buckling. Furthermore, the continuous reversal of strain means that misalignment does not tend to be self-correcting as the test proceeds, as it would be in tensile loading, but rather that damage concentrates in those regions subjected to the greatest bending strains.

Non-axiality can arise from a number of causes, and these have been discussed comprehensively by Bressers [8], who also describes the method of estimating the resulting bending strains in terms of the magnitude and type of misalignment. The main sources of bending are, briefly:

misalignment of loading bars such that the centre lines, although parallel with each other and with the axis of movement of the actuator or crosshead of the machine, are displaced; angular misalignment of the loading bars that produces bending in the specimen even when the centre lines appear to coincide at the point of contact between specimen and loading bar; inadequacies in the specimen seating due to poor preparation of the loading surfaces or the presence of dirt or oxide; lack of concentricity in the specimen or incorrect angular relationships between load-bearing surfaces and the specimen axis.

The alignment of the loading system should be checked using a dummy specimen equipped with suitably disposed strain-gauges. Another method is to use a split specimen, and to measure the extent of any displacement between the mating halves when brought into contact. When excessive non-axiality is encountered the source must be identified and eliminated before proceeding with the test programme.

6.2.2.2. Compatibility Between Control Signal and Machine Response

Axial loading should achieve uniformity of the applied strain across the specimen. However, in order to ensure that the strain is also applied with precision at the point of reversal, it is necessary that the dynamic response of the testing machine be compatible with the rates of strain demanded by the control signal. This points to a need for dynamic calibration of load cells in addition to a means of checking that the error between control and feedback signals is within prescribed limits. In this context, a low limit on the hysteresis characteristic of the strain transducer is also necessary. Furthermore, there arises a requirement for calibration of the extensometer under conditions similar to those experienced in use, i.e. including both strain reversal and through-zero strain. As yet, no standard exists to cover this situation.

Computer-controlled testing machines are increasingly being used for high temperature fatigue testing. There is a strong temptation to assume that the computer is producing the correct control signal, but this may not be so, and errors in the program are less obvious than incorrectly set dials or potentiometers would be in analogue control systems. This emphasises the importance of calibrating the control program and checking it before use.

6.2.2.3. Waveform, Frequency and Strain Rate

The waveform of the control signal can significantly influence measured performance in terms of both stress–strain behaviour and endurance. This is because of the time-dependent nature of deformation and other damaging processes, e.g. oxidation, at elevated temperatures, and the most obvious examples of the effects of waveform are the substantial differences in stress–strain hysteresis loop shape and endurance between tests with and without dwell periods. However, even in the absence of major effects due to dwell periods, the form of the test cycle is of significance. For example, the extent of the time-dependent strain, which generally has a different, usually more damaging, effect on high temperature fatigue life than time-independent strain, becomes greater as the available time increases, i.e. at lower rates of strain. Indeed, the fact that the mechanism of failure may change from transgranular at high strain rates to intergranular at low strain rates, or when the cycles include dwell periods, emphasises the importance of specifying limits on waveform and strain rate.

As a consequence of these effects, the continuous variation in strain rate associated with a sinusoidal waveform can cloud subsequent interpretation and comparison of results. Furthermore, a number of levels of strain range is generally required in a test programme, and this means that average strain rates would not be the same for all tests. Such a waveform must therefore be considered unsuitable for the performance of standard high temperature fatigue tests, and a triangular waveform, such that total strain rate remains constant throughout the tensile and compressive-going parts of the cycle, is most commonly used. Clearly, unless the test is specifically aimed at determining the effects of incorporating more time-dependent strain in one part of the cycle than the other, then the tensile and compressive-going strain rates need to be equal.

The time-dependence of strain at high temperatures means that fatigue properties (both stress–strain behaviour and endurance) are dependent on frequency. The effects of oxidation on the test material may add to this frequency-dependence (particularly of endurance, although carburisation and similar effects could also affect stress–strain behaviour). When attempting to characterise fatigue life as a function of the applied strain amplitude, a series of tests over a range of strain amplitudes is required. This raises the question as to whether frequency or strain rate should be kept constant in a standard test programme. Keeping frequency constant would have the result that tests at higher strain amplitudes would be conducted at higher strain rates. Because of the significant effect of strain rate on stress–strain behaviour and life, constant strain rate is generally preferred and should be taken as standard practice.

It is less easy to decide whether the strain rate itself should be standardised. If the test were to become the determination of an accepted material property, similar to tensile strength or the high cycle fatigue limit, then definition of a standard strain rate (or limitation on the permissible range of strain rate) would be necessary. However, no such property is yet recognised and tests are often carried out to simulate, in a controlled and simplified manner, a strain-load–time pattern that is relevant to service conditions. For example, the test programme may be designed to establish the response of a material to combined creep and fatigue, i.e. using fatigue cycles incorporating creep dwell periods. In addition to determining the effects of duration of dwell periods and whether these are in tension, compression, or both, there is almost always a comparison made between endurance with and without dwell periods. In other words, the property evaluation procedure contains a subprogramme involving the determination of LCF life. Although the strain rates for this subprogramme may be selected to reflect the service application, they generally tend to be relatively high in order that time-dependent strains are minimised. For example, strain rates are typically of the order $10^{-3}\,s^{-1}$, and the majority of testing in the USA is at $4 \times 10^{-3}\,s^{-1}$. Clearly, the higher the temperature the greater the strain rate needed to minimise time-dependent effects (but more sophisticated methods of control and recording become necessary for strain rates greater than about $10^{-2}\,s^{-1}$). Consequently, there does seem to be a developing need for the definition of minimum values of strain rates for the determination of low-cycle fatigue performance, albeit in conjunction with specific ranges of temperature and, possibly, class of material.

Materials which fail by transgranular cracking over a wide range of waveforms and strain rates, for example 9Cr–1Mo steel which shows no creep–fatigue interaction, do not require the specification on mechanistic grounds of a minimum strain rate for determining endurance. Even so, for economy, a relatively high strain rate would be desirable. However, for materials which do show a creep–fatigue interaction, a realistic minimum strain rate for endurance testing would be a value comfortably above that at which the failure mechanism changes from intergranular to transgranular as strain rate is increased. A limit on the permitted range of strain rate is also required for the consistent and comparable measurement of stress–strain behaviour.

6.2.2.4. Choice of Control Parameter

Most of the foregoing has tacitly assumed that tests are carried out under strain control, but fatigue tests at high temperature are sometimes performed at constant load range. Indeed, for long endurances, where there are considerable difficulties in accurately measuring and controlling the low levels of strain that may be required (usually associated with lives in excess of 10^5 cycles), strain control is hardly practical. For these low-amplitude tests it appears that the results obtained under load control can be correlated with those obtained under strain control, provided that the calculation of the strain amplitude of the load-controlled tests is based on cyclically stabilised stress–strain behaviour [10].

When testing under conditions involving rather higher strain amplitudes, the differences between load and strain control become more important. In particular, the final stage of crack development is usually rapid when testing at a constant load range, since the tensile stress on the net section increases as the crack extends. As a consequence, the difference between the number of cycles to the first indication of a crack and cycles to complete separation of the specimen is relatively small. On the other hand, tests at constant strain amplitude frequently show a much greater difference between the life to first indication of cracking and that to complete separation. The criterion for determining failure (see Section 6.2.3.2 below) therefore becomes of considerable importance when comparing results from load-limited with strain-limited tests. However, if life to first indication of cracking is compared then, provided that the strain rates used in each test are high enough to avoid time-dependent strain, similar results can be obtained when expressed as a function of inelastic strain range [11].

Nevertheless, for a load-controlled test at constant loading rate the strain rate will not be constant, and will in fact increase rapidly once the elastic limit is exceeded. The practical implications of this depend upon the magnitude of any time-dependent strain involved and thus on the strain rates used in comparable tests under load or strain control at the same frequency. Where the strain rates in either test are in the regime where significant time-dependent strains occur then load-controlled tests will give different stress–strain behaviour and, probably, different lifetime information from strain-controlled tests. Load-controlled testing also causes strain ratchetting throughout the test. This is particularly evident at high strains, and complicates the subsequent interpretation and comparison of results. In general, strain control is to be preferred for a standard test in spite of the somewhat greater cost and difficulty.

6.2.2.5. Limits on Temperature Control and Measurement

In considering the requirements for measurement and control of temperature it is necessary to address the effects on material properties on the one hand, and on load and strain measurement or control on the other.

In the absence of dwell periods, the sensitivity of LCF life to temperature changes is dependent on the choice of the parameter against which endurance is assessed. When life is expressed by the Coffin–Manson equation the influence of temperature can be relatively small. This is because, in the absence of significant changes in material ductility, the major effect of temperature is to change the proportion of plastic to elastic strain in a given total strain range. Since the extent of plastic strain in the cycle is the principal life-limiting parameter, a plot of life as a function of plastic strain range generally shows a relatively low sensitivity of endurance to temperature [12] (see also Fig. 1 of Chapter 2). However, this may not be true if comparisons are made over a wide range of temperature such that significant differences in ductility or failure mechanism occur. Conversely, plotting life as a function of total strain does not necessarily show the same degree of temperature independence, and when life is plotted as a function of stress range considerable effects of temperature are evident, as shown schematically in Fig. 3. This is a consequence of the sensitivity of tensile properties, notably the elastic limit and work-hardening characteristics, to temperature.

In creep–fatigue testing, however, the influence of temperature on properties is great since time-dependent, thermally activated strain mechanisms are now included. The requirement for good temperature control is particularly important when the creep strain is developed as a

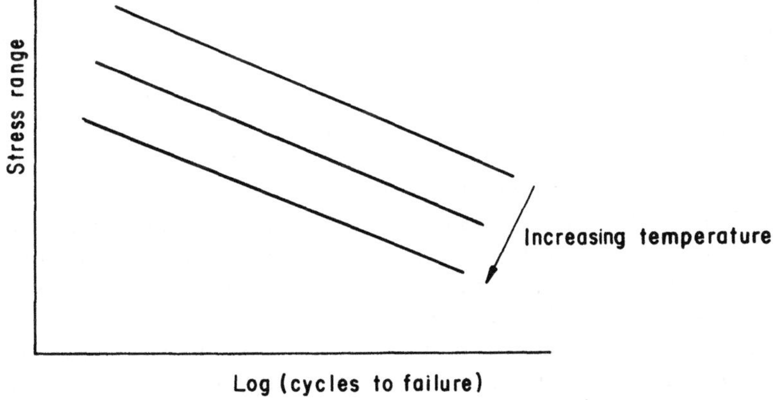

FIG. 3. Schematic showing the effect of temperature on endurance as a function of stress range.

consequence of stress relaxation during a dwell at constant total strain. Variations in temperature produce expansion and contraction in the specimen and the loading train and, because the strain control system acts to maintain a constant strain during the dwell period, the stress may not decrease smoothly during relaxation but can vary by considerable amounts. Clearly, this situation is undesirable because of the high sensitivity of creep rate to stress level. Temperature variations therefore result in inconsistencies in the amount of creep strain from cycle to cycle, and in the rate of creep strain during the dwell period. Furthermore, the damaging effect of creep strain is dependent upon the strain rate. These factors can make considerable difficulties in obtaining consistent and reliable data and in applying the results for the purposes of design and life prediction. It therefore becomes especially important to limit temperature variations as much as possible during such tests.

Differences in the actual temperature of the specimen and that intended can arise from several sources, as follows: inaccuracies in thermocouples leading to uncertainty in knowledge of the actual temperature of the specimen at any one point; uneven temperature distribution both along and through the specimen; variation with time as a result of drift in thermocouple output or response of the temperature control system; short-term variation or cycling of temperature due to the particular characteristics of the control and heating systems; cycling of temperature due to the thermoelastic effect during strain cycling. An additional source of temperature cycling can occur if induction heating is used when testing ferromagnetic materials. This is due to the change in magnetic characteristics as the material is strained, i.e. the magnetomechanical effect, and it may be advisable to avoid this method of heating when testing ferritic steels, for example [13].

In general, inaccuracies in properly constructed and fitted Pt-13%Rh vs. Pt thermocouples should result in an initial uncertainty of $<1°C$ at temperatures up to 1000°C [14]. Drift in output would be expected to be within $\pm 1°C$ over a period of 1000 h [15], and the time stability of good, modern temperature control systems is such as to increase this figure by no more than 0·5°C. Provided some care is taken in matching the control system and heating system, then cyclic variation in temperature arising from this source can also be kept within limits of less than 0·5°C. The indicated temperature gradients along and across the specimen can be relatively easily kept at less than $\pm 1°C$ of the average indicated temperature in most cases. These limits are less than those given in BS 3500 [16] for creep and rupture testing, and this is because gauge lengths are generally short compared with standard creep testpieces. The contribution to temperature variation arising from the thermoelastic effect is dependent on frequency and strain amplitude, and can result in the temperature changing throughout the cycle by as much as 2°C. The conclusion from considering all these sources of uncertainty, drift, short-time cyclic variation and gradient is that an achievable standard of temperature measurement and control is ±5°C peak-to-peak over the test duration with the average temperature typically within ±3°C of the nominal value.

Another aspect of temperature control that should be considered is the effect on the extensometer of short-term variations in the ambient temperature. Depending upon the type of extensometer, these can produce quite large deviations in the level of strain output and can thus affect control and measurement. The type of extensometer that uses metallic limbs to transfer the displacement at the specimen to a transducer that is outside a furnace is particularly susceptible to the effect of draughts. This is because conduction ensures that the portions of the limbs outside the furnace are at temperatures that are considerably above ambient, with the consequence that the cooling effect of a draught can result in significant contractions that are not necessarily equal in each limb. Other effects are those due to the temperature dependence of the transducer itself but these are usually not large. In general, these factors point to a need to site the testing machine in a draught-free, temperature-controlled environment.

6.2.3. Data Recording and Analysis

6.2.3.1. Factors Affecting Accuracy of Recording Methods

Clearly, standardisation of load and extension measurements is relatively straightforward, since the load cell and extensometers can be calibrated according to relevant standards. This implies typical accuracies for peak-value measurements of load to within 1% and strain to within 2%. However, it is also necessary to pay some attention to the equipment used for recording test data and to specify some minimum requirements of performance. It is important that the response characteristics of the recording equipment are compatible with those of the control system, particularly with respect to speed of response. In most cases the rates of straining used in high temperature fatigue testing, at least under strain control, are sufficiently low (generally $<10^{-3}\,s^{-1}$) that a good quality *X–Y* recorder will be satisfactory in this respect for the recording of load–extension hysteresis loops, such as shown in idealised form in Fig. 4.

Load and strain as a function of time are generally required in tests incorporating dwell periods and, again, most modern chart recorders are

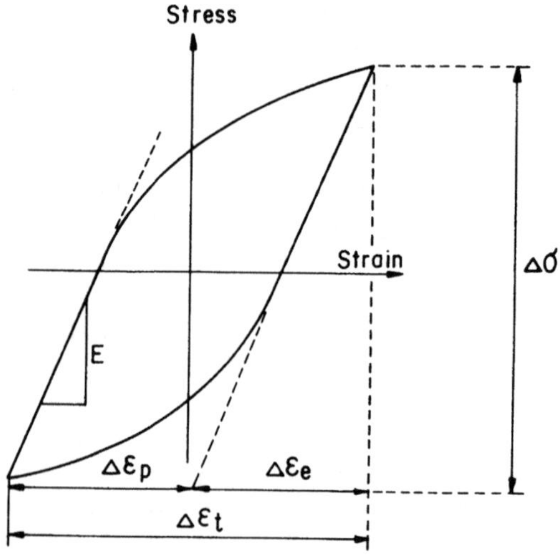

FIG. 4. Ideal stress–strain hysteresis loop.

capable of giving accurate records at these rates of strain, except possibly where detailed information is required in circumstances when the rate of change is particularly high, for example, during the stress relaxation that occurs immediately upon commencement of a dwell period at constant strain. For such specialised purposes, and for tests at high strain rates or frequencies, the use of an oscilloscope in conjunction with a transient recorder may be necessary.

Digital recording of data using dedicated computers is commonly used because the data are then available in a form suitable for computer processing and analysis. This does not necessarily mean dispensing with *X–Y* and chart recorders, and these are frequently used in addition to the computer in order to provide some insurance against loss of data from the computer or storage device, and also to give a convenient record of the status of the tests at intervals throughout life. Standardisation in this case should again be concerned with ensuring that the rate of data capture is sufficiently rapid and that, in turn, sufficient 'bits' of information are stored for each cycle recorded in order to fully characterise the test parameters. In general, a level of 400 bits (200 load and 200 extension values) per cycle is adequate for a low-cycle fatigue test, but for a creep–fatigue test additional information needs to be recorded, the extent of which depends upon the duration and stress or strain level of the dwell periods.

6.2.3.2. *Definition of Failure: Use of Data Records*

The data records store information about stress–strain behaviour and facilitate assessment of the influence of repeated cyclic strain, thereby providing valuable information for design purposes. However, the other objective of the test is to determine endurance, and the data records are particularly useful here since failure is not necessarily considered to be separation of the specimen into two pieces and, in practice, several other definitions are used. The choice of a particular definition of failure is usually linked to the service application of the material but is also frequently influenced by experimental convenience, and in any particular case, the quantitative expression of the definitions is often arbitrary. This appears to be an area particularly in need of standardisation, as attested to by the variety of failure criteria in use, described below.

Criteria for failure, apart from fracture into two pieces, are based on the concept of the development of a crack to a given size. Generally, the presence of the crack is inferred from the consequent change in compliance as manifested in differences in the apparent stress–strain behaviour between tension and compression. In practice, allowance must be made for real changes in stress–strain behaviour of the material before it can be concluded that the failure criterion has been achieved. Therefore, the test must be taken to a point which is clearly beyond that at which 'failure' would be deemed to have occurred, and the data records must then be used to establish the defined failure point retrospectively. The definitions commonly used at present include the attainment of: (a) a chosen percentage decrease in the value of the maximum stress; (b) in the case of fully reversed straining (or loading), a drop in the maximum stress (or increase in the maximum strain) that is not matched by a corresponding drop in the minimum stress (or increase in minimum strain); (c) a chosen percentage deviation of the curve of maximum stress versus cycles from the extrapolated cyclic hardening or softening behaviour; (d) the earliest identification of the formation of a cusp in the compressive portion of the hysteresis loop. Another method that has recently been proposed involves monitoring the elastic stiffness from the linear portion of the hysteresis loop, since this decreases rapidly after crack initiation [17].

It is clear that considerable scope for discrepency in the determination of lifetime would arise from the application of different failure criteria to otherwise similar tests. Even if the same basic definition of failure is used, the results are sensitive to the choice of the particular value of the percentage change in the chosen parameter. The criterion that is probably among the more sensitive to the onset of cracking and that most

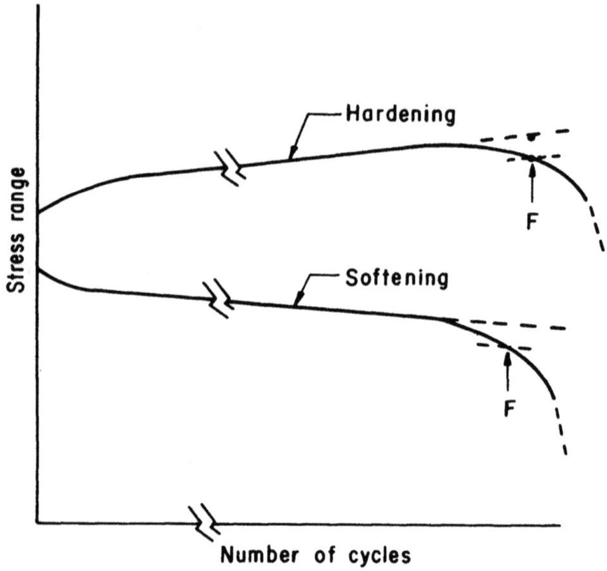

FIG. 5. Identification of failure (F) according to comparison of actual stress range with that obtained from extrapolated cyclic hardening or softening behaviour.

conveniently takes account of changes in the stress–strain response of the material due to strain hardening or softening is that contained in definition (c) above, and which is shown in schematic form in Fig. 5. The critical level of deviation can be taken to be as low as 1 or 2%, using a retrospective examination of the record.

It has been reported [18] that the proportional decrease in tensile load in cases in which the stress–strain behaviour is stable corresponds to the fraction of the area of cross section of the specimen occupied by the crack. It was also shown in this work that the identification of a cusp cannot be achieved with confidence until the cracked area has reached between 10% and 20% of the cross section of the specimen, so that this criterion is relatively insensitive with respect to the identification of crack initiation.

Having argued in favour of the adoption of the percentage deviation of the tensile behaviour from that extrapolated from the general strain hardening or softening, it then remains to propose a standard value for that percentage. Again, particular circumstances may influence the value required for any given application. A value as low as 1% is probably at the limit of detection, but may be necessary if the component were considered to have failed as soon as a low-cycle fatigue crack originated, as for example in an aero-engine rotor blade where crack growth resulting from high-frequency vibrations would very soon fracture a cracked blade. The implications for testing are that the strain control and extensometer system need to be particularly accurate and sensitive, and that a very stable specimen temperature is required for this level to be determined within the normal range of confidence limits. A less demanding value would be 50%, but such a high value may be considered to give a misleadingly long endurance and it would be valuable if a compromise figure could be agreed that would give a realistic, reproducible and relatively easily determined failure criterion, e.g. 5%. If such a figure were adopted as a standard value, which should be reported for all test programmes, it would provide a valuable point of cross reference between laboratories and materials. Obviously, this would not preclude the retrospective identification and use of other failure criteria as well, provided that the test was continued for a sufficient number of cycles.

It is worth noting that the adoption of fixed percentage changes for any of these failure criteria will only be truly consistent for specimens of the same gauge dimensions. This is because initiation *per se* is not actually identified but the attainment of the failure criterion involves growth of the crack to a size sufficient to produce the specified change in compliance or load drop. For example, a 5% drop in load in a specimen of 20 mm diameter implies a crack of considerably greater depth than in a specimen of 5 mm diameter. The endurance in the former specimen therefore includes a greater number of cycles of crack growth.

An analogous case appears to apply for differences in gauge length, a longer specimen exhibiting a lower sensitivity to the presence of a crack of a given size, other things being equal [19]. The situation is further affected by the mode of cracking in as much that multiple crack initiation, for example, at a number of surface grain boundaries, gives a different maximum crack depth from single crack initiation for the same measured change in compliance. These factors emphasise the importance of specifying limits on specimen dimensions and using a failure criterion that involves minimal crack growth. Indeed, the importance of the latter to subsequent interpretation and use of data to assess predictive methods may be considerable. It has been shown [20] that, in MAR M509 alloy, whereas Miner's life fraction rule successfully described LCF data for life to crack initiation (corresponding to a crack depth of 0·2 to 0·3 mm in a specimen of 8 mm diameter) it resulted in large overestimation of life to fracture.

Ideally, the lifetime data required for the purposes of design and life estimation are thus seen to be (a) life to crack initiation under specified conditions on standard specimens and (b) crack growth rate measurements on specimens, validated according to fracture mechanics concepts using standard procedures [21].

6.2.3.3. Determining the Plastic Strain Range

The parameter most commonly used to correlate endurances in low-cycle fatigue at high temperatures is the inelastic strain range. In the absence of time-dependent inelastic strain this is often referred to as 'plastic' strain, and its magnitude is determined from the hysteresis loop. However, there are different views as to how this should be done. The situation is further complicated by semantic arguments over the term 'plastic strain' in cycles where time-dependent strain is also present.

Leaving the semantic aspects for the moment and considering a cycle without dwell periods, how should the inelastic strain range be determined? For an ideal, fully reversed hysteresis loop, such as shown in Fig. 4, there would be no difficulty. The required strain range is the width of the hysteresis loop at zero stress, which in this case also happens to be the mean stress. This can be measured and, within the limits of measurement accuracy, will give the same result as would be obtained by subtracting from the measured total strain range, $\Delta\varepsilon_t$, the elastic strain range determined from the measured modulus of elasticity, E, and the measured stress range, $\Delta\sigma$, i.e. using the equation:

$$\Delta\varepsilon_{IN} = \Delta\varepsilon_t - \frac{\Delta\sigma}{E} \tag{2}$$

where $\Delta\varepsilon_{IN}$ is the inelastic strain range. The difficulty arises when hysteresis loops exhibit significant departure from the ideal shape, as is frequently the case in tests at high temperature, and they have the shapes shown in Fig. 6 or Fig. 7. Hysteresis loops of the form shown in Fig. 6 occur when the Bauschinger effect, which results in the flow stress in tension being reduced in magnitude as a result of the previous compressive plastic strain (and *vice versa*), is of sufficient magnitude that tensile plastic strain occurs while the material is still subject to an applied compressive load (and *vice versa*) (see, for example Ref. 22 and Chapter 2). In these circumstances, the loop width at zero stress no longer gives the correct value for plastic strain range and eqn (2) must be used to determine this parameter.

The situation can be more complicated since, in practice, hysteresis loops frequently have the appearance shown in Fig. 7. In this case, the region in which strain is directly proportional to stress is virtually non-existent, and

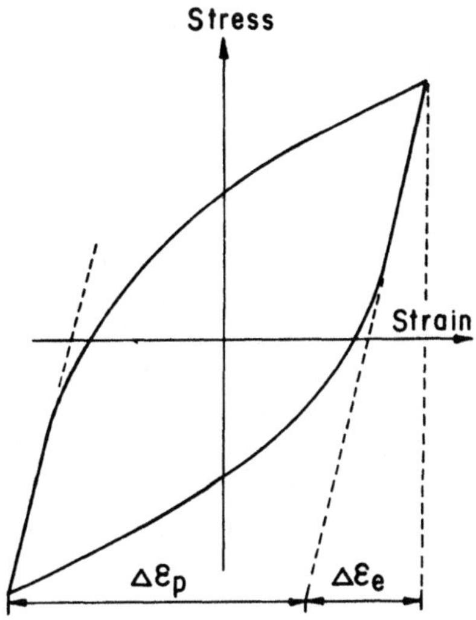

FIG. 6. Bauschinger effect leading to plastic strain being greater than the loop width at zero stress.

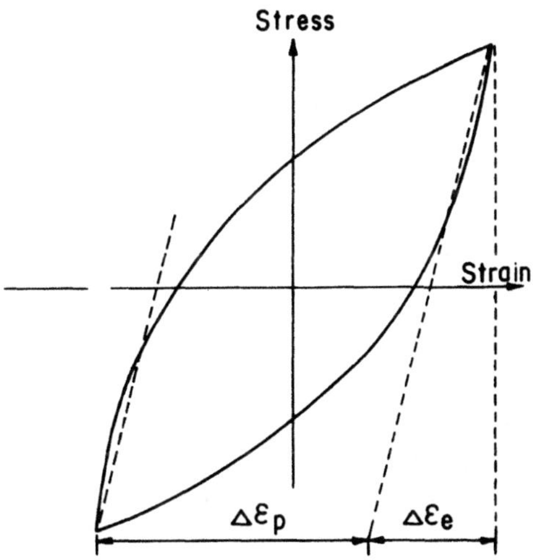

FIG. 7. Non-ideal loop showing apparent absence of linear elastic region.

it is difficult to determine from the loop the exact value of the elastic modulus. The reason for this continuous curvature of the stress–strain path is not always clear. Sometimes it can be shown to be due to lack of compatibility between the speed of response of the recording equipment and that of the control system. However, this is not always so, and in many cases the effect appears to be associated with a real material response, and it has been proposed that anelastic recovery may be responsible [23]. There seems to be a need for more work to fully understand the effects involved in this phenomenon, but in the meantime a practical solution to the problem of determining plastic strain range is to take the value of elastic modulus as that which is measured during the first quarter-cycle, i.e. the initial loading of the test. This is then used as indicated in Fig. 7 to determine the elastic strain and hence the plastic strain range. However, the method will not be accurate in circumstances in which the modulus changes during the course of the test, and in such cases it will be necessary to determine the modulus at intervals, perhaps in a separate test, in order to evaluate the correct value of plastic strain range.

When dwell periods are incorporated in the cycle, it is important to note that the elastic strain is not necessarily dependent on the total stress range,

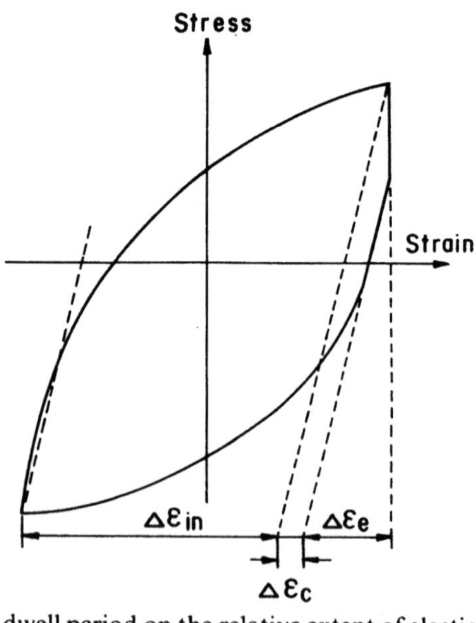

FIG. 8. Effect of a dwell period on the relative extent of elastic and inelastic strain.

since it is the stress at the end of the dwell period that determines the elastic strain, as shown in Fig. 8. It is also worth noting that the stress–strain path following a dwell period frequently differs from that following reversal without a dwell period. It generally exhibits a straight portion even when the other part of the cycle shows a curved unloading line. This is further evidence to support the idea that anelastic recovery plays a role in determining the shape of the stress–strain path following reversal.

Let us return to the semantic difficulty. This arises largely because of the adoption by a number of researchers, initially in the USA, of the term 'plastic' to mean time-independent inelastic strain, whereas it was originally defined as strain that is not recovered on unloading and would therefore include both time-independent and time-dependent inelastic strain. Although the use of different words to describe the same thing may not matter in practice, provided that the term is clearly defined in each case, it is an undesirable feature and provides another example to support the case for standardisation. Clearly, this question of definition of the terms used to describe inelastic strain will need to be resolved if international standards are to be agreed.

6.2.4. Post-test Examination

In most cases, those responsible for high temperature fatigue test programmes will ensure that the specimens are examined after test. There is valuable information to be gleaned: the site of crack nucleation; whether the crack path is predominantly transgranular or intergranular; the extent of oxidation and whether it appears to have influenced crack initiation and development; fundamental microstructural information on grain size and structure; and whether defects or manufacturing flaws that may invalidate the result are present. Such examination also enables correlation of crack depth with tensile load drop.

The acquisition and reporting of this information considerably aids interpretation of results, indicates possible design implications and ultimately contributes to the improved understanding of the subject (and in turn to the development of better testing methods). This is a strong case, therefore, for arguing that a post-test examination of specimens and the reporting of the results should be included in any standard for high temperature fatigue testing. However, the extra work involved will inevitably increase the cost of the test programme and, in order to avoid expense in routine materials property assessment, it may be better that this item should be an optional, although recommended, part of the procedure.

6.3. DISCUSSION

6.3.1. Quantification of the Effects of Differences in Test Procedures

Most of the foregoing, by implication and by intention, has been concerned with strain-controlled axially loaded tests. Attention has been drawn to differences in practice and discrepancies arising when comparing results from nominally similar tests. It is worth considering whether these factors may be quantified and if other forms of test may have relative advantages or disadvantages.

There may be a systematic difference between sets of results such as would be obtained from choice of different failure criteria, or an increase in the scatter of a set of results as would, for example, be caused by a high level of uncertainty in the measurement of strain. In Table 1 an attempt has been made to relate the predominant type of error to the causative factor, and also to indicate the interactions or secondary influences that contribute to or modify the resultant discrepancy.

If we accept that the suggested limits on accuracy of the measured parameters of strain, load and temperature are achieved, then the main areas of uncertainty concern the effects of specimen form, surface finish, failure criterion and misalignment. As yet, there is little published evidence to help in quantifying the influences of these factors on life or, in the cases of specimen form and misalignment, on stress–strain behaviour. A limited comparison of LCF data for IN738LC alloy available from five laboratories [25] indicated that endurance as a function of inelastic strain range could vary by more than an order of magnitude at low values of inelastic strain range (about 0·03%), although the variation decreased with increasing strain, being about a factor of two for inelastic strain ranges greater than about 0·5%. The requirement here is for systematic investigation with the objective of establishing just what degree of uniformity is required to reduce differences in results to acceptable levels.

A 'round robin' programme of tests to investigate the extent of laboratory-to-laboratory variation in the determination of low-cycle fatigue endurance and stress–strain behaviour is currently in progress, involving a number of organisations in Europe. The programme specifies the test conditions, and limits on measurement and alignment. The test materials will be from common casts and will be characterised in respect of microstructure and tensile properties prior to distribution to the testing laboratories. The main variables are therefore specimen shape and method of preparation, although in the case of the latter a minimum standard of surface finish is required. Full information on the changes in load during the strain-controlled tests will be obtained, and the tests will be taken to fracture or very close to fracture, so that it will be possible to apply a number of different failure criteria to each test retrospectively. In this way, it will be possible to establish quantitatively the extent to which specimen form and choice of failure criterion affect results.

It is known that the characteristics of the material, particularly the cyclic hardening or softening behaviour, will influence the way in which specimen shape and failure criterion affect life. Consequently, the programme will address this aspect by including three types of material classified according to their cyclic stress–strain behaviour as hardening, softening or stable.

This programme will therefore provide the kind of information that is necessary before decisions regarding further restrictions on specimen design (or designs) can be made with confidence. The requirements for specimens for determining endurance may be rather more stringent than for specimens used in tests for establishing the stress–strain behaviour of the material.

There is also a need for further investigation of the effects of misalignment on LCF property determination. As noted previously, even when the bending strain complies with the recommended limit, significant reductions in life are possible. However, the achievement of better alignment limits is likely to involve significantly greater expense and effort on preparation of specimens, alignment of machines and checking the specimen/machine system prior to test.

It is therefore necessary for work to be carried out to establish the feasibility (and cost) of obtaining lower bending strains before specifying these as a matter of routine. There is a corresponding need for a systematic investigation of the effects of misalignments within the recommended limits on representative materials, in order to define the extent of uncertainty in property measurements arising from this source.

6.3.2. Other Test Methods: Correlation with a 'Standard' Method

If sufficient progress can be made to produce a generally acceptable standard procedure for fatigue testing at high temperatures under axial strain control in the near future, an assumption that does not appear unduly optimistic in view of the good groundwork represented by the American standard for ambient temperature low-cycle fatigue, then the question arises as to the value of other test procedures and of the results of tests carried out over the years using 'non-standard' procedures.

Other test procedures may include, for example, those in which the strain is produced by controlled displacement of the machine crosshead and those

TABLE 1

POSSIBLE SOURCES OF INCONSISTENCY IN DATA FROM LOW-CYCLE FATIGUE TESTS AT ELEVATED TEMPERATURES

Source	*Significant factors*	*Dependent on:*	*Interacting with or affected by:*	*Properties affected (A = endurance: B = stress–strain behaviour)*	*Predominant type of error (1 = systematic; 2 = scatter)*
Specimen shape/extensometer					
(a) Smooth/axial side contact	Slipping at contact	Contact pressure	Material (hardness, oxidation response)	B, A	2
	Notching at contact	Contact pressure	Temperature	A	1
(b) Ridged/axial	Stress concentration	Profile of ridge	Material (cyclic stress–strain characteristics)	A	1
	Gauge length error (small)	Profile of ridge		B	1
(c) Smooth/axial shoulder-mounted	Gauge length error (large)	Blend radius	Material Strain type (elastic, plastic, creep)	B	1
(d) Hourglass/diametral	Small volume of material		Material (homogeneity, anisotropy, grain size)	A, B	2
	Notching at contact	Contact pressure	Material (hardness)	A, B	1
	Oxide at contact	Temperature	Material (oxidation response)	B, A	1

	Conversion from diametral to axial strain	Poisson's ratio	Material (anisotropy, grain size)	B	1
Specimen preparation	Surface finish	Final stage	Temperature (oxidation)	A	2,(1)
	Working of near-surface layer	Machine/grinding schedule	Material	A	1
Axiality of loading	System alignment Concentricity of specimen	Bending strain		B, A	2
Failure criterion	Load drop Loop shape	Sensitivity of detection	Material (cyclic stress–strain behaviour)	A	1
	Separation (fracture)	Final stages of test	Crack growth rate and specimen shape		
Control of test conditions	Temperature	Measurement and calibration		B, (A)	2
	Strain or load range			B, A	2
	Waveform	Strain rate		B, A	1
Other procedures during testing or setting up	Plastic strain during setting up	Excessive preload	Material	A, B	2
	Time at temperature prior to starting test	Test temperature, heating rate, time to reach stable temperature	Initial heat treatment of material	A, B	2

involving cyclic bending of strip specimens [25]. The results from these are not as easily interpreted as those from a direct strain-controlled test, but the tests have advantages under specific circumstances. For example, controlled displacement tests avoid the need for extensometers on the specimen and permit the use of small and therefore relatively simple-to-use environmental chambers surrounding the specimen [26]; cyclic bend tests provide a reliable and economical means of testing for very long durations [27]. In the case of displacement control, the strain in the specimen has to be calculated from the crosshead displacement allowing for the effect of deformation outside the gauge portion, i.e. in the loading train and, in particular, in the shoulder region of the specimen.

The strain distribution in cyclic bend tests is non-uniform through the thickness of the strip specimen, and this leads to some uncertainty in interpretation of the data. In both cases, however, it should be possible to compare the results with those obtained from a series of tests under standard conditions in order to provide a cross-reference or calibration by which comparisons with other results can be made. There is evidence that such an approach could be successful in comparing bend test data with uniaxial strain data [27]. In the same way, results from tests using, for example, non-standard specimens or failure criteria could also be calibrated against a standard test to provide a standardised database for low-cycle fatigue properties at high temperatures. The information arising from the 'round robin' test programme referred to earlier should provide at least some indication as to whether such an approach is feasible.

Consideration of possible future developments in test procedures suggests that another, more sophisticated, form of test may be required because of the significant influence of strain rate when time-dependent strains occur. Although at present the majority, by far, of strain-controlled tests are carried out at constant total strain rate, the strain rate that is fundamentally important appears to be the inelastic strain rate (see Fig. 14 of Chapter 2). It can therefore be envisaged that control of inelastic strain rate may be necessary, at least in some circumstances. This implies the need for computer-controlled testing, since on-line calculation and control of inelastic strain would be required. Tests of this sort therefore sound expensive but the development of computer-controlled testing machines has been rapid in recent years and they are increasingly used for high temperature fatigue testing. With the continuation of the existing trend towards cheap, large-capacity computing power, this type of test would not present prohibitive economical or technical difficulties and would be within the capability of a number of specialised test facilities currently in use.

Nevertheless, the aspect of diminishing returns must be borne in mind when considering the value of increased sophistication and precision of test procedures. This is particularly so when acquiring long-term data in the regime that combines low strain amplitude with long dwell times or low strain rates. These data are of very considerable practical significance in, for example, the design or life assessment of high temperature components used in the generation of electricity from both nuclear and fossil energy sources. There is a good case to be made for the use of relatively simple test procedures that require only basic and cheap test machines, thereby enabling large quantities of long-term data to be obtained within realistic budgets.

These considerations emphasise the value of a standard test procedure that can be used, as indicated above, to 'calibrate' such simple tests. However, this approach requires that the processes governing high temperature fatigue behaviour are well understood, and that departures from 'standard' procedures can be accounted for with confidence. This, in turn, indicates the need for the parallel development of at least some extremely precise and highly flexible test machines to generate the necessary information.

6.4. CONCLUDING REMARKS

The present position concerning the development of standard procedures for high temperature fatigue testing can be summarised as follows.

There is an increasing need for this type of test, and more organisations are producing data for design and quality control purposes. In many cases there are differences in approach forced by the application and by, for example, availability and section size of material. However, there are other arbitrary differences in procedures and inconsistencies in the definition of terms.

Where the choice of procedure is determined by the application, factors such as choice of failure criterion and selection of specimen type often differ from one laboratory to another, and may also depend upon whether endurance or stress–strain data are required. The argument for determining endurance to a standard failure criterion using standard specimens, in addition to any other criterion more specifically related to the application, is strong and the procedure would not be unduly onerous.

Other differences in approach are usually dictated by material size or characteristics. For example, if the cross section of the sample is particularly small then, to avoid buckling problems, the gauge section may be too small for axial strain measurement and diametral methods may be needed. The investigation of the behaviour of a limited region of material, as in a weld, may necessitate an hourglass specimen. The sensitivity of a material to specimen preparation procedures may dictate the choice of either grinding or machining in particular cases. It is therefore difficult to include every type of test in a standard procedure, although any standard test procedure would certainly provide useful guidance on several aspects of 'non-standard' tests.

Aspects of the test procedure that have up to now been optional include the quality of surface finish, whether the initial load is in tension or compression, the schedule for heating up the specimen before the test, the rate of strain and the waveform. In general, these aspects could be standardised without undue difficulty.

Standardisation of high temperature fatigue testing is currently a topic of a relatively high level of activity. The ASTM standard [2], on recommended procedures for low-cycle fatigue under axial strain control, includes an appendix that indicates precautionary measures for testing at high temperatures. This standard is currently being redrafted in order to incorporate these into the main body of the document.

More recently, there has been a significant level of activity in the UK, originating from the UK High Temperature Mechanical Testing Committee whose secretariat is at the National Physical Laboratory. The Low Cycle Fatigue Working Party of this committee has established the extent of consensus in the UK and has produced a recommended code of practice [28]. Although this is not yet under consideration by the British Standards Institute (BSI) for adoption as a standard, it would form a useful basis to such a document and in the meantime provides a convenient and valuable source of reference and guidance. The BSI has recently organised the preparation of a draft test method for low-cycle fatigue at ambient temperature and it is likely that a corresponding standard for elevated temperatures will be drafted in due course.

The effects of specimen form and failure criterion are currently under investigation through the European 'round robin' programme described previously, and the results will show whether further restrictions should be placed on the form of specimen for a standard test.

There are other areas needing further attention for improved standard procedures to be recommended with confidence. In particular, systematic information is needed on the effects of bending strains at levels below the present recommended value of $\pm 5\%$ of the applied axial strain range; on the influence of surface finish as a function of test temperature; and, importantly, further elucidation of the effects associated with the non-ideal character of the stress–strain hysteresis loops observed at high temperature.

Finally, there has been considerable progress towards the production of standard procedures for fatigue testing at high temperatures since a previous summary of this subject [29], and there has been a good deal of effective and constructive communication within the UK. The production of a standard by BSI is probable and, eventually, with the forging of International contacts in the standards field, especially within Europe and with the USA and Japan, an International Standard will be a realisable goal.

ACKNOWLEDGEMENT

The author wishes to thank ERA Technology Ltd for permission to publish and for the use of typing and drawing facilities.

REFERENCES

1. ASME, 1979, *ASME Boiler and Pressure Vessel Code, Case N47-17, Class 1 components in elevated temperature service*, Section III, Division 1, New York, ASME.
2. ASTM, 1980, *Recommended practice for constant amplitude low cycle fatigue testing*, E606-80, Philadelphia, ASTM.
3. Hales, R., 1983, Fatigue testing methods at elevated temperature, *Fatigue at high temperature*, ed. R. P. Skelton, London, Elsevier Applied Science, pp. 63–96.
4. Massarelli, L. and Marchionni, M., 1977, High temperature low cycle fatigue behaviour of nickel-base superalloys; influence of testing frequency and stress or strain waveform, Final Report to COST 50, October 1977. Laboratorio per la Technologia dei Materiali Metallici non Tradizionali del CNR, Milan.
5. Day, M. F. and Harrison, G. F., 1982, Design and calibration of extensometers and transducers, *Measurement of high temperature mechanical properties of materials*, ed. M. S. Loveday, M. F. Day and B. F. Dyson, London, HMSO, pp. 225–40.
6. Thomas, G. B., 1985, Axial extensometry for ridged specimens, *Techniques for high temperature fatigue testing*, ed. G. Sumner and V. B. Livesey, London, Elsevier Applied Science, pp. 45–56.

7. Sumner, G., 1985, Heating methods and grips, *Techniques for high temperature fatigue testing*, op. cit. [6], pp. 71–96.
8. Bressers, J., 1982, Axiality of loading, *Measurement of high temperature mechanical properties of materials*, op. cit. [5], pp. 278–95.
9. ASTM, 1984, *Verification of specimen alignment under tensile loading*, E1012, Philadelphia, ASTM.
10. Soo, P. and Chow, J. G. Y., 1978, Correlation of low cycle and high cycle fatigue data for solution-annealed Incoloy 800, *Alloy 800, Proc. Petten Int. Conf.*, Amsterdam, North Holland, pp. 169–74.
11. Day, M. F. and Thomas, G. B., 1979, Microstructural assessment of fractional life approach to low cycle fatigue at high temperatures, *Metal Sci.*, **13**, 25–33.
12. Thomas, G. B. and Varma, R. K., 1984, Assessment of LCF and creep–fatigue behaviour of Ni–Cr–base superalloys, Final Report, COST 50, Round III, NPL Report No DMA (A) 90.
13. Hufnagel, R. J., Schmale, D. T. and Jones, W. B., 1984, Elevated temperature mechanical testing of metals, *Indust. Engng Chem. Prod. Res. Dev.*, **23**, 144–9.
14. Colclough, A. R., 1982, Methods of practical thermometry in the range 0 to 3000°C—a survey, *Measurement of high temperature mechanical properties of materials*, op. cit. [5], pp. 58–90.
15. Desvaux, M. P. E., 1982, The practical realisation of temperature measurement standards, *Measurement of high temperature mechanical properties of materials*, op. cit. [5], pp. 91–112.
16. BSI, 1969, *Methods for creep and rupture testing of metals*, BS 3500.
17. Bhanu Sankara Rao, K., Valsan, M., Sandhya, R., Ray, S. K., Mannan, S. L. and Rodriguez, P., 1985, On the failure condition in strain-controlled low cycle fatigue, *Int. J. Fatigue*, **7**(3), 141–7.
18. Raynor, D. and Skelton, R. P., 1985, The onset of cracking and failure criteria in high strain fatigue, *Techniques for high temperature fatigue testing*, op. cit. [6], pp. 143–66.
19. Skelton, R. P., 1986, Personal communication.
20. Remy, L., Reger, M., Reuchet, J. and Rezai-Aria, F., 1982, An operational definition of life to crack initiation in high temperature fatigue, *High temperature alloys for gas turbines 1982*, ed. R. Brunetaud *et al.*, Dordrecht, Holland, D. Reidel, pp. 619–32.
21. ASTM, 1983, *Method of measurement of fatigue crack growth rate*, ASTM E647-83T.
22. Challenger, K. D. and Vining, P. G., 1983, Sub-structure and back stress changes resulting from the cyclic loading of $2\frac{1}{4}$Cr–1Mo steel at 755 K, *Mat. Sci. Engng*, **58**, 257–67.
23. Gostelow, C. R., 1985, Strain measurements in high strength anistropic materials, *Techniques for high temperature fatigue testing*, op. cit. [6], pp. 57–70.
24. Thomas, G. B., Bressers, J. and Raynor, D., 1982, Low cycle fatigue and life prediction methods, *High temperature alloys for gas turbines*, op. cit. [20], pp. 291–317.
25. Forrest, P. G. and Penfold, A. B., 1961, New approach to thermal fatigue testing, *Engineering*, **192**, 522–3.
26. Skelton, R. P., 1985, High strain fatigue testing in inert environments, *Techniques for high temperature fatigue testing*, op. cit. [6], pp. 117–41.

27. Batte, A. D., Dawson, R. A. T., Evans, R. B. and Thomas, G., 1985, Reverse-bend testing to obtain long-term endurance data, *Techniques for high temperature fatigue testing*, op. cit. [6], pp. 97–115.
28. UK High Temperature Mechanical Testing Committee, 1986, *A code of practice for constant amplitude low cycle fatigue testing at elevated temperature* (UKHTMTC/LCF/2/86, ISBN 0-946754-05-5), National Physical Laboratory.
29. Thomas, G. B., 1982, Standards for high temperature fatigue, *Measurement of high temperature mechanical properties of materials*, op. cit. [5], pp. 32–41.

7

The Relation Between Laboratory Specimen and the Practical Case

R. P. SKELTON

Central Electricity Research Laboratories, Leatherhead, Surrey, UK

SUMMARY

The smooth cylindrical fatigue specimen provides the starting point for most investigations in high strain fatigue. If it is to be used for a determination of cyclic stress–strain properties or a metallurgical investigation of microstructural changes or for crack morphology studies at elevated temperatures, one need enquire no further. However, if specimen endurance is to be related to component behaviour then size effects become important. Further, the traditional laboratory simplification of testing between fixed strain limits warrants closer examination when certain stresses encountered in service are being simulated.

This final chapter thus very briefly attempts to distinguish between specimen and structural (i.e. component) behaviour, and ends with a short discussion on cyclic inelastic analyses, design codes and their role in lifetime prediction.

7.1. INTRODUCTION

When it is required to determine the high temperature mechanical properties of a metal or alloy, a sample of that material is taken, manufactured into a specimen, and certain tests are performed on it. It is implicit that material response is uniform between the strain-measuring points. The resulting behaviour (such as the stress–strain relation, or secondary creep rate, or cavitation-controlled ductility) is then taken to be representative of that which would occur in the bulk of a component fabricated from the same material and subjected to the same stress levels.

In the previous chapter it has been shown that the traditional high-strain fatigue test is likely to become more closely controlled in laboratory testing, recommendations being given on specimen finish, shape and strain measuring methods. It is as well to enquire whether the 'specimen' is being used for a material property determination or is modelling a practical situation. One of the aims of this final chapter is therefore to take a broader view of the push–pull test with respect to the component it simulates and the role it plays in the overall strategy of design and lifetime assessment at elevated temperature.

7.2. DEFECT-FREE AND CRACKED SPECIMENS

The start-up and shutdown cycles which give rise to reverse plasticity have been discussed several times in this book. For a really thick structure (say 150 mm for a steam chest) the initiation of a fatigue crack in a typical laboratory specimen (say 10 mm diameter) after a given number of cycles would be interpreted as those expected in service provided (a) the strain range at the component surface was reproduced accurately, including if necessary periods of large strain (major cycles) and small strain (minor cycles), and (b) the frequency or dwell time effect had been properly simulated. Although crack propagation is not the subject of interest (neither in the specimen nor in the component) for this chapter, in many cases the complete failure of the testpiece would be accepted as the criterion for initiation in the surface layers of the service item.

When, on the other hand, the component is thin (for example a fuel element can of 0·5 mm thickness in a nuclear reactor) the crack initiation criterion in the same 10 mm diameter specimen must be chosen with great care, as there is obviously a danger of overestimating the likely cycles experienced in service [1]. When total strain range is the governing parameter for growth there is not always a one-to-one correspondence in crack advance rate between specimen and component [2] because of differences in load drop for the same crack depth (compliance). In general, this will depend on the ratios a/w and g/w, where w is the specimen width or diameter, a is the crack depth, and g is the gauge length. For $a/w > 0{\cdot}2$, for example, the specimen has become a *structure* (see Chapter 1). It reverts to becoming a specimen if the cyclic crack growth rate is described by a normalising parameter such as stress intensity factor range ΔK, which involves both gross stress and crack depth. Such considerations lie outside our present scope.

7.3. CYCLE OF EVENTS

The role of the push–pull specimen, although multifarious, plays only a part in the sequence of events leading to a lifetime assessment of a critical component in service. This book has shown that its primary purpose is to generate total endurance data. (Entering an agreed failure curve at given strains in order to sum the damage accrued by various service cycles is quite often the easiest part of an assessment.) It may also be used to provide cyclic stress–strain data, as has been discussed at length in Chapter 2. Finally, if some crack growth per service cycle can be tolerated, then that same push–pull specimen can be used to furnish crack growth data [1, 2].

If the component is simple, and the surface strain range can be calculated easily, then the three parameters outlined above can be used together with metallographic evidence (Chapters 3 and 4) to estimate the cyclic lifetime (Chapter 5). For thick components, valuable additional data can be supplied by the *thermal shock* test on suitable billets. This models service behaviour more accurately than an isothermal direct stress test because (a) the temperature is cycled, (b) there is a simulative strain gradient induced with each transient, and (c) the surface stress state is equibiaxial. Further, a strain-concentrating feature such as a circumferential groove may be incorporated and approximate methods used to calculate the strain enhancement [3].

When the service component, or indeed the laboratory specimen, is complicated in shape, then finite element analyses are required to calculate the strain range at specific locations. For thermal phenomena this process requires a knowledge of the heat transfer coefficient, the physical constants of the alloy (expansion coefficient, Young's modulus etc.) and their variation with temperature, and cyclic stress–strain data (constitutive equations). The ensuing calculations are costly and time-consuming, and are only performed when lifetime cannot be guaranteed by simpler methods (see Section 7.5).

If whole-component (structure) testing is impracticable and push–pull specimen testing cannot match the plant conditions, then a compromise may be sought in a *features test*. One example of this is a bellows [4]. Another example selected almost at random from the literature is given in Fig. 1 [5]. Suppose it were required to determine the fatigue properties of the thin-section joint X, and that a bending deformation mode was suspected in service. Push–pull testing could not do justice to the zone in question and so a cantilever-type fatigue test on a section through the tube and sleeve could thus be conceived. This, together with a subsequent

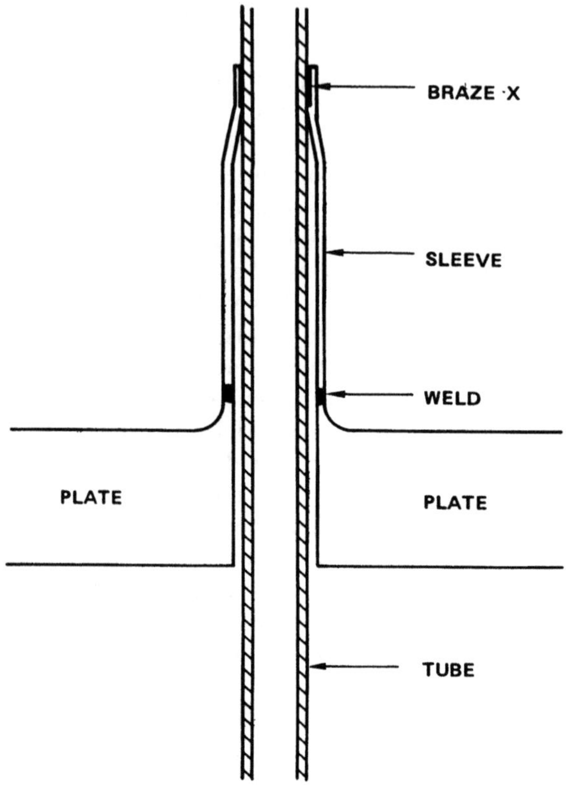

FIG. 1. Example of a 'feature' involving several members of different thickness (after Ref. 5). (Reproduced with permission of Thomas Telford Publications.)

metallographic examination of the crack path, would give a more realistic estimate of likely service cycles.

Fatigue tests on service-exposed material, whether in the form of a specimen or feature, complete the final link in our chain of events. Their purpose is to validate a previous lifetime assessment or to guarantee a further period of operation beyond the original design life.

7.4. PUSH–PULL SIMULATION OF STRUCTURAL BEHAVIOUR

In this section we examine the extent to which the service situation may be modelled in the laboratory. Thermomechanical testing (in which temperature and applied strain are varied in a systematic manner on push–pull specimens) is omitted since it is discussed in detail in Chapter 3.

7.4.1. Bending

Unless specimen alignment is deliberately offset in some way, the bending mode cannot be incorporated in a push–pull test. Instead, the four point reverse bend test may be chosen. Deformation over a constant radius ensures that the surface fibre strain can be matched to that in direct stressing, and so there is a close correspondence between crack initiation cycles in both types of specimen. Despite the presence of a strain gradient across the section, total endurances are also in agreement [6]. The system possesses excellent reliability for long-term tests, although hardening and softening are more difficult to interpret since the bending stress is required to be calculated.

7.4.2. Strain Concentration

Consider a cylindrical specimen containing a circumferential groove, the whole undergoing high strain fatigue. This system borders on a 'features' test since the plastic enclave at the root of the groove is surrounded by material which either deforms elastically or, with increasing remote stress, σ, and strain, ε_t, elasto-plastically. The local strain $\varepsilon_{t\max}$ may in certain circumstances be measured, but the local stress $\sigma_{\max}$ must be calculated, for example by the Neuber relation [7]:

$$\frac{\sigma_{\max}}{\sigma}\frac{\varepsilon_{t\max}}{\varepsilon_t} = K_t^2 \tag{1}$$

where K_t is the elastic stress concentration factor.

Wetzel [8] has shown how a plain push–pull specimen may be used to model material behaviour in the enclave itself. If deformation in the *grooved* specimen is conducted at a constant elastic strain range remotely, then eqn (1) becomes:

$$\sigma_{\max}\varepsilon_{t\max} = K_t^2\varepsilon_t^2 E = \text{constant} \tag{2}$$

where E is Young's modulus. So, for a given value of ε_t and K_t in eqn (2), 'control hyperbolas' with their origins at their respective loop tips may be constructed as shown in Fig. 2, and the hysteresis loop of a *plain* specimen constrained to move between them. The original experiments were performed manually [8] but nowadays computer control is available [9]. The example of Fig. 2 is one of cyclic hardening behaviour in the enclave

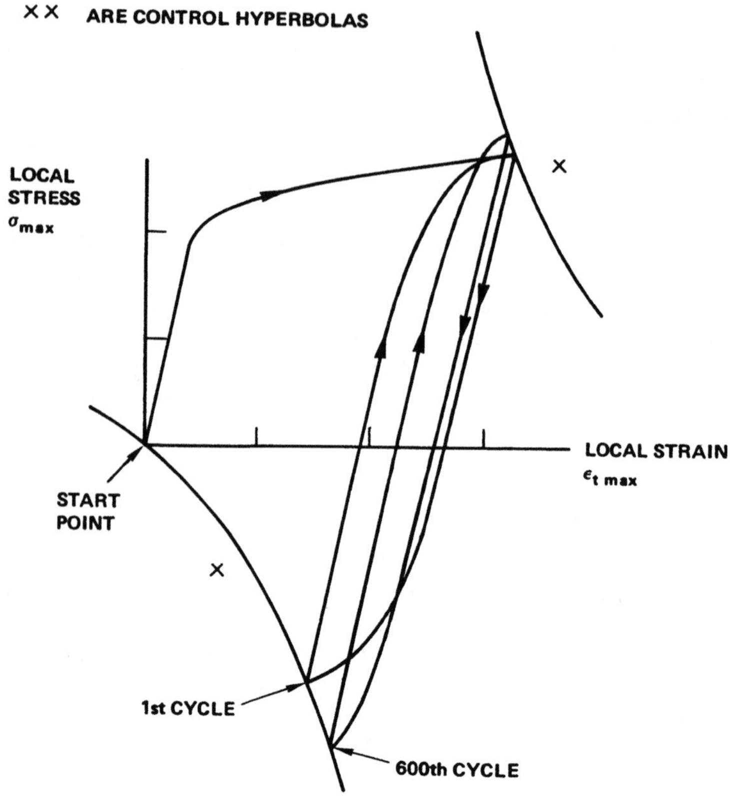

FIG. 2. Using a push–pull specimen to simulate notch-root behaviour (after Ref. 8). (Copyright ASTM. Reproduced with permission.)

and, of course, will be different (cycle by cycle) from that observed in total or plastic strain control (see Fig. 4 of Chapter 2).

For research purposes, a similar set of bounding curves could be set up for constant energy of deformation (area of hysteresis loop; see Chapter 2), although this situation does not possess a direct service analogy. The experiments of Wetzel [8] were conducted to verify the conditions for crack initiation at a notch root.

7.4.3. Welds and Composite Materials

A weld may be regarded as a *metallurgical* concentration, since not only do the cyclic stress–strain properties differ from parent material, for example softening in austenitic weld metal where parent metal hardens, but the values of Young's modulus are often anisotropic in austenitic materials [10]. In ferritic welds, the weld metal is softer than the parent and the heat-affected zone and fusion-boundary regions are favoured for crack initiation and propagation. The testing of weld metal itself in a push–pull gauge presents no difficulties, but a *weldment* incorporates parent metal on both ends. With longitudinal (axial) extensometry, therefore, suitable correction factors are required since the specimen once again has become a feature (material connected in series). It may be shown that if n is the fractional volume occupied by weld metal in the gauge then:

$$n\varepsilon_{\mathrm{tw}} + (1-n)\varepsilon_{\mathrm{tp}} = \varepsilon_{\mathrm{td}} \tag{3}$$

where the suffixes 'w', 'p' and 'd' refer to weld, parent and duplex (composite) gauge material respectively. Thus, if the stress–strain properties of parent and duplex material are known, $\varepsilon_{\mathrm{tp}}$ and $\varepsilon_{\mathrm{td}}$ may be substituted from the Ramberg–Osgood equation (eqn (2) of Chapter 2) to give a stress–strain relation for the central weld region. Diametral extensometry in this case would alternatively be valuable for probing different areas.

However, it would not be helpful in the instance of specimens which have a carburised case and an uncarburised core (parallel arrangement). The corresponding relation becomes

$$n\sigma_{\mathrm{c}} + (1-n)\sigma_{\mathrm{u}} = \sigma_{\mathrm{d}} \tag{4}$$

where the suffixes 'c', 'u' and 'd' refer to carburised, uncarburised and duplex material. A stress–strain relation for the carburised layer itself may then be derived [11].

7.4.4. Elastic Follow-up

Lee [4] has discussed instances in thin-section AGR components (e.g. retention loops which connect cover plates, or bellows convolutions) in which differential temperatures dictate that the feature is under displacement control. Stresses may thus be classed as secondary (see Section 7.4.6), yet during relaxation at 650°C over several hundred hours some extension occurs at the critical location. This is termed *incremental behaviour* or *elastic follow-up* and, in the laboratory, is likened to a plain specimen connected to a spring S in Fig. 3. During true stress relaxation, the total strain across T_1T_2 is kept constant and so relaxation strain $\varepsilon_{\mathrm{pr}}$ is given by

$$\varepsilon_{\mathrm{pr}} = -(\sigma_0 - \sigma_{\mathrm{f}})/E \tag{5}$$

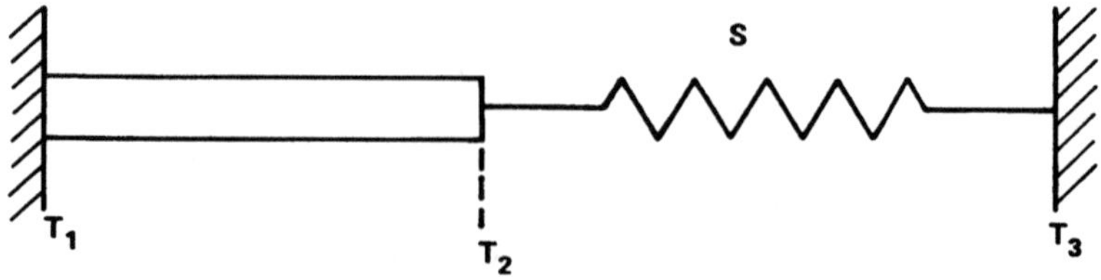

FIG. 3. During elastic follow-up, the specimen T_1T_2 is supposed to be in series with a spring S of variable stiffness.

(see eqn (4) and Fig. 2(d) of Chapter 2). When follow-up is present, the total displacement across T_1T_3 is kept constant and the relaxation strain ε'_{pr} is now given by:

$$\varepsilon'_{pr} = Z(\sigma_0 - \sigma_f)/E \tag{6}$$

where $\varepsilon'_{pr} > \varepsilon_{pr}$. A typical value of Z in service is 3·5 [4]. The term E/Z may be defined as an effective modulus $E^* < E$, and follow-up may be regarded as an exchange of plastic strain for elastic strain in the whole system, not just the gauge length. The value of Z is determined by the 'spring constant' S in Fig. 3 and in the laboratory may be modelled either by computer control, or by deliberately positioning extensometry away from the gauge section.

In the bellows convolution crest considered by Lee [4], strain enhancement was calculated as follows. In the stress–strain coordinate system of Fig. 4 the point A was arrived at by elastic means. The slope AB was defined by the factor Z discussed above, intersecting the (stabilised) cyclic stress–strain curve at B′. Next, a stress level σ_{rH} was chosen below which relaxation could not occur during the specified time interval, so that point B'' was fixed. Upon unloading from B'' elastically into compression a mirror stress–strain curve to OB′ was constructed, taking O′ as the new origin. From the parallelogram OAB″CO the intersection point D was found. The process was then repeated from D to obtain the point B'_2. The final steady loop is thus $B'_2B''DB'_2$, the total strain range of which is $\Delta\varepsilon$ and the creep range $\Delta\varepsilon_{cr}$. It should be noted that the cyclic yield points in tension and compression have been assumed to be equal.

A very similar construction may be performed when point A lies below the conventional yield stress σ_{yT} [4]. Elastic follow-up may still occur but, provided that the compression excursion does not exceed σ_{yC}, shakedown to elastic action is guaranteed. For the component considered, this occurred at a limiting total strain range $\Delta\varepsilon$ of 0·1%.

This method of intersecting an appropriate stress–strain curve resembles the Neuber construction discussed in Chapter 2, i.e. an elastic solution is first calculated and then afterwards modified.

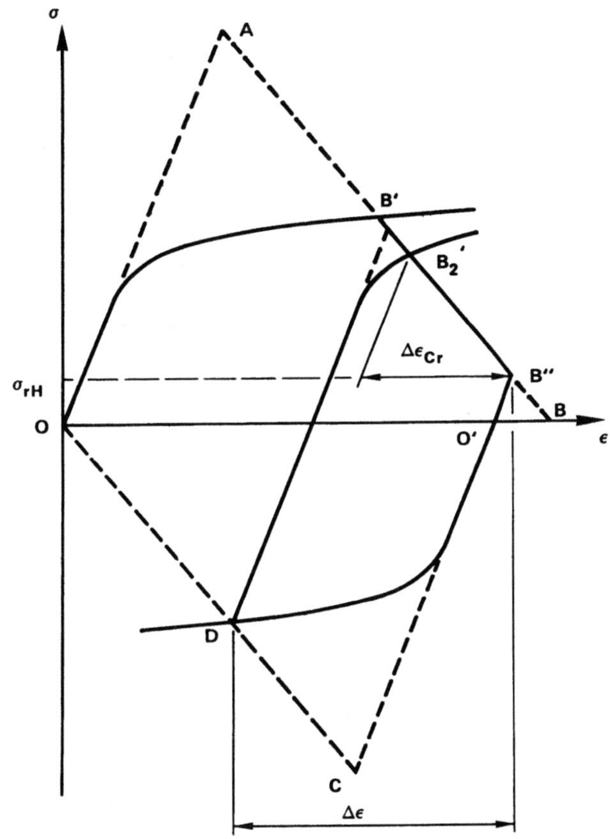

FIG. 4. Method of determining the steady-state hysteresis loop when follow-up is present (after Ref. 4). (Reproduced with permission of Thomas Telford Publications.)

7.4.5. Constrained Versus Contained Yielding

If a component of uniform cross section is prevented from expanding by constraints A and B in Fig. 5(a), then hysteresis behaviour is adequately modelled in the laboratory by choosing an appropriate temperature and replacing the expansion strain $\alpha\Delta T$ (where α is the coefficient of expansion and ΔT is the change in temperature) by a mechanical displacement, that is, a total strain range ε_t. Similarly, if the component is thick and only the surface undergoes the same change ΔT (Fig. 5(b)), it is customary to model this case also by testing between constant strain limits ε_t, thus simulating

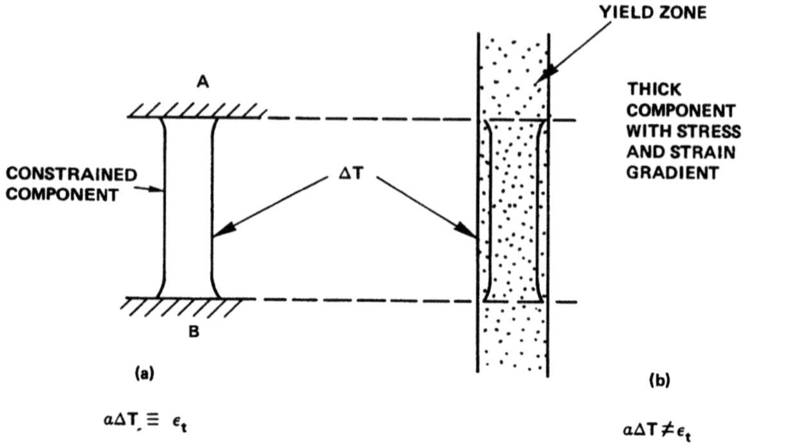

FIG. 5. Comparison of (a) externally constrained and (b) internally constrained ('contained') deformation.

the residual/steady load stress states during heating and cooling, as discussed in Chapter 3. However, Chell has pointed out [12] that because of the steep stress gradient the Neuber relation (eqns (1) and (2)) applies in this case even though no stress concentrator is present. It is merely sufficient that small-scale yielding is contained by the remaining structure. Accordingly, if in Fig. 6 we take OP as an elastic solution, the actual intersection with the equibiaxial cyclic stress–strain curve is not along PQ (strain invariance) but is obtained by putting $K_t = 1$ in eqn (1) and constructing the hyperbola PR. Thus fully reversed cyclic modelling for this case should strictly be conducted between 'control hyperbolas' as shown in Fig. 2, although the steepness indicates only a small departure from total strain control.

7.4.6. Primary and Secondary Stress Combination

It is important to distinguish between two types of stresses. Thermal transients, the effects of which at the surface are modelled as described in the last section, are examples of *secondary* stresses, i.e. they cause no net force across the component and so are self-equilibrated. Thus an induced tensile stress at the surface due to cooling must be balanced elsewhere in the section by a compressive stress. On the other hand, *primary* stresses arise from mechanical sources such as end loads (which may be displacement-controlled) and hoop stresses due to internal pressure. It is thus seen that either the secondary or primary members are simulated by the push–pull

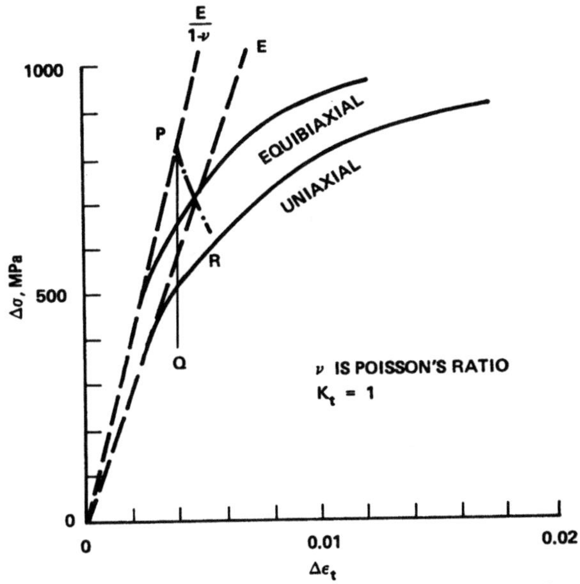

FIG. 6. Calculation of the surface strain range during thermal quenching of a plain thick steel cylinder (see Chapter 1 of the earlier volume for details).

specimen according to whether the component itself is slender (as shown in Fig. 5(a)) or thick (Fig. 5(b)). What happens when we wish to model the effect of both stress types at once?

Since the mean stress in continuous cycling high-strain fatigue is always zero, or very nearly so, any attempt to represent the primary stress by an appropriate increase in the tensile strain limit, for example, will simply result in a new loop which settles down to zero mean stress, and a correspondingly greater plastic strain range. We have only succeeded in increasing the secondary stress range. Similarly, reverting to load control will not suffice, since Fig. 13(b) of Chapter 2 shows that even when the load limits are equal, ratchetting may ensue and this would be exacerbated if the tension value were increased by an amount equal to the primary stress. Clearly, a type of cycle is required in which the upper limit is a specified load while the lower limit is a specified strain. This type of cycle can be executed by computer or other forms of control.

Chell has argued [12] that, where gradients are involved, the upper limit is not the sum of the primary and secondary stresses since the small-scale yielding criterion, e.g. eqn (1), breaks down if the primary stress spreads the yielded enclave deep into the section. An estimate of the intersection point

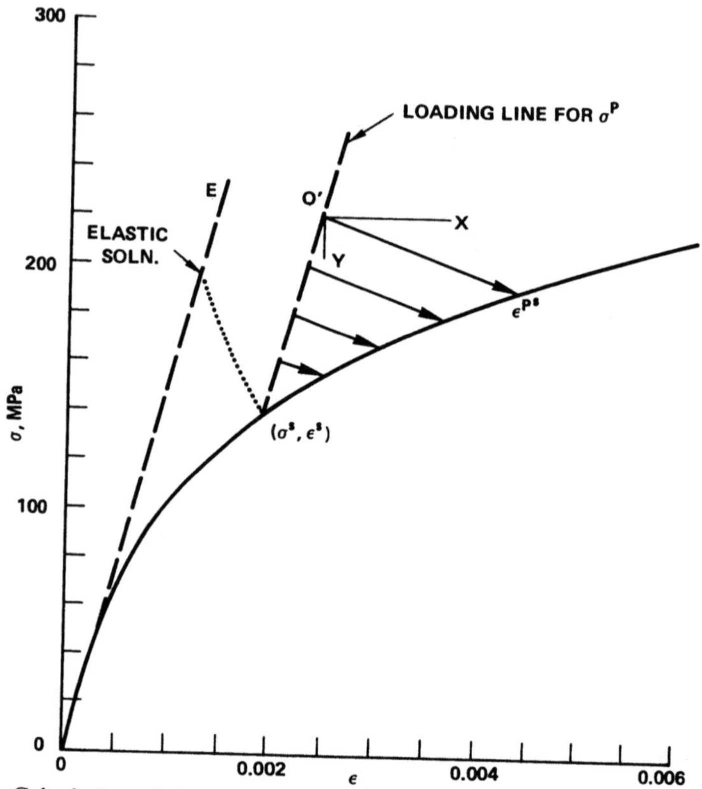

FIG. 7. Calculation of the eventual stress–strain state in the presence of primary and secondary stresses.

on the stress–strain curve is given by means of the following bounding theorem [12]:

$$\varepsilon^{\mathrm{ps}} = \varepsilon f(\sigma^{\mathrm{p}} + \sigma^*) + \frac{(\sigma^{\mathrm{s}} + \sigma^*)}{2\sigma^*}[\varepsilon^{\mathrm{s}} - \varepsilon f(\sigma^*)] \tag{7}$$

where $\varepsilon^{\mathrm{ps}}$ is the required strain due to primary and secondary stresses, σ^{p} is the primary stress, and ε^{s} is the secondary strain, while σ^* is a dummy variable chosen to minimise the right-hand side of eqn (7). This may be solved by using the Ramberg–Osgood equation (eqn (2) of Chapter 2) so that the first term in eqn (7) becomes

$$\left(\frac{\sigma^{\mathrm{p}} + \sigma^*}{E}\right) + \left(\frac{\sigma^{\mathrm{p}} + \sigma^*}{A}\right)^{1/\beta}$$

the first term in square brackets is

$$\frac{\sigma^s}{E} + \left(\frac{\sigma^s}{A}\right)^{1/\beta}$$

where σ^s is the secondary stress, while the final term is

$$\frac{\sigma^*}{E} + \left(\frac{\sigma^*}{A}\right)^{1/\beta}$$

Referring to Fig. 7, the point $(\sigma^s, \varepsilon^s)$ is arrived at by first applying the Neuber equation (1) to the elastically calculated secondary stress. Then, for selected values of primary stress σ^p (shown at 20 MPa intervals in Fig. 7) the corresponding values of ε^{ps} are found from eqn (7). It is seen that the final values are each bounded by a constant stress solution (along O′X) and a strain-invariant solution (along O′Y).

As the primary stress increases with respect to a repeatedly applied secondary stress, ratchetting can indeed occur [13]. Incremental strain per cycle is given by a relation of the form $[\sigma_y/E](\overline{\sigma^s}\overline{\sigma^p} - 1)$, where σ_y is yield stress and $\overline{\sigma^s}\overline{\sigma^p}$ are normalised to that yield stress: this behaviour could also be modelled by suitable specimen control.

7.5. MODELLING HYSTERESIS LOOP BEHAVIOUR

The study of the evaluation of stress–strain fields and their evolution in structures is known as *cyclic inelastic analysis*. The calculations must be pursued until stress redistribution has occurred and stabilisation reached in each volume element of the structure. Unless we are lucky enough for the problem to be reduced to single-specimen behaviour, more complicated finite element calculations requiring constitutive equations are employed. However, less complex methods are available which may save a costly and time-consuming undertaking.

For the case of continuous cycling, data can be generated readily in the laboratory, and in design calculations the monotonic, 10th cycle (for example) or stabilised cyclic curve can be entered. Where prolonged dwell times are involved, laboratory data are not likely to be available, and so an estimate of the shape of the loop may be made by (a) assuming monotonic behaviour in the compression-going direction after the tensile dwell (because softening has occurred), (b) isotropic hardening during the tension-going phase (to maximise the peak tension stress), and (c) a suitable relaxation law during dwell [14] (see also Chapter 5). A kinematic hardening rule, as shown in Fig. 23(b) of Chapter 2, may alternatively be used. In either case, the stabilised plastic strain range is quickly established.

If the design calculation requires more accurate cycle-by-cycle data, then these may be supplied by appropriate computer simulation of specimen behaviour, for example using the Ramberg–Osgood law and empirical hardening or softening relations such as eqns (34)–(39) of Chapter 2. Alternatively, visco-elastic laws and the Bauschinger effect may describe the process [15]. In either case, the resulting gradual change in stress range and plastic strain range are recorded.

Turning to behaviour in a structure, simplified methods, such as use of the Neuber rule, direct entry into the stabilised stress–strain curve, or a check for ratchetting may preclude further costly calculations. Specialised thin-shell programs may be used. Goodman and Hutton describe [16] a thin shell/flange feature which has a 675°C temperature on one side and is at 494°C on the other during the steady state. When the upper temperature is rapidly reduced, the thin shell immediately adjusts but the flange region lags behind. Reversed thermal stresses over 300 cycles with 700 h dwell periods were expected. First-cycle results were validated against known

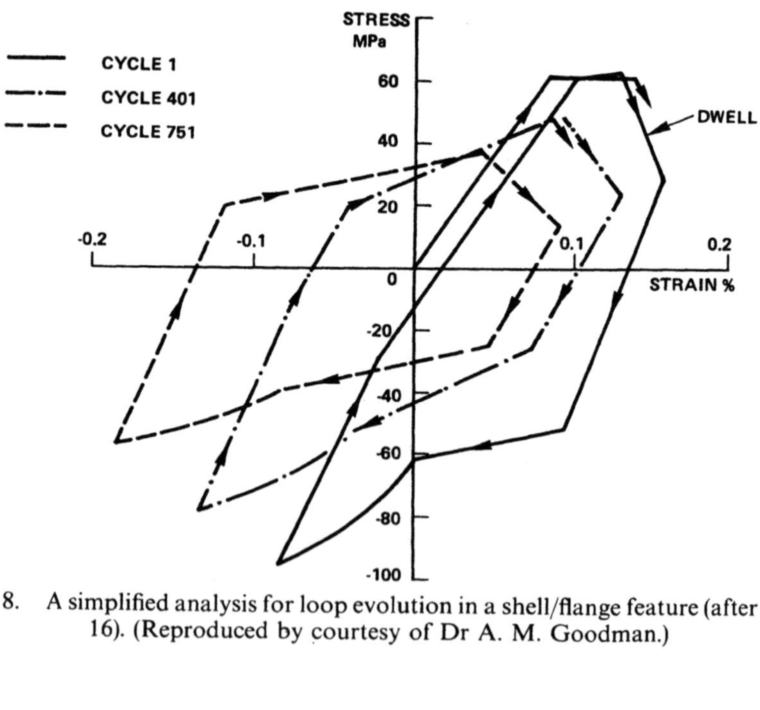

FIG. 8. A simplified analysis for loop evolution in a shell/flange feature (after Ref. 16). (Reproduced by courtesy of Dr A. M. Goodman.)

finite element solutions but subsequent behaviour was predicted on the simplified program. Figure 8 shows the loop evolution [16]. The authors were able to show that the region of maximum damage moved from a bend in the shell to a location nearer the flange.

Further discussion of finite element techniques lies beyond the scope of this closing chapter. Chaboche [17] points out that structural stabilisation depends on (a) the hardening/softening properties of the material itself, (b) the adoption of a mean stress level which is related to (a) and also the loading condition, and lastly (c) stress redistribution induced by plastic flow, related to component geometry and loading conditions. An example using the French model EVPCYCL [17] is given in Fig. 9, which shows the calculated hysteresis loop development at a notch root. Similar loops have been calculated for thermal cycling of thick cylinders [18]. Many other instances can be quoted, for example the program CREPLACYL for

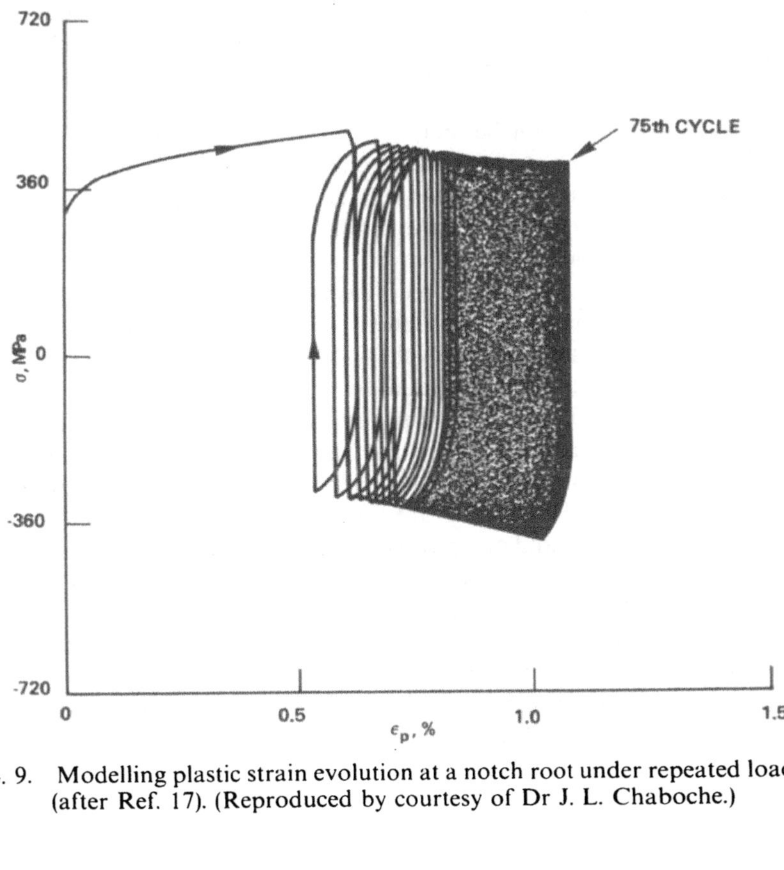

FIG. 9. Modelling plastic strain evolution at a notch root under repeated loading (after Ref. 17). (Reproduced by courtesy of Dr J. L. Chaboche.)

determining the elastic–plastic stress distribution in a thick 316 boiler stop valve in supercritical steam plant which was run up and shut down from 600°C [19]. The CEGB program BERSAFE [20] for calculation of reversed loading requires a series of tip-to-tip parameters, i.e. stress values and corresponding plastic strain ranges, and incorporates a kinematic hardening rule.

7.6. DESIGN CODES

In the field of design and lifetime assessment of plant components, one purpose of a Code is to save the designer from having to make judgements on metallurgical or mechanical properties data. It is assumed that this has already been drawn together (from the results of laboratory tests) in the construction of recommended assessment curves, e.g. fatigue endurance. However, a return to the original data, in order to determine the degree of scatter, or variation in properties from one material to another, can be justified in a sensitivity analysis which is undertaken when the initial lifetime conditions cannot be met.

A design route must strive to be simple. In the first instance it must be conservative, and subsequent tests related to a specific application may highlight that conservatism. For example, in the feature considered by Goodman and Hutton [16], application of the inelastic assessment route of Code Case N-47 [21] led to a creep-damage summation of 22 (when a factor of unity is sought!). Consequently, component tests were set up whereby a heating and cooling cycle was employed. A reduction of the dwell time to 6 h was compensated by a slight increase in maximum temperature. After 750 cycles were performed in the rig, a metallographic examination of the critical location showed no sign of distress. Similarly, the lifetime assessment rules in N-47 regarding thin-section components operating in the AGR system are quite severe, and by undertaking hardness, stress–strain and creep tests on Alloy 800 at 650°C, Lee [4] was able to demonstrate some benefit in life for a cold-worked bellows compared with that of a similar annealed unit.

An interesting aspect of the latter work was that the inverse of the cyclic hardening exponent in the Ramberg–Osgood law was found to be equal to the index in the Norton creep law. Analysis shows [22] that in this case the enhancement factor Z in Fig. 4 is given by $Z = (1/\beta + 2)/3$, so for a typical value of $\beta = 0{\cdot}125$, $Z = 3{\cdot}3$. This is discussed further below.

Appendix T of ASME Code Case N-47 has received widespread

attention from both the design/assessment and the materials testing point of view. Its application is optional for components which see mostly secondary stresses, for example in the fast breeder reactor, but alterations to the basic structure of the Code have been proposed in several countries [23] in areas where it was felt to be deficient. The Code itself is continually under review and the latest edition includes a fatigue assessment curve for the ferritic $2\frac{1}{4}$Cr–1Mo steel at elevated temperature when, for many years, only an austenitic failure curve had been available.

The Code contains an elastic route which is highly conservative, and if this cannot be satisfied one enters an inelastic route which requires a better estimate of plastic and creep strains but, in token of this, is less pessimistic. Rules are given for checking against ratchetting, but the model on which these are based is a cylinder subjected to a primary stress together with a self-equilibrating secondary (thermal) stress acting across the wall. However, as has been shown, certain displacement-controlled stresses (another example is a nozzle-to-shell intersection) can give rise to elastic follow-up, which is a form of ratchetting. These secondary stresses must therefore either be reclassified as primary in nature, and therefore not allowed to relax, or must be computed as has been done in Fig. 4. The first assumption can incur severe penalties because both primary and secondary stresses are summed in a damage calculation.

This phenomenon of elastic follow-up is recognised in a Japanese development of the Code [24], with the proviso that stresses must be reclassified as primary if the total calculated strain (depending on Z) exceeds 0·5%. A distinction is also made between cyclic stress–strain properties and monotonic values according to the length of the dwell. The mode of assessing creep and fatigue damage has also been changed (e.g. continuous cycling fatigue damage can be found for three different strain rates). Similarly, the French Code RCC-MR [25] introduces cyclic stress–strain curves in an appendix of materials properties, proposes a new ratchetting condition based on an equivalent primary stress and an 'efficiency diagram', and re-expresses the Neuber relation in the presence of primary and secondary loadings (see Section 7.4.6). Within the UK, there has been a proposal to re-configure the ASME N-47 creep–fatigue interaction diagram for austenitic steels [26] (see Chapter 5), and for thick-section ferritic materials undergoing thermal fatigue, a CEGB assessment route has been produced [27] comprising (a) stress analysis, (b) damage summation for crack initiation criteria, and (c) the subsequent growth of short and long cracks. All this activity is evidence of a healthy communication between engineers and stress analysts on the one hand and materials scientists and metallurgists on the other. In the experimental field, the features test in particular has done much to reduce the areas of misunderstanding.

ACKNOWLEDGEMENTS

The author would like to thank several colleagues in the CEGB, UKAEA and NNC for discussions which have enabled this short overview to be written.

REFERENCES

1. Skelton, R. P., 1985, High strain fatigue testing at elevated temperature: a review, *High Temp. Technol.*, **3**, 179–93.
2. Skelton, R. P., 1986, The application of small specimen crack growth data to engineering components at high temperature: a review, Symposium on Low Cycle Fatigue, Lake George, ASTM STP 942, Philadelphia, ASTM, ed. H. Solomon (to be published).
3. Skelton, R. P. and Miles, L., 1984, Crack propagation in thick cylinders of ½Cr–Mo–V steel during thermal shock, *High Temp. Technol.*, **2**, 23–34.
4. Lee, D. J., 1985, The design of AGR thermal shield details for high temperature applications, *Nucl. Energy*, **24**, 409–17.
5. Nettley, P. and Hamer, A., 1981, UKAEA development programme on structural materials for fast reactors, *Nucl. Energy*, **20**, 71–8.
6. Batte, A. D., Dawson, R. A. T., Evans, R. B. and Thomas, G., 1985, Reverse bend testing to obtain long-term endurance data, *Techniques for high temperature fatigue testing*, London, Elsevier Applied Science, pp. 97–115.
7. Neuber, H., 1961, Theory of stress concentration for shear-strained prismatical bodies with arbitrary non-linear stress–strain law, *Trans. ASME, Ser. E*, **28**, 544–50.
8. Wetzel, R. M., 1968, Smooth specimen simulation of fatigue behaviour of notches, *J. Mat., JMSLA*, **3**, 646–57.
9. Truchon, M., 1982, Application of low cycle fatigue test results to crack initiation from notches, *Low cycle fatigue and life prediction*, STP 770, Philadelphia, ASTM, pp. 254–68.
10. Kimmins, S. T. and Horton, C. A. P., 1982, Anisotropy of elastic and tensile properties in type 17Cr–8Ni–2Mo weld metal, *4th Internat. Conf. Welding in Nuclear Engineering* (Aachen), Dusseldorf, German Welding Society, pp. 280–4.
11. Skelton, R. P., 1986, Crack growth and cyclic stress–strain properties of 9Cr–1Mo steel at elevated temperature, *Int. Spring Meeting, Fatigue at High Temperature*, Paris, Société Française de Metallurgie, pp. 185–203.
12. Chell, G. G., 1986, A *J* estimation procedure for combined mechanical, thermal and residual stresses, *Int. J. Press. Vessels Piping*, **23**, 187–213.

13. Bree, J., 1967, Elastic–plastic behaviour of thin tubes subjected to internal pressure and intermittent high heat fluxes with application to fast-nuclear reactor fuel elements, *J. Strain Anal.*, **2**, 226–38.
14. Hales, R., 1986, The construction of creep–fatigue design curves for type 316 stainless steel for use in conjunction with an elastic assessment procedure, CEGB Report TPRD/B/0753/R86.
15. Suzuki, A., 1986, A modified fraction model for the visco plastic behaviour of metals, *Int. Conf. on Creep*, Tokyo, JSME, pp. 471–6.
16. Goodman, A. M. and Hutton, M. F., 1986, Life assessment of thin axisymmetric structures subjected to combined creep and fatigue, *ibid.*, pp. 515–20.
17. Chaboche, J. L., 1986, Methods for structure calculations under high temperature fatigue: present and future possibilities, *Int. Spring Meeting, Fatigue at High Temperature*, op. cit. [11], pp. 359–84.
18. Burlet, H., Cailletaud, G. and Pineau, A., 1986, Fatigue thermique d'outillage de coulee par centrifugation, op. cit. [11], pp. 457–71.
19. Argo, H. C., DeLong, J. F., Kadoya, Y., Nakamura, M. and Ando, K., 1984, Eddystone experience on long-term exposed 316 ss steam turbine valve components, Paper, 84-JPGC-Pwr-15, New York, ASME, pp. 1–11.
20. Hellen, T. K. and Harper, P. G., 1984, *BERSAFE users guide, Phase III, Level 3*, CEGB document, Berkeley Nuclear Laboratories, UK.
21. Anon., 1985, *ASME Boiler and Pressure Vessel Code, Case N-47-23, Class 1 components in elevated temperature service*, Section III, Division 1, New York, ASME.
22. Nguyen, U. L. and Lee, D. J., 1985, Simplified analysis of a cantilever beam subjected to elastic follow up and cyclic loading, *8th Int. Conf. Struct. Mech. in Reactor Technol.*, Vol. L, Paper L7/7, pp. 365–70.
23. Terzaghi, A., Tunarelli, L. and Wood, D. S., 1984, Overview of Commission of European Communities Activity (Working Group Codes & Standards) in structural analysis and materials relevant to LMFBRs, Paper ASME 84-PVP-103, New York, ASME, pp. 1–6.
24. Anon., 1980, *Structural design guide for Class I components of prototype fast breeder reactor for elevated temperature service*, Power Reactor and Nuclear Fuel Development Corporation, Tokyo.
25. Roche, R., Delbecq, J. M., Drubay, B. and Forgeron, T., 1986, Creep design rules in the French Code RCC-MR, *Int. Conf. on Creep*, Tokyo, JSME, pp. 545–54.
26. Bestwick, R. D. W. and Clayton, A. M., 1985, Design methodology for creep fatigue assessment using creep ductility criteria, *5th Int. Seminar on Inelastic Analysis and Life Prediction in High Temperature Environment* (Post SMIRT discussion), Paris, pp. 55–60.
27. Chell, G. G., Raynor, D. and Skelton, R. P., 1986, A procedure for assessing the lifetimes of ferritic components subjected to high temperature fatigue, CEGB Report TPRD/L/3024/R86.

Index

Alloy instability, 180
Aluminium alloys, 216–17
ASME Code Case N-47, 230, 232–4, 246, 255, 265, 316–17
ASTM E606-80, 265
Asymmetric loading, 235

Back-stress, 33, 57, 58–60, 85, 255
Basquin equation, 266
Bauschinger effect, 27, 57, 286
Bauschinger strain, 33
Bending mode, 305
BERSAFE program, 316
British Standards Institution, 296–7

Cavity
  growth mechanisms, 212
  nucleation
    grain boundary sliding, by, 159–60
    growth, and, 202–6
    map, 208
    matrix deformation, by, 160–2
Coffin–Manson equation, 266. *See also* Manson
Companion specimens, 36
Composite materials, 306–7
Compression, measurements, 7–8
Compressive stress, damage measurement, 238–9
Constant
  plastic strain limits, 50–1
  stress limits, 49–50
  total strain limits, 49
Continuous cycling, 31–3
Crack
  density effects, 149–51
    studies, 129–34
  growth
    stage I, 179
    stage II, 179
  initiation, 179, 302
    titanium alloys, 218–19
Creep
  damage, 114, 232, 250
  ductility, 236
  failure, 250–1
  studies, 15
  testing, 16
Creep–fatigue
  assessment, 18
  behaviour, 234–9
  endurance curve, 253–4
  failure mechanism maps, 240–5
    alternative approaches, 243–5
    verification of model, 243
  interaction, 252
    model, 240–3
CREPLACYL program, 315
Cyclic
  bend tests, 294
  deformation processes, 178
  hardening, 31

Cyclic—*contd.*
  hardening—*contd.*
    exponent, 34, 66–8, 76
  inelastic analysis, 313
  softening, 31
  strength coefficient, 34, 76
  stress–strain curves, 80, 87
    data, presentation methods, 39–48
    properties, 27–112
    relation, 33–4
    test techniques, 36–8
Cylindrical specimens, 270–2

Damage measurement
  quantitative, 237–9
  tensile stress, 237–8
Damage observations, qualitative, 235–7
Data
  recording and analysis, standard test procedures, 281–9
  requirements for standard test procedures, 266–7
Deformation
  behaviour, 11
    titanium alloys, 217–18
  processes, 180–97
    ambient temperature, 181–3
    elevated temperatures, 184–6
    pure metals, 181–6
    solid-solution alloys, 186–7
    two-phase alloys, 187–97
Design codes, 98, 316–18
Diametral control, 36
Differences in test procedures, effects of standard test procedures, 290–1
Dislocation
  bundles, 181
  subcells, 183
Displacement-controlled tests, 13
Downshock events, 114
Dwell
  behaviour, 210, 214
  effects, titanium alloys, 219–20
  periods, 35, 86, 157, 158, 197, 209, 212, 288

Elastic follow-up, 307–8
Elastic laws, 5
Elastic modulus, 7
Elastic–plastic laws, 6–8
Elastic–plastic stress distribution, 316
Elastic stress concentration factor, 89
Elasticity
  modulus, 76
  studies, 5
Endurance
  behaviour, limiting properties, 252–4
  curves, 252
  prediction, 249
Energy changes around hysteresis loop, 51–3
Environment effects, 78–9, 180, 207–9
Equivalent energy method, 89
EVPCYCL program, 315
Extensometer
  method of strain measurement, 269–72
  requirements, 121–3

Failure
  assessment diagrams, 246
  criteria, 229–59
  definition, 283
  identification, 284
Fatigue
  damage, 197–9, 232
    intergranular damage, 201–9
    transgranular damage, 198–201
  mechanism map, 208
  overview of process, 178–80
  testing
    constant strain, 13–14
    elastic regime, 10–11
    elastic–plastic regime, 11–13
    history of, 1–25
    sequence of events, 10–14
    traditional methods, 10–13
  traditional description, 178
Fatigue–creep
  fracture map, 213
  interactions, 209–13

Features test, 2, 303
Feltham equation, 251
Finite element analysis, 303, 315
Frequency–temperature map, 209
Friction stress, 33, 57

Gauge lengths, 8
Gittus equation, 251
Grain boundary sliding, 134
  cavity nucleation by, 159–60

Hardening rule, 87
Hardening/softening relations, 61–5
Hardness changes, 83
High strain fatigue (HSF) tests, 28
  evolutionary stage, 35
  saturation stage, 35
  test parameters, 38
High temperature alloys, 60–1
History
  fatigue testing, of, 1–25
  sequential effects, and, 79–81
Hooke's law, 6
Hourglass specimens, 268–70
Hysteresis loop, 8, 12, 31, 36, 40, 48, 50, 53, 282, 286
  anomalous behaviour, 81
  behaviour modelling, 313–16
  energy changes around, 51–3
  internal, 59
  plastic strain rate variation around, 56
  steady-state, 309

Incremental
  behaviour, 307
  step test, 36
Intergranular damage, 236–7
  versus calculated net damage, 239
Internal
  cracks, 198
  stress, 3, 57–60
Irradiation effects, 77–8
Isotropic hardening, 88

Kinematic hardening, 88

Labyrinth structure, 183
Longitudinal control, 36
Loop inversion technique, 166
Low-cycle fatigue tests. *See* Standard test procedures

Machine modulus, 35
Manson–Coffin relationship, 163. *See also* Coffin
Materials
  behaviour, standard test procedures, 265–6
  science, 3, 17–21
Matrix deformation, cavity nucleation by, 160–2
Maximum
  strain error, 110
  stress error, 111
Metallography, 17–21, 177–228
Microstructural
  damage, 17
  response, 85
Microstructure effects, 192–4
Monotonic data, 45–8
Multiaxial effects, 255
Multiple step test, 36

Neuber equation, 89, 93, 305, 310, 313
Neuber rule, 314
Norton creep, 316
Notch-root behaviour, 306

Offset, 35

Paris equation, 179
Particle dislocation interaction, 187–92
Persistent slip bands (PSBs), 181–4, 186, 187, 189, 190, 199, 200, 216
Plastic strain
  evolution, 315
  hysteresis energy per cycle, 51
  range, 286–9
    versus total strain range, 43–5

Plastic strain—*contd.*
rate variation around hysteresis loop, 56
Poisson's ratio, 5, 36, 269
Primary stresses, 310–13
Prior exposure effect, 79
Push–pull tests, 302, 303
simulation of structural behaviour, 304–13

Ramberg–Osgood equation, 27, 34, 98, 307, 312, 314, 316
Rapid continuous cycling, 234
Ratchetting under load control, 50
Relative energy changes, 38
Residual stress range, 33
Reverse bend test, 305

Safety factors, 21
Saturation stress–strain data, 68–9, 88–94
Secant modulus, 33
Secondary hardening, 64
Secondary stresses, 310–13
SFE, 187
Shakedown behaviour, 48–51
Shell/flange feature, 314
Slow continuous cycling, 235
S–N curves, 11
Specimen
forms, 268–73
grip ends, 272–3
manufacture, 9
tests, 1–4
relation with practice, 301–19
Spectrum loading, 9
Standard test procedures, 261–99
correlation with non-standard procedures, 291–5
data
recording and analysis, 281–9
requirements for, 266–7
effects of differences in test procedures, 290–1
extensometry, 268–73

Standard test procedures—*contd.*
factors affecting reliability of results, 268–89
loading procedures, 274–81
materials behaviour, and, 265–6
need for, 264–5
plastic strain range, 286–9
post-test examination, 289
purpose of, 262
relevance to high temperature fatigue, 262–4
specimen design, 268–73
Step–stress method, 166
Stored energy
per cycle variation with plastic strain, 53–5
per half cycle, 55
values, 52
variation at constant total strain range, 55
Stowell–Hardrath relation, 89
Strain
amplitude, 42
concentration, 305–6
factor, 88–94, 91
distribution, 294
measurement, and specimen form, 268–72
range, 38
partitioning (SRP), 163–6, 245
application, 167–70
basic method, 163–4
simplified methods, 165–6
versus fraction of intergranular cracks, 133, 134
versus life, 145, 170
versus number of cycles to failure, 132
rate, 34, 77
rate–frequency relation, 110
Strain-controlled tests, 13
Strength coefficient, 68
Stress
amplitude, 42
analysis, 3
concentration factor, 88–94
deviation curve, 39
range, 38

Stress—*contd.*
range—*contd.*
versus plastic strain range, 39–41
versus total strain range, 41
relaxation, 32, 212, 251–2, 307
Stress–strain models, 255
fields, 313
laws, 4–8, 12
relationships, 3, 250, 312
Surface cracks, 198
Surface finish test specimen, 272

T-1430 design curve, 247
Tangent modulus, 33, 49
Temperature effects, 69–77
Tensile stress, damage measurement, 237–8
Tensile yield stress, 49
Tension measurements, 6–7
Thermal aging, 256
Thermal fatigue
damage modes, 114
origins of, 230–2
Thermal mechanical creep–fatigue test data, 147–58
austenitic steels, 149–54
failure mechanisms, 149
ferritic steels, 147–9
life prediction, 162–70
nickel-base superalloys, 154–6
summary of general trends in, 156–8
Thermal mechanical fatigue test data, 124–47
austenitic steels, 129–39
classification of failures, 144
comparison of data and other tests, 136–9
crack density studies, 129–34
ferritic steels, 124–9
grain boundary sliding studies, 134
Thermal mechanical fatigue test data—*contd.*
nickel-base superalloys, 139–41
summary of general trends in, 141–7
Thermal mechanical strain cycling, 113–75
extensometer requirements, 121–3
specimen design, 118–19
temperature control, 119–21
testing methodology, 117–24
thermal cycles, 123–4
Thermal shock test, 303
Thermomechanical tests, 81
Time-dependent mechanical deformation, 3
Titanium alloys, 217–20
crack initiation, 218–19
deformation behaviour, 217–18
dwell effects, 219–20
Type II service cycle, 248

Ultimate tensile strength, 4
Upshock events, 114

Vacuum effect, 78
Voids, 198, 209

Welds, 306–7

Yield point, 86
stress, 33, 55, 83
Yielding, constrained versus contained, 309–10
Young's modulus, 7, 33, 41, 69, 96, 166

GPSR Compliance

The European Union's (EU) General Product Safety Regulation (GPSR) is a set of rules that requires consumer products to be safe and our obligations to ensure this.

If you have any concerns about our products, you can contact us on ProductSafety@springernature.com

In case Publisher is established outside the EU, the EU authorized representative is:

Springer Nature Customer Service Center GmbH
Europaplatz 3
69115 Heidelberg, Germany

Batch number: 10607212

Printed by Printforce, the Netherlands